SEED MONEY

SEED MONEY

Monsanto's Past and Our Food Future

Bartow J. Elmore

W. W. NORTON & COMPANY
Independent Publishers Since 1923

For information about permission to reproduce selections from this book, write to Permissions, W. W. Norton & Company, Inc., 500 Fifth Avenue, New York, NY 10110

For information about special discounts for bulk purchases, please contact W. W. Norton Special Sales at specialsales@wwnorton.com or 800-233-4830

Manufacturing by LSC Communications, Harrisonburg
Book design by Lovedog Studio
Production manager: Lauren Abbate

Library of Congress Cataloging-in-Publication Data

Names: Elmore, Bartow J., author.
Title: Seed money : Monsanto's past and our food future / Bartow J. Elmore.
Description: First edition. | New York, NY : W.W. Norton & Company, [2021] |
    Includes bibliographical references and index.
Identifiers: LCCN 2021022107 | ISBN 9781324002048 (hardcover) |
    ISBN 9781324002055 (epub)
Subjects: LCSH: Monsanto Company. | Agricultural chemicals industry—
    United States—History. | Seed industry and trade—United States—History. |
    Plant biotechnology industry—United States—History. |
    Transgenic plants—United States—History. | Agriculture—United States—
    History.
Classification: LCC HD9482.U64 M644 2021 | DDC 338.7/6600973—dc23
LC record available at https://lccn.loc.gov/2021022107

W. W. Norton & Company, Inc., 500 Fifth Avenue, New York, N.Y. 10110
www.wwnorton.com

W. W. Norton & Company Ltd., 15 Carlisle Street, London W1D 3BS

1 2 3 4 5 6 7 8 9 0

For Joya Elmore

In Memory of Susan Elmore

Contents

List of Figures ix

Part I: **SEEDS**

INTRODUCTION
"Don't Do It. Expect Lawsuits." 3

Part II: **ROOTS**

CHAPTER 1
"You Are Getting into Chemistry Now, Senator,
on Which Subject I Am Rather Weak" 21

CHAPTER 2
"A Coal-Tar War" 38

CHAPTER 3
"A Die-Hard Admirer of the Tooth-and-Claw" 53

Part III: **PLANTS**

CHAPTER 4
"Wonderful Stuff, This 2,4,5-T!" 79

CHAPTER 5
"So You See, I Am Prepared to Argue on Either Side" 98

CHAPTER 6
"Sell the Hell out of Them as Long as We Can" 109

CHAPTER 7

"Strategic Exit"   127

CHAPTER 8

"They Can Have My House; I Just Need Thirty Days
  to Get Out"   138

CHAPTER 9

"Trespassing to Get to Our Own Property"   158

CHAPTER 10

"The Only Weed Control You Need"   173

CHAPTER 11

"I Have to Cry for Them"   192

Part IV: **WEEDS**

CHAPTER 12

"Oh Shit, the Margins Were Very, Very, Very Good"   209

CHAPTER 13

"They Are Selling Us a Problem We Don't Have"   241

Part V: **HARVEST**

CONCLUSION

"Malicious Code"   267

Acknowledgments   281

Notes   289

Illustration Credits   375

Index   377

List of Figures

p. 174   *Figure 1:* Genetically engineered (GE) glyphosate-tolerant crop adoption rates in the United States: percentage of planted acres of corn, cotton, and soybean, 1996–2016.

p. 186   *Figure 2:* Rise in the number of weed species resistant to ALS inhibitors, 1982–2000.

p. 219   *Figure 3:* Estimated agricultural use of glyphosate in the United States, 1992 and 2017.

p. 225   *Figure 4:* Rise in the number of glyphosate-resistant weed species, 1996–2019.

p. 230   *Figures 5a and 5b:* Herbicide use trends for soybean, 1992–2016: pounds of glyphosate per acre compared with pounds of all other herbicides per acre (top); pounds of glyphosate compared with pounds of all other herbicides (bottom). Data compiled from Arkansas, Illinois, Iowa, Minnesota, Missouri, Nebraska, and Ohio.

Monsanto made a lot of money from its herbicide Roundup®,
and eventually got into biotech to make its patented Roundup
Ready® seed system. But in less than two decades, that seed
system began to falter. Monsanto said it had the answer:
Xtend® seeds like those seen here.

PART I

SEEDS

"Don't Do It.
Expect Lawsuits."

BLACK SUVS VEERED INTO THE PARKING LOT OF THE RUSH
Hudson Limbaugh Sr. US Courthouse in Cape Girardeau, Missouri.
"It looked like the feds showing up," recalled Bev Randles, an attorney with the Kansas City firm Randles & Splittgerber. Then in her
late forties, Bev was a native of the Show-Me State. She had grown
up on a farm just thirty miles from Cape Girardeau. In a way, coming to the courthouse was coming back home.[1]

Thick fog enveloped the area as more than a dozen dark-suited
corporate attorneys for German chemical and pharmaceutical companies BASF and Bayer filed out of their vehicles and into the sleek
courthouse—a "palace," Bev called it—along the banks of the
Mississippi River. They were there to meet federal judge Stephen
N. Limbaugh Jr., grandson of the famed Missouri attorney for
whom the courthouse was named and first cousin to the conservative
talk show host most people know. This was Limbaugh country, a little over an hour south of St. Louis in the fertile farmland just north
of what Missourians call the "bootheel" of the state.[2]

Something big was going down in this small town, though only a
handful of reporters were there to document what was happening.
It was January 27, 2020, the start of the *Bader Farms v. Monsanto
and BASF* jury trial, and Bev Randles and her husband, Billy, were
getting ready to start the biggest case of their lives.[3]

The Randles were representing Bill Bader, a Missouri peach
farmer who filed this case back in 2016—roughly a year and a half

before Bayer bought Monsanto in a mega-merger that made head-lines across the globe. Around that time, Bev Randles had been run-ning for lieutenant governor on the GOP ticket, hoping to become the first Black politician elected to statewide office, when she vis-ited Bader's farm for a photo op. She had spent enough time around farms to know that Bader's peaches did not look right. They had curled leaves, and many looked like they were dying. When Bader told her that he believed a herbicide sold by BASF and Monsanto was to blame, Bev told him she wanted to help.[4]

The herbicide was dicamba, and it was used to fix a problem created years before. Back in the mid-1990s, Monsanto had intro-duced Roundup Ready technology that made commodity crops—such as soybeans, corn, and cotton—genetically engineered (GE) to be resistant to its blockbuster herbicide, Roundup, which was first commercialized in the 1970s and contained a powerful weed-killing chemical called glyphosate. Farmers had loved the system because it allowed them to spray Roundup throughout the grow-ing season, keeping fields clean of unwanted plants. But within a few years weeds started developing resistance to Monsanto's her-bicide, which is why the company started working feverishly to create crops that would tolerate Roundup *and* dicamba, another powerful herbicide that had been around since the 1960s. In 2007, Monsanto acquired from the University of Nebraska the gene sequence that bestowed dicamba resistance to plants, and eight years later Monsanto commercialized its first dicamba-tolerant seeds—branded Roundup Ready Xtend—beginning with cotton in 2015 and then soybeans in 2016.[5]

But there was a problem, and it was a serious one: dicamba was volatile—much more so than Roundup. When soybean and cot-ton farmers sprayed their fields, this chemical often vaporized—especially in hot temperatures—drifting onto adjacent farms and ecosystems, damaging everything from watermelons to sycamore trees. Farmers without dicamba-tolerant GE crops were incensed, especially fruit farmers like Bill Bader who had no way to avoid damage when farmers nearby sprayed dicamba on their own fields. After all, there was no such thing as a dicamba-tolerant peach tree.[6]

"This is the nastiest litigation I've ever been involved in," Billy Randles said of the trial, which was saying a lot, considering the fact that he had been involved in legal suits where he represented Philip Morris back when cigarette companies were still selling doubt about the link between smoking and cancer. An alumnus of Harvard Law School, Randles had been in practice for thirty years and had even run for governor of Missouri. He was comfortable speaking in front of big audiences, including at the Kansas City church where he moonlighted as a preacher. Nevertheless, Randles and his wife, Bev, had never been lead lawyers for a case this big.[7]

Bayer, now the owner of Monsanto's technology, was clearly sending its top litigation specialists to Cape Girardeau to battle a husband-and-wife team whose lawsuit could do real damage to the firm. Recognizing the stakes, Jan Miller, Bayer/Monsanto's lead attorney, appealed to Judge Limbaugh before the trial to put a gag order in place that would prevent Bader's legal team from talking to the press. Limbaugh honored the request.[8]

Observers in the gallery may well have been surprised by the judge's order, but when Billy Randles took to the podium to deliver his opening statement, it quickly became clear why Miller had made his move. As Bev sat pensively at the plaintiff's desk next to the farmer she had promised to help roughly four years prior, her husband began to lay out a series of internal memoranda and documents that no corporation would want exposed.

"Don't do it. Expect lawsuits," concluded a Monsanto employee in a document summarizing the findings of an advisory panel to Monsanto that Randles cited in the first moments of trial. Monsanto had created the academic review committee in an attempt to get frank feedback about its dicamba system, and the panel had concluded that Monsanto's seeds were going to wreak havoc, especially for "specialty crop" farmers growing fruits and vegetables. Steve Smith, a tomato grower Monsanto had asked to serve on the panel, claimed dicamba-tolerant crops were the "most serious threat to specialty crops of anything I had seen." He fired off a blistering email to Monsanto managers saying, "While I know you are hearing the comments I and others are making, I'm not sure you are HEARING."[9]

Documents showed that Monsanto knew dicamba's tendency to drift off-target could help them make money. After all, vaporized dicamba could also harm soybean and cotton farmers that did not use Monsanto's Roundup Ready Xtend traits. If they wanted to protect their crops from dicamba drift, they were going to have to use the company's new seeds that made crops resistant to both Roundup and dicamba. In 2013, as Monsanto prepared to launch its new product, a company slideshow coached salespeople on how to convince commodity crop growers who were not troubled by Roundup-resistant weeds to buy the new dicamba-tolerant seeds. "Why should I pay for something I don't need?" a farmer might ask. Push "'protection' from your neighbor," one slide said, revealing that the firm's officials were not only aware of the implications of drift for neighboring farms and ecosystems but were thinking of it as an asset that would force farmers to purchase company seeds.[10]

In 2013, Monsanto did not have its own dicamba herbicide. German chemical company BASF was the main distributor of dicamba brands, but the Environmental Protection Agency (EPA) did not grant approval for these BASF-made herbicides to be used on dicamba-tolerant crops during the hot growing season, in part because of concerns about volatility. Not until 2017 would Monsanto introduce an EPA-approved dicamba formulation, called XtendiMax® with VaporGrip®, which it claimed was much less volatile than older dicamba brands.

So when Monsanto first sold its dicamba-tolerant cotton seeds in 2015 and soybeans in 2016, there was no EPA-approved dicamba herbicide that could be used on these crops during the growing season. Monsanto put pink labels on its seed bags warning farmers not to spray dicamba over their dicamba-tolerant crops, but internal communications revealed that company employees knew what was going to happen. "I . . . get to work with a group of renegades that . . . thinks one sticker is going to keep us out of jail," said an Xtend team member in a 2015 email. "Let's face reality," said Boyd Carey, Monsanto's point person for dicamba complaints, "Regardless of whether it's legal or not there will probably be guys who spray dicamba." BASF had little doubt: "Dicamba demand spike with DT [dicamba-tolerant] traits" read the sales report in 2016.[11]

The eight jurors in the case, mainly working-class people from Missouri, were taking it all in. As Billy Randles said in closing statements, they were now the "most informed people in the world" on the dicamba issue, having "seen company documents no one else has ever seen."[12]

And that included records showing how Monsanto tried to block university access to XtendiMax data. Internal correspondence from 2015 showed that the firm had decided to "pull back some of this academic testing with Xtend and XtendiMax formulations to ensure that these formulations keep a 'clean' slate" when going through EPA review. Carey later testified that preventing university weed scientists from analyzing the volatility of a herbicide was extremely rare, maybe a once-in-thirty-year occurrence. Publicly, the company claimed that this was because "of the difficulty in producing quantities that would allow for broad testing." But internally, that logic was considered a joke. "Ha ha ha," laughed a Monsanto official in a 2015 email, "Difficulty in producing enough product for field testing. Ha ha ha. Bullshit."[13]

Plaintiff Bill Bader looked mad. The only man before the judge without a tie, he sat and listened as Randles read confidential company correspondence that showed how the firm planned to deal with his complaints. Bader had called Monsanto when his problems started getting bad in 2015 and 2016, but the company refused to send anyone to come look at what was happening on his farm.[14]

It was all part of the plan. "Do not visit a driftee inquiry if the driftee is not a [Monsanto] customer," said Carey in a 2017 directive marked highly confidential. Because the company was publicly denying that dicamba drift was really a problem, Carey was careful to write: "Note 'driftee' is meant as an internal term only." It was now all so clear to Bill Bader. Monsanto never intended to respond to his requests for help.[15]

Randles went on, citing another confidential email in which a Monsanto official joked that a "decent lawyer will have a field day with Mr. Bader" if his case went to trial. The plan was to "point the finger at disease," which is exactly what Monsanto's team did, focusing on root rot and other pests on Bader's farm as key culprits causing peach damage. "Deny, deny, deny," Carey exclaimed in another Monsanto document. The company had made a policy never to admit that it had a serious drift problem.[16]

But BASF, Monsanto's partner in this dicamba system, knew how bad things were. Randles put a 2016 company report before the jury: "There must be a huge cloud of dicamba blanketing the Missouri Bootheel," the report stated. "That ticking time bomb finally exploded. The scope of damage is on a massive scale, and fingers are pointing in all directions from grower to grower."[17]

Even as the damage got worse, Monsanto kept promoting its Roundup Ready Xtend system. As Randles showed, the firm continued to treat the drift problem as a way to sell seeds. John Cantwell, a Monsanto employee, outlined the strategy in a confidential email: "I think we can significantly grow business and have a positive effect on the outcome of 2017 if we reach out to the driftee people. . . . Most driftee people were interested in the technology and can be . . . turned into new users."[18]

There was a lot of money to be made in all this. A 2017 Monsanto sales meeting ended with plaudits: "Xtend-deli-icious, invigorating, success, Xtendiful, cha ching."[19]

* * * *

MILES AWAY FROM LIMBAUGH country, in the San Francisco Bay Area, retired schoolyard groundskeeper Dewayne "Lee" Johnson

was dying. For years, Johnson had sprayed Monsanto's glyphosate on weeds for the Benicia Unified School District just north of
Berkeley, and on one occasion, a hose exploded, soaking him in
Monsanto's herbicide. In August 2018, Johnson won a landmark
case against Monsanto in which a jury determined that Johnson's
chronic Roundup exposure was a "substantial contributing factor"
to his non-Hodgkin's lymphoma. A California appeals court later
reduced the $289 million verdict to $20.5 million but did not over-
turn the lower court's ruling. Thousands of people from all across
the country filed similar litigation. By 2020, there were more than
120,000 lawsuits underway or set to be filed.[20]

For Johnson, the outcome was bittersweet. A few years earlier, he
broke down and cried when he told his young boys that he was diagnosed with cancer. His physicians told him he probably would not
make it through 2020. No amount of money would change this fate.
But Johnson was hopeful that his case would help others by exposing problems with the world's most widely used herbicide.[21]

During the trial, Johnson's attorneys revealed Monsanto papers
that had never before been available to the public. The most damaging documents showed that despite Monsanto officials' claims to
the contrary, the firm could not prove its Roundup formulations—
now used on more than 90 percent of all soybeans and corn grown
in the United States and sprayed over millions of acres of farmland
worldwide—were harmless. "You cannot say that Roundup is not
a carcinogen," Monsanto toxicologist Donna Farmer said in a 2003
email, "we have not done the necessary testing on the formulation
to make this statement." Seven years later, the situation was still the
same. "With regards to the carcinogenicity of our formulations,"
another Monsanto scientist said internally in 2010, "we don't have
such testing on them directly."[22]

These in-house memos said internally what other scientists were
saying publicly: that despite the copious amount of research done
on the health effects of glyphosate, many studies did not consider
the chemicals called "surfactants" that helped Roundup penetrate

plants. This was concerning, especially considering that Monsanto's own scientists had determined that surfactants (such as polyoxyethylene tallow amine) "are able to increase glyphosate absorption through the skin." Dr. William Heydens, a Monsanto toxicologist that worked closely with Roundup, had wanted to avoid research that would involve surfactants because "of the potential for this work to blow Roundup risk evaluations (getting a much higher dermal penetration than we've ever seen before)."[23]

In 2014, almost twenty years after Roundup Ready technology had been introduced, the World Health Organization's International Agency for Research on Cancer (IARC) announced that it would review glyphosate in its next round of cancer studies. In internal correspondence, Monsanto officials admitted that there was "vulnerability in the area of epidemiology" and "potential vulnerabilities in the other areas that IARC will consider." "More than just pure bad luck is working against glyphosate."[24]

What Monsanto scientists refused to publicly admit was that evidence of a problem went back decades. In 1999, a Monsanto-funded study conducted by University of Wales professor James Parry showed that "glyphosate is capable of producing genotoxicity," or damage to genetic material, in lab animals. Monsanto was clearly not happy with the finding. Dr. Heydens said that he wanted to "find/develop someone" who would be willing to offer a more positive assessment of glyphosate. "Parry is not currently such a person, and it would take quite some time and $$$/studies to get him there."[25]

When the IARC announced in March 2015 that it had found enough evidence to classify Roundup as a probable human carcinogen, Monsanto set out to "ghost write" articles to try and save its signature herbicide, but not everyone was on board with this. When Monsanto told an ex-employee and consultant for the firm that the company was going to keep his name off a glyphosate paper he worked on, he said "I can't be a part of deceptive authorship on a presentation or publication. . . . We call that ghost writing and it is unethical." But others had no problem with the plan. "Ghost wrote cancer review paper Greim et al.," explained Monsanto's Dr. David Saltmiras in a summary report of his 2015 accomplishments.[26]

After IARC's ruling, the EPA quickly moved to conduct a reevaluation of Roundup's active ingredient, issuing a finding in September 2016 that glyphosate was "not likely to be carcinogenic to humans." But in 2019, the Centers for Disease Control and Prevention (CDC) did its own review and delivered more nuanced findings. In a section of an issue paper titled "cancer effects," the CDC noted that "meta-analyses reported positive associations between glyphosate use and selected lymphohematopoietic cancers." The CDC also cited several studies that "reported risk ratios greater than 1 for associations between glyphosate exposure and risk of non-Hodgkin's lymphoma or multiple myeloma." In 2020, there was no definitive proof that Roundup caused California groundskeeper Lee Johnson's lymphoma or the cancers of other Roundup litigants, but there were clearly a lot of unanswered questions.[27]

In the end, the jury in the Johnson case did not need irrefutable scientific evidence of the link between glyphosate and cancer. They awarded damages to Johnson despite the discrepancies between the IARC's and the EPA's findings. One juror admitted that the way Monsanto meddled with science through ghost writing was something that really bothered him: "They were protecting a product that was very important to the corporation's bottom line."[28]

And it continued to be so. Despite the jury verdict, Roundup remained an EPA-approved product, and growers continued to spray it on hundreds of millions of farm acres. In 2019, President Donald Trump's EPA appointee Andrew Wheeler reaffirmed his agency's approval of glyphosate, saying, "EPA has found no risks to public health from the current registered uses of glyphosate." USDA secretary and former Georgia governor Sonny Perdue chimed in: "USDA applauds the EPA's decision. . . . If we are going to feed 10 billion people by 2050, we are going to need all the tools at our disposal, which includes the use of glyphosate."[29]

✳   ✳   ✳   ✳

ON THE OTHER SIDE of the world, in the elegant Centec Tower in downtown Ho Chi Minh City, Vietnamese salespeople were hard at work promoting Monsanto's Roundup Ready system. In 2014, Monsanto heard the good news that the Vietnamese government had approved the use of genetically engineered seeds in Vietnamese fields. A few months later, Monsanto celebrated the first harvest of Roundup Ready corn in Vietnam's history. These plants promised big profits. It was yet another triumph in Monsanto's concerted quest to expand its seed empire into developing countries around the world, one that executives might have celebrated over expensive cocktails at the trendy rooftop bar upstairs.[30]

To accomplish this feat, this seed seller had to overcome its historic image as a harbinger of death. During the Vietnam War, Monsanto had been the largest producer of Agent Orange, which had destroyed millions of acres of lush, tropical forest and riddled communities throughout the country with serious health problems. As Monsanto's corn crops rose from the ground in 2015, the US government was still trying to deal with that toxic history, channeling hundreds of millions of American taxpayer dollars toward an expensive cleanup program for hot spots still contaminated by Agent Orange spills that occurred more than four decades earlier. Few Americans knew this was going on. Even fewer knew that Monsanto spent no money on these remediation efforts, though some Vietnamese people had tried hard to force the company to bear responsibility for its role in causing these pollution problems.[31]

A few blocks away from the Centec Tower, Vietnamese citizens and foreign tourists can still visit the War Remnants Museum and read allegations of human deformity caused by Agent Orange. A blind man tapping keys on an electronic piano welcomes guests as they enter the facility, with a sign adjacent to him implying that American herbicide campaigns during the Vietnam War caused his impairment. Upstairs, visitors find a house of horrors: a whole room devoted to suspected Agent Orange tragedies, including deformed fetuses soaking in formaldehyde and gruesome pictures of disfigured men, women, and children purportedly ravaged by America's chemical storm. Museum curators call war culprits by name, specifically mentioning Monsanto in photographs and display captions.

Back up the street, phones ring as Monsanto salespeople pitch seeds that will bring tremendous volumes of another herbicide, Roundup, to Vietnam. Dicamba will surely follow.

* * * *

AGENT ORANGE. ROUNDUP. DICAMBA. The histories of Monsanto's herbicides are as entangled with one another as they are with the evolutionary pedigrees of the weeds they were brought in to kill. These man-made chemicals are more than a half-century old and yet, as these stories reveal, we live with all of their legacies today. It is hard to escape them. In Vietnam, dioxin contamination remains below air bases in Bien Hoa and beyond. 2,4-D, one of the active ingredients in Agent Orange, is still used to kill weeds on farms across the world. Glyphosate is so widely used today in the United States that many processed foods, including Cheerios, have trace amounts of this chemical. And dicamba drift has meant that even farmers who sought to avoid buying Monsanto's Xtend seeds have had little choice but to join the GE revolution.[32]

This is not the way it was supposed to be. Agricultural chemicals were supposed to offer liberation—at least that is what the advertisements said. When Monsanto introduced Roundup Ready seeds in the 1990s, the company told farmers it would be the "system that sets you free." Agent Orange too was supposed to be a tool of emancipation, helping save American and South Vietnamese lives by clearing jungle havens of Communist insurgents threatening democratic governments in Southeast Asia. Bayer recently explained that its Roundup Ready Xtend system was designed to give farmers the "freedom" to "get the weed control you need." "Farmers should have choices," the tagline promised.[33]

But told they were buying cutting-edge innovations, Monsanto's clients were in fact purchasing tools recycled from a time long ago. Dicamba, now sprayed on Monsanto's Xtend crops, is older than the chemicals it has been brought in to aid. As it turns out, the future of agriculture is in fact heavily dependent on a distant chemical past that helped create some of today's most pressing farming problems.

This is the story of how we got here, to a time where many farmers and consumers feel they have few options but to buy "solutions" from the very firm that helped create the problems they now struggle to solve. *Seed Money* traces Monsanto's remarkable journey from making DDT to restructuring DNA, examining how it gained such outsized influence on our food system, becoming by the twenty-first century the largest seller of seeds in the world. Monsanto's name may now be history, its brand subsumed under the Bayer banner, but that history is making the future of food, which is why it is important to uncover the roots of the system we live with today.[34]

Monsanto earned the seed money to finance a seed empire at a time when most American chemical companies were dependent on German and Swiss firms for supplies. When John Queeny created Monsanto in 1901, he was on a quest to free the American economy from the stranglehold of European chemical concerns. He looked toward natural resources that the United States harbored in abundance, especially coal and petroleum, seeking to turn those mineral deposits into synthetic products that would drive the global economy.

But in that pursuit of freedom—noble as it may have been—Monsanto, along with its many chemical competitors in the United States, helped to create a new chemical economy dangerously reliant on fossil remains buried deep in the ground. Despite the magic that chemical firms spoke of—a world in which new plastics and pesticides, fabrics and fibers seemed to appear out of thin air—the truth was this seemingly limitless bounty ultimately came from resources that were finite even in the vast geographic expanse of the United States—coal, oil, gas, phosphate, and other mined minerals.

This book takes readers to small towns—Nitro, West Virginia, Anniston, Alabama, and Soda Springs, Idaho—where fossil remains, scavenged from the earth, became the chemicals on which we depend. *Seed Money* also explores rural fields in Ohio, Vietnam, Brazil, and beyond, where farmers labor to produce the food we eat. This is a story, in other words, that connects the histories of people who work in plants with the histories of people who work with plants. Their lives are here presented side by side. Some of these individuals never saw this book in print, their lives taken as they worked dangerous jobs. This book is written to honor them and the testimonials they left behind that may well help us avoid the hazards of a chemical age that wrecked their bodies and still threaten ours.[35]

Which is why *Seed Money* is also for the living, particularly employees inside Bayer. This is not another Frankenfood book that promotes conspiracy theories about GE technology or tries to cast Monsanto as a company filled only with wrongdoers. Because of its toxic legacies, Monsanto has consistently ended up on lists of the most hated firms in America, with some people dubbing the company "Monsatan." Such labels reduce the complexity of the human story behind this brand. Over the years, there were people in Monsanto that made unethical decisions that resulted in devastating and far-reaching ecological and human health costs. Those stories are in these pages. But there were also people who worked from within Monsanto—people like Bob Shapiro, Ernest Jaworski, and John Franz—hoping to change the world for the better. Today, inside Bayer's corporate laboratories, people commit themselves to the creation of new drugs that might help cure cancers—cancers like the one that imperiled Lee Johnson's life when he took his case to trial. Others are genetically engineering crops to be drought resistant, many believing, earnestly, that these efforts will feed the 10 billion people USDA secretary Sonny Perdue invoked in his defense of glyphosate.[36]

There is no reason to doubt the sincerity with which many Bayer scientists and researchers pursue their work to feed the world and heal the sick. Many carry out their tasks with dignity and distinction with an eye toward the public good. But if these well-meaning men and women fail to look up from lab microscopes and widen the aperture to take stock of the history in which they are embedded, they may fail to see the harvest these seeds might bear.[37]

It has now been twenty-five years since Monsanto sparked the GE seed boom that forever changed our food system, which makes it a fitting time to look to history, to take stock of where we have been before we figure out where we are going.

The journey begins in a small factory on South Second Street in St. Louis, Missouri, along the banks of the Mississippi River just an hour north of Cape Girardeau, where Monsanto's founder, John Queeny, was tinkering with chemical equipment bought second-hand. He was working day and night, determined to make it as a self-employed businessman and driven by visions of freeing Americans from their dependence on a German firm called Bayer.

This family portrait shows Olga Monsanto with her husband,
John, and her son and daughter, Edgar and Olguita.
According to one source, John Queeny named his company
after Olga, in part because he was still working at the Meyer
Brothers drug company and wanted to avoid the confusion
of holding that post and operating a chemical concern
that would bear his last name.

PART II

ROOTS

"You Are Getting into Chemistry Now, Senator, on Which Subject I Am Rather Weak"

ON A COLD DAY IN FEBRUARY 1902, A BOURBON-FILLED OLD fashioned may well have been waiting for John Queeny halfway down the bar at Tony Faust's center city restaurant in St. Louis. At some point, this apparently became the custom. Queeny—a mustached man standing roughly six feet tall, then in his forties— liked to take his trusted colleague Louis Veillon, a Swiss chemist, to Faust's for lunch, where they could discuss the building of their chemical plant. On this particular day, Queeny brought a special treat: a small sample of white powder.[1]

Veillon thought these lunches a bit odd, in part because he often found himself drinking booze alone. Arriving at their table, Queeny typically urged Veillon to purchase a beer, before excusing himself so that he could use the restroom. Upon his return, Queeny would ask the waiter for hot coffee while Veillon drank his alcoholic brew.[2]

Frustrated by his lunchmate's habits, one day Veillon decided to follow Queeny to the toilet, staying a few paces back so as not to be detected. To his amazement, he discovered that Queeny's path included a detour to the bar. There, Veillon explained, "I saw him drinking an old fashioned." That was how Queeny got his midday fix.[3]

It all made sense. Those who knew Queeny were well aware that he could be a serious drinker, sometimes even to the point of having to be "carted home" from conventions, but Queeny knew this was something he had to hide, probably because his teetotaler wife, Olga Monsanto, detested it so much. Queeny's propensity for drink extended to the office, where he allegedly took little swigs from a "medicine bottle" placed above the fireplace near his desk.[4]

But today, Queeny could be excused for enjoying a midday drink. There was cause for celebration because Veillon and Queeny had just discovered that they had made their first batch of saccharin, an artificial sweetener. Allegedly it was the waiter who confirmed the good news, guzzling a glass of water infused with Queeny's white powder and declaring it sweet. Veillon and Queeny had not been sure, their senses dulled by inundation with saccharin particles floating in the air back at their plant.[5]

The precise details are lost to history, but legend has it that Veillon raised a ruckus, stripping the tablecloth from under their plates and waving it over his head, like a prizefighter who had just won a bout. They had done it—they had finally done it. This was a very sweet beginning.[6]

✳   ✳   ✳   ✳

QUEENY WAS AN OLDER man now, and the desire to do something big, to have the independence to run his own shop, must have eaten at this midwesterner, described by those who knew him well as "irascible," "tenacious," a "bulldog," a real "table-pounder." "He had a relentless stirring within him," his son once remarked.[7]

From an early age, Queeny threw his heart into salesmanship, not chemistry, though this professional path was one born of necessity. He was the eldest son in a household of five children and grew up in Chicago. His mother, Sarah Flaherty, and his father, John, had come to the United States from County Galway, and by the time they had their first child in 1859, the elder John Queeny had secured a steady job as a building contractor. The family made much of its money from rents accrued from housing investments purchased via family savings. Then, in 1871, tragedy struck. The Great Chicago Fire, a conflagration that destroyed more than three square miles of the city, also consumed the family's investments. The younger Queeny was forced to find work.[8]

He was just twelve years old, with six years of public schooling to his name and little vision for what he wanted to do. But in 1871 that was really beside the point. His family needed money—badly. Years later, John Queeny explained why he ended up at the wholesale drug firm of Tolman & King, saying, "I took my first job as an office boy in a drug concern at $2.50 per week because that was the first one I happened to run across." Such was the contingency of life that drove Queeny into the drug business.[9]

By all accounts, he was a hard worker. His first job was as a "runner for shorts," which meant he saddled up in a horse-drawn wagon and called on other wholesalers to secure small amounts of drugs needed to fill Tolman & King requirements. Queeny's employer peddled everything from common chemicals—such as baking powder, the painkiller morphine, and the antimalarial drug quinine—to wild patent medicines and quackery cure-alls. Take "Brunker's Carminative Balsam," the "champion of remedies for babies fretting, teething, summer complaint flux of cholera infantum, or for adults for dysentery flux, cholera morbus, Asiatic cholera, congestion of the stomach or hemorrhage of the stomach and bowels." In an era in which there was no Pure Food and Drug Act, such were the fantastical claims drug companies could make on potions and elixirs.[10]

Not everyone bought into the hype, including John A. King, Queeny's boss. When asked by the *Chicago Tribune* if he sold patent medicines, he replied, "Of course, but do not recommend them. I guess the most of them are bad. They guarantee to bring a dead man to life almost, and people believe it. I do not know what they are made of. I have no doubt that some of them would destroy an iron kettle."[11]

Dangerous wares, but Queeny moved them fast, speeding up deliveries by getting to know the traffic cops at major intersections. "Queeny wore out more wagons than any" other employee, his boss allegedly exclaimed. It was a good gig and it brought in much needed cash. For about a decade, Queeny rode his circuit in Chicago before setting off to take a job in New Orleans with another drug concern, I. L. Lyons & Company. Both Tolman & King and Lyons & Company were wholesalers largely dependent upon other drugmakers to fill their inventories.[12]

This was the dawn of the chemical age, a time when big German firms such as BASF, Bayer, and Hoechst—collectively known as the "Big Three"—were moving beyond patent medicines derived from exotic plants and leading the way in synthetic organic chemistry. (Organic here simply refers to chemicals synthesized from carbon-based resources.) Most of these companies started out producing dyes but diversified by the 1880s and 1890s into blockbuster drugs derived from coal tar—the thick, black by-product left over in the furnace conversion of coal to coke (coke being coal without its impurities). Geography mattered. The Big Three benefited from their location near premier research universities as well as their position along the Rhine River, which served as a coal conduit in the nineteenth century. From these fossil remains Bayer produced aspirin while Hoechst developed Novocain, powerful drugs designed to alleviate the aches and pains of people living in a rapidly industrializing world.[13]

In Switzerland, smaller firms, such as Ciba, Geigy, and Sandoz, competed with their large German rivals, though much of their raw materials—an estimated 80 percent—came from German suppliers. In Great Britain, firms such as Nobel Industries—named after Swedish chemist Alfred Nobel, who patented dynamite technology—developed chemical production operations, though never on the scale seen on the Rhine.[14]

America had its own budding chemical industry, but it lagged far behind Europe in synthetic organic chemistry. DuPont, founded in 1802 by E. I. du Pont de Nemours of Brandywine Creek, Delaware, still focused mainly on gunpowder and explosives production and did not enter into diversified chemical sales until the twentieth century. Herbert Dow's Midland, Michigan, chemical company, established in 1897, became a leader in the production of chlorine and bromine from salt brine in the late 1890s but was a slow entrant into organic chemistry. Pfizer, founded in 1849 by German immigrants Charles Pfizer and Charles Erhart in Brooklyn, became a leading producer of pharmaceutical products in the United States, though the firm focused on producing chemicals such as citric acid from unripe fruit (later via fermentation of mold) rather than investing initially in coal-tar chemistry. Colonel Eli Lilly's business founded in Indianapolis in 1876 and Dr. Wallace Abbott's Chicago laboratories established in 1888 both sourced chemicals from the big players in Europe. Merck, a German firm that opened an office in New York in 1887, remained, up to World War II, one of the largest drug suppliers in the United States.[15]

Such was the world Queeny saw as a young salesman, one in which America was largely reliant upon foreign suppliers for organic chemicals derived from coal tar that were becoming increasingly central to the global economy.

✳   ✳   ✳   ✳

QUEENY SPENT THE BETTER part of a decade in New Orleans, purchasing drugs and patent medicines for I. L. Lyons, many of which promised to liberate Americans from pernicious dependencies and addictions. "Drunkenness or the Liquor Habit, Positively Cured by administering Dr. Haines' Golden Specific," read the Gilded Age advertisement by I. L. Lyons in the 1890s. This elixir, contents not disclosed, promised to "effect a permanent and speedy cure, whether the patient is a moderate drinker or an alcoholic wreck." Bottled spring water from Choctaw County, Alabama, held similar magic according to the drug company Queeny worked for, curing "cutaneous afflictions, alcoholism, and female complaints."[16]

But if Lyons claimed to curb addictions with some products, the company clearly exacerbated those same addictions with others.

"Boneset Bourbon Tonic," for example, was mainly "pure old Kentucky Whisky" but billed as an "antidote to malaria" and a cure-all that "relieves indigestion, restores appetite, and is especially adapted to the feebleness of overworked clergymen, delicate women, and old people." A similar story played out with morphine. America's drug firms offered up potions to end the menace of "morphinism," even as these same houses continued to sell morphine-infused patent medicines to calm people's nerves. Lyons and others made money selling cures to problems the drug business helped create.[17]

In 1891, Queeny, now in his early thirties, left New Orleans and soon became a buyer for St. Louis–based Meyer Brothers Drug Company, one of the largest drug wholesalers in the country. In 1894, he took an offer in New York as a sales manager at the American subsidiary of the German chemical firm Merck. To this point, he still had virtually no experience with chemical manufacture, knowing the drug industry only from sales and purchasing positions.[18]

John Queeny's time with Merck was relatively short and uneventful, save his courtship with his bride, Olga Monsanto, whose maiden name would become his brand. Olga came from high society, described by some as a bit pretentious, with heritage that extended back to elite families in both Spain and Germany. (According to company historians, her grandfather was Don Emmanuel Mendez de Monsanto, a nobleman knighted by Queen Isabella, who first came to the Spanish West Indies in the 1830s to run a sugar plantation in what is today Puerto Rico.) After a Catholic wedding at St. Paul's Church attended by some six hundred guests—by all accounts a lavish affair written up in the *New York Tribune* and *Merck's Market Report*—Olga and John traveled to St. Louis so John could return to Meyer Brothers Drug Company.[19]

That is when Queeny began to take on more responsibility. In 1897, Olga bore a son, Edgar, and two years later, the day John Queeny turned forty, his daughter Olguita arrived. There were now two children to take care of in the Queeny household as well as a woman of wealth who expected certain standards of living.[20]

In this moment, John Queeny used all his savings, some $6,000, to build a sulfuric acid plant in East St. Louis on the Illinois side of the Mississippi River. Sulfuric acid was one of the basic "heavy" or bulk chemicals used by many industries, from fertilizer manufacture to oil refining and pharmaceuticals, and it promised to be a lucrative business, especially given that the Mississippi River offered an artery into the nearby sulfur mines of Louisiana, which featured some of the richest deposits in the world. Hedging his bets, he negotiated with his employer to hang on to his job, just in case the gig did not pan out.[21]

It didn't. The day the sulfur refinery opened for business, Queeny was allegedly at his desk at Meyer Brothers when he got a phone call from the plant operator across the river. "Nothing can save it," screamed a man on the other end of the line who watched helplessly as a fire engulfed the plant. Company lore holds that Queeny then put down the phone and calmly went back to work, later returning home to a dinner party where he made no mention of the great loss he had borne that day.[22]

Whether this story is true is impossible to confirm, but what is clear is that no mention of this fire ever made it into the newspapers. In fact, during the late 1890s, it was Olga Queeny, not John Queeny, who more commonly made the press because of her profession. She was a gifted pianist, a talent she passed on to her son, Edgar, and became a preeminent member in the Music Teacher's National Association. The only other substantive media mention of John Queeny, besides the announcement of his marriage to Olga and the death of his father in 1898, was a story detailing how Queeny had caught a "porch-climber" snooping around southside homes in St. Louis. As far as his chemical business was concerned, there simply was no concern.[23]

There he was: forty years of age, his savings spent, his father deceased, with two kids back at home both under the age of three. Maybe this is what drove him to the saloon many mornings, where he downed nickel beers and sandwiches with his boss, Carl Meyer.[24]

*   *   *   *

THOUGH QUEENY HAD SUFFERED through some business failures, being a buyer for one of the biggest wholesale pharmaceutical firms in the country was no small job, and by 1899, Queeny began to make a name for himself in the National Wholesale Druggists Association, serving as the chairman of the Committee on Adulterations.[25]

Today, Monsanto is well known for fighting federal regulations that threatened its business, but in 1899 the founder of this firm believed that regulation was the only thing that would save his industry. As chairman of the Committee on Adulterations, he fought hard for a pure food and drug law that he believed would help root out corrupt companies selling contaminated products or patent medicines of dubious efficacy, thereby restoring the public's faith in drug companies.[26]

Many in the drug business opposed regulation, but Queeny clearly felt that without the government's imprimatur, his industry would lose the public's faith. It is an "absolute necessity" he argued in an 1899 minority report issued by his committee, that druggists have "a national law prohibiting the sale of impure and adulterated drugs and chemicals for medicinal uses." He could not "understand how anyone in the drug business who has the welfare of humanity at heart, no matter how well-informed he may be," could oppose "having an additional safeguard for the purity of the goods he handles." How sincere Queeny was in making this statement is hard to tell. After all, filling the Meyer Brothers catalog, he bought many questionable nostrums whose medicinal qualities were dubious at best, including cod liver oil concoctions billed as "brain food" for "brain workers" that could cure various ailments, from tuberculosis to "general debility." Nevertheless, there is no doubt he was a leading voice fighting for a pure food and drug law that he thought would bring a measure of respectability to his craft.[27]

To make his case, Queeny pointed to agriculture and the seed business. He noted that the US Department of Agriculture was "taking the laudable interest in seeing that farmers are able to obtain pure seeds of prime quantity." Why should drug companies not welcome similar regulatory oversight? The stakes were much higher in the drug business: "Seeds poor in quality, or adulterated, cause a loss only to farmers, while inferior drugs cause untold suffering and loss of life."[28]

The committee chairman faced fierce resistance from many association members. A Minneapolis druggist wrote Queeny, saying he could not support a drug law "as long as political bums and leg-pullers are the executive officers." Others made similar protests.[29]

But Queeny also had fans. One of his biggest boosters was none other than Harvey Wiley, the charismatic if controversial head of the USDA's Bureau of Chemistry and a rising celebrity in the battle for pure food. He applauded Queeny's efforts to expose the "fraud practiced in the drug trade." He urged Queeny and his fellow advocates to lobby Congress for change, which Queeny did, even testifying in favor of the Pure Food and Drug Act on the floor of the US House of Representatives.[30]

But that was in the future. Queeny, for the moment, remained in his post at Meyer Brothers, though his longing for a business and brand to call his own remained strong. At the turn of the twentieth century, he once again made a go for independence. And this time he picked a product that he thought could only be a winner: saccharin, an artificial sweetener.

* * * *

FIRST DISCOVERED IN 1879 by Constantin Fahlberg, a German graduate student studying in the United States, saccharin was a white powder 300 to 500 times sweeter than sugar. Chemically speaking, it was essentially benzoic sulfimide distilled from coal tar. In the 1880s, some believed that its synthesis from coal portended devilish harm to human health, but by 1893, when it appeared at Chicago's World Columbian Exposition, it was clear there was a big American market for it.[31]

Queeny knew what he was doing when it came to exploiting consumer markets. He had spent his whole life crisscrossing the country,
jotting down the names of new drug companies featured on billboards that whistled by his railcar. He knew that saccharin was a
hot item that could make him big money. At the time, a new soft
drink industry was booming. Dr Pepper (1885), Coca-Cola (1886),
and Pepsi-Cola (1898), among many other brands, became household names all across America, and each of these firms needed lots
of sugar to make its products. By the 1910s, Coca-Cola boasted that
it was the largest consumer of sugar on the planet, using some 100
million pounds of sugar per year. That was a lot of sweetener, and it
represented a huge cost to soft drink firms. If they could find a way to
reduce that expense, say by using domestically produced saccharin,
it would be an economic boon. The trick was not having to pay the
duty rate for saccharin. The Dingley Tariff of 1897 imposed a high
levy on both imported saccharin and sugar, but the saccharin impost
essentially doubled the cost of this chemical. By avoiding this duty,
Queeny's saccharin would be significantly cheaper as a sweetening
agent for soft drinks than tariff-protected sugar sold in the United
States, because even though the per-pound price for Monsanto's saccharin ($2.50 in 1902) was orders more expensive than sugar (which
cost just a few cents per pound), the fact that saccharin packed such
sweetening power meant buyers would still see savings when they
switched to the man-made chemical if it was sold at duty-free prices.[32]

The problem was, Queeny didn't know what he was doing when
it came to manufacturing products. He was not a chemist or even
an inventor and he had never actually made a chemical compound
before. In fact, throughout his life, he confessed his lack of scientific
training. In a 1924 US Senate hearing, Queeny quipped: "You are
getting into chemistry now, Senator, on which subject I am rather
weak." And even if he had the chemical training to launch his business, another problem was that a handful of German and European
companies—Merck among them—made and controlled the intermediate chemicals he needed to make a finished product.[33]

The good news was that he had a buyer. In 1901, Jacob Baur, owner of the Liquid Carbonic Company in Chicago, a well-known supplier of carbonic acid and other ingredients to the soft drink industry, said that he would lend Queeny $3,500 to help get a saccharin plant built in St. Louis. Queeny had supplied Baur over the years with Epsom salts, and they had developed a strong relationship. In addition to investing in Queeny's business, Baur also agreed to purchase Queeny's initial artificial sweetener output. This was going to be a big contract, some 8,000 pounds of saccharin per year sold at $2.50 a pound.[34]

If Queeny and Baur's deal worked out, saccharin from St. Louis would end up in thousands of soft drink bottles across the country, and most American consumers would not be the wiser. At that time, there were no labeling requirements forcing bottlers to explain what was in their products, so there was no way, other than a particularly well-trained tongue, for most people to know what sweetened their drink. What Queeny realized was that he did not necessarily have to convince the world that saccharin was a good thing; he just had to convince a few powerful businessmen to purchase his stuff, and that would change everything. Queeny focused on selling to industry, not consumers. It was a foundational sales strategy—one Monsanto eventually bragged about in its mid-century branding slogan, "Serving Industry . . . Which Serves Mankind."[35]

But who would actually make sweet powder for Queeny? He needed someone who knew how to run a chemical plant. He did not want to have another fiasco like the sulfur factory fire a few years back.

That was when he learned of Louis Veillon, a twenty-something Swiss chemist with a newly received PhD from Zurich Polytechnic who was working for the Sandoz Chemical Company of Basel, Switzerland. Sandoz, a firm Queeny had gotten to know as a Meyer Brothers' buyer, was one of the chief manufacturers outside of Germany that made the intermediate chemicals needed to synthesize saccharin. Queeny approached the firm's officials and asked them if they would be willing to lend him young talent to get his plant

started. In exchange, Queeny agreed to buy his ingredients from
the Swiss firm.[36]

Sandoz accepted the deal, and Veillon sailed for America, arriving
on New Year's Eve in 1900 at the St. Louis railroad terminal sport-
ing a fur coat and a Henri Quatre beard. According to Veillon, his
appearance disturbed Queeny, who wondered whether this young
man, who spoke only broken English, would "be ready to rollup [sic]
his shirt sleeves and get down to real work." But the next day, the
two were hard at it, scouring the city for the best place to build their
factory, and before the day was done, they decided to purchase space
in an old Diamond Match Company warehouse on South Second
Street that still had lots of matches stacked along its walls. Given
Queeny's previous run-ins with disastrous fires, one has to wonder
whether he considered the dangers of another conflagration devouring his dreams once again in this house of flames.[37]

By all accounts, this first factory was an extremely rudimentary operation. Queeny purchased almost everything—a boiler,
steam engine, pipes, filters, and wooden tanks—secondhand. The
most expensive piece of equipment was a centrifuge costing $1,000,
imported from Switzerland. Queeny's operation, like most other
businesses in the American chemical industry, was wholly dependent upon essential brainpower, machinery, and chemicals imported
from abroad.[38]

Queeny and Veillon were able to start making chemicals in a matter of months. It was February of 1902 when Veillon and Queeny
enjoyed their celebratory saccharin lunch at Faust's. Queeny must
have been proud standing in front of the Monsanto Chemical
Works, a firm with assets of nearly $14,000. This was his business,
his brand.[39]

Queeny's affection for Olga no doubt shaped his decision to
name his firm after his wife's maiden name, but there were likely
other, more practical considerations that influenced the choice as
well. Because Queeny still held his post at Meyer Brothers at the
time of incorporation, some thought it would look bad if he used

his surname for his startup. Over the next three years, Queeny shuf-fled back and forth between his desk at Meyer Brothers and his new plant on South Second Street, so to avoid confusion, some argued, it made sense to go with Monsanto instead of Queeny.[40]

Whatever the case may be, it is clear that the women in Queeny's life played an essential role in the early success of his chemical busi-ness, and not just in terms of branding. No one was as critical as Olga's mother, German-born Emma Cleeves, who served as a kind of interpreter, helping Veillon and Queeny work through details of their operation. Veillon was more comfortable in German, Emma's native tongue, than he was in English, a language he learned largely through a few courses in school and chit-chat with his Scottish brother-in-law. "There were a lot of technical things that Queeny wanted to know from me, which I know in German," Veillon said, but Queeny did not speak German, or any foreign language for that matter. Emma translated Veillon's comments to Queeny late into the evening at the family dining room table. Back at the Monsanto Chemical Works, men largely ran things. The first three people Veil-lon hired to help work his factory were all men, and as late as 1916, only 4 women out of 250 employees worked on the South Second Street shop floor, which was typical in the chemical industry as a whole. Nevertheless, women like Emma, often marginalized in com-pany histories, were critical to the success of Queeny's business.[41]

A celebratory plant christening in February 1902, in which Queeny handed out cigars and $2 bonuses to employees, was the high point in a year that otherwise ended dismally for Queeny. On April 5, Veillon jotted down in his diary, "Money is getting scarce," adding a note a few days later: "Money still scarcer, salary not payed [sic] in full." By the end of the year, the company was already roughly $1,500 in the red, and at the start of 1903, Veillon expressed doubts they would be able to keep their plant running: "The question arises of stopping manufacture of S. buying it from competitors."[42]

The problem was the Germans. Fearing a new entry into their market, the big German saccharin producers were putting the squeeze on Monsanto, drastically dropping wholesale prices from about $4.50 a pound to less than a dollar. Fiery Queeny, desperate not to fail again, did everything he could not to go bankrupt, liquidating his horse and carriage and mortgaging his life insurance to raise money. He was just trying to outlast the onslaught from abroad.[43]

Then, there was good news. "Whole output of S for 1903 sold in advance," Veillon wrote in his diary. Queeny had found a buyer—a big buyer. It was the Coca-Cola Company of Atlanta, Georgia. To this day, Coca-Cola makes no mention of the fact that it at one time sweetened its sacrosanct secret formula with saccharin in an effort to keep down costs. But the Coca-Cola deal was big news at Monsanto. With the Coke contract, Queeny was able to cut losses to just $63 by 1904.[44]

But the company was still not making a profit. Monsanto needed a new product that could bring in more revenue.

The answer was caffeine. While the saccharin struggle continued, Queeny looked to diversify and reached out to another Swiss chemist from Veillon's alma mater, Gaston DuBois, who had intimate knowledge of caffeine manufacture. Queeny had first met DuBois while touring chemical plants in Europe in 1903. Ralph Wright, the son of one of Queeny's wealthy friends, had recommended DuBois as a translator. Wright, whose father some dubbed the "financial angel" of Monsanto because of his early investments in the firm, had studied with DuBois in Zurich and knew he would be a good guide for Queeny. DuBois impressed Monsanto's founder, who soon offered him a salary of $75 a month to come to St. Louis, and soon thereafter, DuBois helped launch caffeine production at Monsanto.[45]

It was a fascinating process. At base, it involved treating "tea waste"—essentially tea stems and other "sweepings" from the tea trade—with a solvent, typically "refined gas oil," to extract out the small percentage of caffeine contained within. After Monsanto extracted caffeine, spent tea sweepings sometimes ended up as "fertilizer" in the gardens of employees (an early Monsanto contribution to agriculture), but much of it was bundled with chemical by-products and simply, as one Monsanto employee put it, "dumped in the river." From the beginning, then, the Mississippi River served as a kind of garbage disposal for the Monsanto enterprise.[46]

"Contract closed with Coca-Cola for 1905 output," Veillon cheerfully scribbled in his diary in July 1904. He may have been referring to another saccharin deal, but if he meant the new caffeine contract, he had reason to be cheerful. Once again, the Atlanta firm saved Monsanto, becoming its chief caffeine buyer. But Coke, always wary of talking about its ingredients and secret formula, made no public pronouncement of its partnership. In fact, years later, when Edgar Queeny recounted his firm's early history, his interviewer made a special note: "Don't stress Coca-Cola, says Queeny."[47]

But how could he not? The Coke contract literally pulled the firm out of the red and into the black, with Monsanto posting its first profit, $10,600, in 1905. That year, Monsanto made more than $63,000 from its caffeine sales, almost twice as much as it earned from saccharin, whose sales totaled a little over $33,000. Three years later, with caffeine selling for $4.10 per pound in Meyer Brothers listings, Coca-Cola agreed to buy all of Monsanto's caffeine production for the next three years. By 1910, Monsanto made a $1 profit on every pound of caffeine it sold.[48]

And there were other new products Monsanto hoped to turn into moneymakers. In December 1904, Veillon and DuBois finished a processing unit capable of synthesizing vanillin, a flavoring compound. It was a complicated process that involved importing cloves from the islands of Zanzibar in eastern Africa and extracting out an aromatic chemical associated with vanilla flavor. Naturally, all of this was expensive, and in June 1905, Veillon reported: "We are selling Vanillin at considerable loss, competition severe." Again, the Germans were to blame, working through American partners, such as American Condiments Co., to drop vanillin prices from $2.50 to $1.25 per pound.[49]

Queeny was frustrated. He wanted to be liberated from the stranglehold of Europe, and the only way to do that was to find a way to start synthesizing in-house the raw materials he needed. Importing tea waste from Asia or cloves from Zanzibar was costly and risky, and Queeny still had no means of producing the intermediates he needed for saccharin. It was time to find salvation through synthesis.

That is when he reached out to Jules Bebie, the third of the Swiss chemists that would later become known as the Swiss "Triumvirate" at Monsanto. Bebie had gone to school with Veillon when he was a teenager and served in the same mandatory military training program. He had also been classmates with the other two members of the Triumvirate at Zurich Polytechnic. But unlike DuBois and Veillon, Bebie had experience with producing saccharin "from the bottom up." When Veillon accepted the job to come work at Monsanto, Bebie had taken a post at Zimmerman Chemical Works in Brugg, Switzerland, in order to be closer to his mother, who was in ill-health. There he had learned the details of how to synthesize the intermediate chemicals needed to make saccharin.[50]

To get Bebie to come to the United States, Queeny had to pay him a hefty salary, $175 per month. "Dr. Bebie cable accepting our offer," Veillon wrote in his diary in July 1905, adding the lament, "salary . . . more [than] mine." It was the price Queeny felt he had to pay. He hoped Bebie would be the critical team member that could help him gain freedom from the European cartels, and in the coming months, Bebie worked hard to achieve that objective, building the machinery needed for bottom-up synthesis of saccharin. By 1906, Bebie successfully synthesized an essential compound needed to make saccharin that Monsanto once imported from overseas. At the same time, he began to manufacture phenacetin, an anti-inflammatory, pain-relieving drug popular at the time.[51]

In 1907, Queeny decided to leave his job at Meyer Brothers. Forty-eight years old, he was officially on his own. The year prior he had paid out the first dividend to investors, who now held more than 600 shares in his company. Having started with $5,000, Queeny now ran an enterprise with a net worth of $75,000. The firm was still heavily reliant on caffeine contracts with Coca-Cola, which one Monsanto chemist dubbed the company's "milk cow," and Veillon continued to write shaky notes in his diary in 1907—"Financial difficulties continuing, more capital needed"—but this was to be expected for a growing company making major investments. A few years later, a fire at the nearby Diamond Match Company warehouse—remarkably sparked by the collapse of a water tower serving the building's sprinkler system—brought flames close to Queeny's shop, but the fire was extinguished before it could do serious damage to the chemical factory. He was making so much money from his Coke contract that he purchased the Diamond Match property and expanded his business. This time fire fueled his commercial ambitions rather than quashed them. Queeny must have poured himself a celebratory drink.[52]

Then all hell broke loose.

CHAPTER 2

"A Coal-Tar War"

THE YEAR 1911 WAS A NIGHTMARE FOR QUEENY. FIRST THERE was Harvey Wiley, head of the USDA's Bureau of Chemistry who had once written to John Queeny thanking him for his support of the Pure Food and Drug Act. Now, using that very law, Wiley led the agency's recently created "poison squad" to ban the sale of saccharin in the United States. In 1911, he was largely successful, the USDA issuing Food Inspection Decision 135, which prohibited the interstate and international trade in "foods containing saccharin." Simultaneously, Wiley championed the 1911 *United States v. Forty Barrels and Twenty Kegs of Coca-Cola* case, in which the USDA argued that Coca-Cola was an adulterated product because, among other things, it included caffeine that was "added" to the beverage to induce addiction. Of course, much of the caffeine in question came from Monsanto, though Coca-Cola also sourced the white powder from other suppliers, including Schaefer Alkaloid Works of Maywood, New Jersey. Finally, there was talk of reducing tariffs on imported chemicals, meaning foreign competitors would now be able to offer US consumers their products at attractive prices, which meant heightened competition for Monsanto.[1]

These were dark clouds hanging over a young and vulnerable Monsanto. Queeny, now fifty-two, had worked so hard to get to this point. He had to fight back.

The Coca-Cola trial was a big deal because caffeine made Monsanto profitable. At that time, caffeine was by far the firm's best-selling product, earning nearly seven times as much revenue as saccharin sales. No other product the company sold—phenacetin, vanillin, a sedative called chloral hydrate—even came close to bringing in the kind of cash caffeine did. Queeny could not afford to lose this contract, which is why he headed to Chattanooga to testify in the Coke case in March.[2]

The trial was a spectacle. Harvey Wiley, now a national celebrity, the "crusading chemist," brought along his new bride, Anna Kelton, and the press followed them around, reporting on their "honeymoon," which included stops at Lookout Mountain and Chickamauga Civil War battlefield. Asa Candler, the now-famous baron of Coca-Cola, delivered passionate testimony asserting the healthfulness of his drink, which was now sold in every state in the country. Internationally renowned scientists testified about the effects of caffeine on the human body, with government witnesses reporting the death of frogs and rabbits given steady doses of caffeine. This was absurd, protested Coca-Cola's lawyers, claiming the caffeine content given to these lab animals was more than any human could ever put down from drinking Coca-Cola. But the rebuttals did little to quell newspapers seeking sensational news. Jarring headlines ran throughout the country—"The Caffeine in Eight Coca-Cola's Would Kill," "Dangerous 'Soft Drinks,'" "Coca-Cola Contains Caffeine, Deadly to Interior Organisms."[3]

And there were other riveting revelations. Observers learned about Coca-Cola's secret formula and its cocaine connection. In a remarkable day of testimony, the government made clear that around 1903, Coca-Cola had contracted with the Schaefer Alkaloid Works to remove a trace amount of cocaine, about "1.400 grain to the ounce," from the firm's signature drink. John S. Candler, Asa Candler's brother and counsel for Coca-Cola, explained that the firm had reached out to Dr. Louis Schaefer who created a "decocainized" version of Merchandise #5, a secret ingredient that consisted of coca leaf flavoring extract (sans cocaine after 1903) and kola nut powder (with trace amounts of caffeine). The press was amazed. Not

only was Coca-Cola admitting that its drink had once contained a
taboo narcotic, it was also confessing that its supposedly sacrosanct
secret formula had in fact been altered.[4]

But caffeine, not cocaine, was the main government target in
this Coke case. Government lawyers argued that caffeine was an
"added" ingredient mixed with the goal of stimulating addiction.
They trucked in medical professionals, including Nashville physician John Witherspoon, who said he treated people that drank
"eight, ten, fifteen or twenty" Coca-Cola's a day. They were like
"morphine habitués," Witherspoon testified, unable to control their
Coca-Cola cravings. And no one was safe. Memphis doctor Louis
Le Roy claimed on the stand that he too had become a Coca-Cola
addict, consuming "half a dozen or so bottles a day." He just could
not "leave it alone."[5]

In the end, neither Coca-Cola nor Queeny ever had to prove that
caffeine added to soft drinks was harmless because presiding judge
Edward T. Sanford never allowed the jury to take up this issue. In the
fourth week of trial, Coke's legal team approached the bench arguing that the case should be dismissed because the plaintiffs had never
proven that caffeine was actually an "added" ingredient in Coca-
Cola. The argument resonated with Sanford, who issued instructions to the jury to find in favor of Coca-Cola. In those instructions,
Sanford indirectly alluded to Queeny's testimony, saying, "A natural
article of food, for example, coffee, cannot be deemed adulterated,
even although the average cup contains a larger amount of caffeine
than an ordinary drink of Coca-Cola . . . since such caffeine . . . is
one of the essential ingredients naturally and normally entering into
its composition." So it was with the Atlanta soft drink: "Coca-Cola
without caffeine would not be Coca-Cola as it is known to the public . . . and if it were sold as 'Coca-Cola' without containing caffeine
the public buying it under this name would be in fact deceived."[6]

This was a pivotal victory for Monsanto in an era in which chemical companies were beginning to change the composition of the
foods Americans ate. Sanford had essentially naturalized a product

that was in fact made up of a host of secret ingredients, including saccharin, that consumers knew little about. There would be no investigation into potential health problems caused by caffeine. On April 6, the Coke case closed.[7]

But the government's legal team was not done. They filed appeals and took the *Forty Barrels* case all the way to the US Supreme Court, where, in 1916, Chief Justice Charles Evans Hughes gave the court's opinion overturning Sanford's ruling. The whole point of the Pure Food and Drug Act, Hughes wrote, was "to protect the public from lurking dangers caused by the introduction of harmful ingredients." Sanford's decision reduced this law "to an absurdity. Manufacturers would be free, for example, to put arsenic or strychnine . . . into compound articles of food, provided [they] were made according to formula and sold under some fanciful name which would be distinctive." Hughes remanded the case to the lower court charging that a jury should finally take up the issue of whether caffeine was in fact harmful to human health.[8]

But that case never happened. In 1918, Coca-Cola brokered a deal with the government to reduce the caffeine content of its beverage in return for a cessation of litigation on the matter. Once again, a decision made behind closed doors between powerful men transformed the soft drink industry, and American consumers were none the wiser. As far as Queeny was concerned, this was good news. Coca-Cola was growing so fast that the caffeine reduction would do little to hurt Monsanto's business. Queeny kept his caffeine plant humming, churning out white powder for Coca-Cola for decades to come.[9]

* * * *

THE CAFFEINE CRISIS WAS AVERTED, but two more battles loomed, including the troublesome saccharin ban. On this front, Queeny sought out one of the most powerful men in the world, President Theodore Roosevelt, to help him crush Wiley's crusade. "We understand that you have been a constant user of Saccharin for a number of years," he wrote President Roosevelt in July 1911, just a few months after the Coke case closed. Would he be willing to speak in favor of the substance? Four days later, Roosevelt wrote back, saying, "I do not care to have this letter published," before adding, "I always completely disagreed with Mr. Wiley about saccharin, both as to the label and as to its being deleterious. . . . I have used it myself for many years as a substitute for sugar in tea and coffee without feeling the slightest bad effects."[10]

In November, Queeny sent Monsanto's attorney, Warwick Hough, to Congress, along with Roosevelt's letter—the one the president had not wanted published. Hough showed those gathered the president's rejection of Wiley's findings and sought to quiet critics who pointed to Germany's restriction of saccharin as proof that it was an injurious substance. Hough explained that such restrictions had nothing to do with health concerns, but rather were the result of the German government's dependence on the beet sugar industry for tax revenues. And the same thing was going on in the United States. As Hough explained in his multiple appeals to the government, Food Inspection Decision 135 was merely the "machinations of the Sugar Trust."[11]

Here was one of the earliest instances in which Monsanto took a position regarding American agriculture. Back in the 1900s, farmers, specifically tariff-protected US sugar growers, were Monsanto's enemies. Queeny and Hough argued that these agrarian interests exploited American consumers, especially women who paid duty-elevated prices for sugar. There was an irony here because the tariff sugar growers fought for created the price differential that made Queeny's saccharin attractive to buyers. But that is not how Monsanto saw things. "The individual consumer," Warwick Hough argued, should have "the right and opportunity to make his choice" about how to sweeten food. That is what the chemical industry could offer: freedom of choice. Of course, consumers had never really made the conscious decision to switch to saccharin. Nor would they be made aware of the caffeine reduction in Coca-Cola in 1918. In both cases, businessmen made these decisions without them knowing.[12]

Monsanto did have some allies in the agricultural sector, especially in the tobacco fields of the American South because the 1911 restrictions exempted chewing tobacco sweetened with saccharin. The USDA's regulation focused mainly on foods and beverages and hinged on the argument that saccharin "injuriously" affected the "quality and strength" of a given food product. Because tobacco was not a food, it essentially fell outside the bounds of this law. Along with diabetic goods, tobacco remained a big market for Monsanto, and before and after World War I, tobacco farmers remained boosters for Queeny's business.[13]

Nevertheless, Queeny cast agricultural lobbyists associated with the "Sugar Trust" as a cabal committed to undermining American freedom. "Stand up for your rights," he told saccharin users in 1912. "No man with good red blood in his veins or any particle of the spirit of our fore fathers in his soul will quietly submit" to this "unlawful regulation . . . the only result of which would be to swell the coffers of an already gorged Trust." As Queeny explained, Monsanto chemicals were weapons in a war against monopoly power. The saccharin fight was a fight for liberty from "the sugar octopus." It was a battle akin to the contests that led to the Magna Carta, Queeny insisted.[14]

Monsanto—a company that, decades later, pitched chemicals as the best way to produce more food to feed a hungry world—crafted the exact opposite argument in the 1910s. "The people of this country are suffering more from an excess of food values than from the paucity of food values," Warwick Hough exclaimed in 1912. Saccharin, in other words, was a chemical corrective to a food system that had grown gluttonous because of subsidies and tariffs. Americans were already well fed; chemicals could help keep them lean. That was Monsanto's message in the Progressive Era.[15]

But try as he might to break the saccharin ban in the 1910s, Queeny was ultimately unsuccessful. He continued to battle for nearly a decade, spending hundreds of thousands of dollars in legal fees, before the USDA finally lifted its restrictions in 1925. The issue remained a constant nuisance, even as he diversified into other chemicals.[16]

And on top of all this, there was the tariff problem. In 1913, the Underwood Tariff did not offer adequate protections for Monsanto. Duties on imports remained low, part of a Democratic effort to reduce prices for goods coming from overseas.[17]

Without strong tariff protections, Queeny had to shut down his chloral hydrate process, first begun in 1908. Chloral hydrate was a popular sedative and a promising product line for Monsanto, but Queeny simply could not compete with German manufacturers without stiff duties shielding his business. He grumbled, but to no avail. Foreign competition came back in full force. The future did not look bright.[18]

❋   ❋   ❋   ❋

AND THEN, QUEENY LUCKED OUT. The assassination of Archduke Franz Ferdinand in June 1914 sent the world spiraling into war, and within months trade lines with Europe were severed. German chemical firms no longer had access to American markets.

Many times, Queeny had weathered the withering assaults of overseas oligopolies. Weak tariffs had done little to keep firms like Bayer and Merck from dominating smaller American chemical companies. War, it seemed, offered Queeny the opportunity he had long hoped for to find liberation from foreign foes.[19]

And yet, the Great War was also bad news for Monsanto because it meant that the firm no longer had a supply of the raw materials it needed to make many of its products. Despite the best efforts of Jules Bebie, Gaston DuBois, and Louis Veillon, the Swiss Triumvirate had never figured out how to make key chemicals from scratch at competitive prices. Yes, Bebie had succeeded in initiating a bottom-up process for making some essential compounds needed for saccharin synthesis at Monsanto in 1906, but within a few years the firm had turned back to importing basic chemical building blocks from foreign suppliers because it was cheaper to do so. By 1914, they were still almost wholly reliant on Europe for key inputs, which now seemed a world away as World War I commenced.[20]

"That was when research really started at Monsanto," Gaston DuBois said. The situation was serious. "We had practically no raw materials," DuBois remembers, the company even importing its glassware. The firm was going to have to spend a lot of money on new equipment and machinery in order to start making intermediates from domestically produced coal tar.[21]

Queeny really did not want to do that. He was the kind of guy who kept old mail to use as scrap paper so he could save a few cents. "What the hell are you doing with six brooms?" he once screamed at an employee who clearly did not understand Queeny's keen interest in thrift. It took the persistent pleading of DuBois, Veillon, and Bebie to move Queeny to invest, which he finally did, seeing no other options but to try and make the chemicals he needed from the bottom up.[22]

So began a hectic and fast-paced process to break Monsanto's dependence on Europe, once and for all. It involved a lot of crude tinkering in lieu of careful craftsmanship, a hurriedness that would soon result in human casualties. But Monsanto's chemists had no time to waste, even if they had little idea of how they were going to make this all happen. The Swiss Triumvirate took to the library, looking for any scrap of information they could find that could help them figure out how the Germans had made their chemicals from coal tar. Queeny charged the Europeans with sabotage, claiming that when they secured certain textbooks, "the leaves referring to coal-tar derivatives *had been cut out!*" It was a time of grueling work, Louis Veillon saying years later that he didn't take a single day off during the war.[23]

Grinding labor paid off. In 1915, after months of experiments, DuBois finally figured out a way to efficiently distill a critical intermediate needed for saccharin production—the compound *ortho*-toluenesulfonamide—from coal tar. This was a big deal. Despite the USDA saccharin restrictions, Monsanto continued to find markets for its product, especially as wartime sugar rationing forced companies and consumers the world over to find new ways to sweeten their foods and drinks. Saccharin prices exploded, reaching nearly $45 a pound by 1917. Throughout World War I, the company continued to sell saccharin in US markets, in part because it could service exempted industries—tobacco for example—and in part because the USDA "refrained from making a prosecution under its own regulation," as Queeny put it, except in a few select cases in specific jurisdictions. And beyond US borders, Monsanto found buyers overseas, including in China, where the company began selling saccharin in 1918. All told, between 1916 and 1919, Monsanto nearly tripled its saccharin sales.[24]

One down, but there were many more chemicals to go, perhaps the most important of which was phenacetin, a fever-reducing drug seen as the best line of defense against influenza. "Not one man in the United States knew how to make it" without importing intermediate compounds from Europe, Queeny claimed, and national stocks were perilously low. What would happen if the flu broke out in the middle of the war? For the first time, Monsanto claimed that its chemicals were critical to America's national security. By 1916, the Swiss chemists got to work building an apparatus that could convert coal tar derivatives into this drug.[25]

That's when workers started dying.

"They had literally bled to death without a wound," Queeny said when describing the fateful demise of three workers tasked with the phenacetin assignment. There was no exaggeration here. The blood cells in these men's bodies had literally "broken down into a watery serum." And there were more than twenty other men assigned to this processing line who were, according to Queeny, "in almost a dying condition." Monsanto's chemists looked on, "completely baffled" by what they were witnessing. This moment, Queeny said years later, "was the hardest thing I ever had to go through."[26]

The problem should have been easy to spot. Low-wage laborers assigned to the grunt work of handling compounds were literally awash in contaminants. As Queeny explained, "Chemicals would slop over containers and get on the men's shoes and clothing." These reports were common at Monsanto. There was the joke about Queeny's Black washerwoman, who often complained about chemical stains on clothes, and Louis Veillon recounted the story of a worker who turned a Turkish bath red when he rinsed off his body after work. Such was the state of affairs at a firm that was racing to redeem debts it had incurred in the process of trying to make all this work. There was no time to carefully control chemical contamination that spread well beyond the firm into baths and washbasins tended by people who never entered Monsanto's factory. Looking back, Queeny claimed that everything changed in this moment. "The remedy was easy," he said, "We installed shower baths and, when the men finished their work, they were bathed and treated under scrutiny of physicians, and they were provided with new underwear, socks, and shirts *every day*." Problem solved.[27]

But it wasn't that easy. Problems persisted because Queeny, ever concerned about costs, was slow to hire experts who could really understand how chemical exposures were affecting workers. As one Monsanto executive explained: "The technical staff did not include anyone identifiable as having training or experience in biology or chemical structure–biological activity relationships until 1927." In 1915, Queeny did hire a local physician to start monitoring worker health, but Dr. Demko, paid on a "retainer basis," knew little about the chemical compounds that were swishing, slopping, and splattering throughout the Monsanto facility. How could he? Even the chemists in charge were buried in library books trying to understand the basics.[28]

The harsh realities of life at Monsanto could be cruel for laborers, especially Black men. "Put the n—— outside," barked the head of Monsanto's medical department referring to a worker, now deceased, that had just been brought in from across the street. The scene astonished Bert Langreck, a Monsanto employee who came to the firm in 1917 and recounted what he had witnessed. That day, Langreck had been talking to the Black man, who was working as an electrician, when a blast knocked Langreck unconscious. Langreck came to and headed to the infirmary, where he saw the electrician's body being taken to the curb.[29]

As Langreck's story reveals, Monsanto's quest for freedom from foreign control depended upon the labor of those whose liberties were severely restricted in Progressive Era America. Black men and immigrant laborers did the dirty and dangerous jobs that kept Queeny's firm in business. "Most employees spoke not a word of English," explained one Monsanto executive, with workers hailing from Italy, Poland, Germany, Hungary, among other places. Queeny may have been seeking to break free from European companies, but European immigrants were nevertheless essential to his firm's early growth. A company chronicler noted that Queeny "didn't speak the languages necessary to be pals" with his workers. In 1916, Monsanto reported to the United States government that 40 percent of its almost-all-male labor force was "non-American."[30]

And with wartime labor shortages, the company also drew heavily, if reluctantly, on the Black community. Veillon said Monsanto had to hire Black workers, "and I was blamed for the reduction of real estate on Second Street [because Black workers] come in there where the white people had been living." The backlash Veillon described is not surprising. At this time, many Black workers were migrating from the American South to St. Louis in search of factory jobs during World War I. In response to the growing influx of Black people, white city residents passed the nation's first popular referendum codifying residential segregation based on race in 1916, and in 1917 white workers killed dozens, maybe even hundreds (the historical record is unclear), of Black citizens in the East St. Louis Massacre. This was a time, in other words, of tremendous racial strife in St. Louis. And yet, firms like Monsanto would never have survived were it not for the labor of those considered by white residents to be outcasts. These laborers' second-class status on South Second Street pulled them into the risky work of chemical creation.

In 1917, Monsanto was once again in expansion mode, purchasing the Commercial Acid Company of East St. Louis just across the Mississippi River. Queeny made the purchase in an effort to ensure Monsanto could make its own sulfuric acid, derived from sulfur ore mined from rich deposits further downstream in Louisiana. At this new factory, which became known as Plant B, Monsanto would also soon begin the manufacture of chlorine from salt brine brought in from the salt wells in Kansas and mines in Louisiana.[31]

Sulfur deposits, salt wells, coal mines—the strange thing about Monsanto's newfound independence was that it was reliant on a few, key raw materials buried deep in the ground. At the time, Queeny did not necessarily see it that way. Speaking of Monsanto's easy access to the coal-tar by-products of the LacLede Gas Company, operating in South St. Louis, he said that there was "an abundance of coal tar the chemical derivatives of which are numbered by the hundreds." A company publication noted that St. Louis "is particularly fortunate for geographical reasons" given the firm's close "proximity to the great Illinois coal fields." And as for salt and sulfur, those were right within reach of St. Louis as well. The future seemed limitless, and in a way, it was. Over the course of the next several decades, Monsanto and other chemical firms produced a dizzying array of chemicals from these resources, including new drugs and the building blocks of plastics.[32]

And, yet, at base, all of Monsanto's remarkable compounds stemmed from a handful of finite resources. And coal tar, more than any other substance, was the most valuable.[33]

During World War I, Monsanto claimed its commitment to coal was actually a commitment to environmental conservation. And there was some truth to this claim. After all, US companies producing coke that was used to make steel had, for years, simply treated the dark, gooey tar produced at their plants as a worthless waste product unworthy of reclamation. The chemical industry, Monsanto argued, could play a key role in recycling a resource that was simply being thrown away. In 1916, the firm estimated that only 25 percent of American coke ovens captured coal tar for reuse. This was unconscionable: "The result has been a shameful and needless waste of nature's stores that should be a cause of humiliation to any country." In classic conservationist speak of the time, company leaders urged the US government to force industry to make wise use of this by-product and ensure that "a large, natural resource will be conserved."[34]

It seemed coke companies got the message. By 1918, the US Tariff Commission reported that coal-tar reclamation had more than doubled since 1913. This stream of black fluid served as the lifeblood of the Monsanto empire for years to come.[35]

"The present European war has been referred to as a chemical war," a Monsanto man exclaimed, but "it could just as appropriately be spoken of as a coal-tar war." Without ample supplies of this stuff, there would be no American chemical industry: "Coal-tar is the junk pile of the chemical manufacturer." Here was a kind of scavenger capitalism, the nascent years of a new chemical-based economy, born of wartime necessity, that fed on the fossils of plants and animals long dead.[36]

✳   ✳   ✳   ✳

BY WAR'S END, MONSANTO, now straddling both sides of the Mississippi River, had branched out into synthesis of many chemicals. In addition to sulfuric acid, the company produced coumarin (1914), another vanilla flavoring compound; phthalic anhydride (1918), an intermediate for the production of a laxative; and phenol (1916), a versatile chemical that during the war was used to make germicides as well as explosives and years later would become a key constituent of early plastics. In 1917, Bayer's patent on acetylsalicylic acid, commonly known as aspirin, expired, and Monsanto quickly added the profitable chemical to its list. As with phenacetin, this drug became popular among Americans infected with the Spanish flu who were trying to reduce their fevers.[37]

In 1920, Monsanto stretched across the Atlantic, making its first capital investment outside the United States via purchase of R. Graesser Limited, a chemical firm located in Cefn Mawr near Ruabon, Wales. Queeny made the investment in Great Britain in part because he wanted to take advantage of tariff protections in the United Kingdom that insulated British companies from foreign competition. But there was another reason why Graesser looked so attractive: Ruabon sat adjacent to the rich Welsh coal fields of Denbighshire. Global conquest demanded access to this critical raw material needed for chemical manufacture.[38]

This international growth was expensive, and Queeny financed much of the expansion with bank loans, which proved problematic when the red-hot wartime economy cooled. In 1920, Queeny owed about $1.2 million to creditors and another $2 million to investors that had purchased corporate bonds.[39]

It was a terrible time to be in debt. The postwar recession was bad enough, but doubly disturbing was the renewed competition from German chemical companies, which now reopened their doors for business in the United States. In 1921, Monsanto posted a loss of $132,000 for the year. Paying back creditors seemed near impossible. Queeny was not doing well. "I know that at the end of the war he was a very bitterly disappointed man," Edgar Queeny said of his father. In 1922, John Queeny issued lamentations: "You and I are witnessing . . . one of the most unsettled and depressing periods that has ever thrown its blanket on the world. All posterity will point to the aftermath of the great War as a period when chaos ruled." For several months, Queeny had no choice but to shut down Plant B in East St. Louis, having no money to keep the plant running, and in 1921, he had to mothball an employee dental program he had just launched a few months earlier. He saw it as too costly. Then, in 1923, the firm's largest creditor, National City Bank, struck the biggest blow, demanding that a National representative take over leadership of the firm. These were some of the "darkest and most humiliating days for the company," recalled one firm official.[40]

But by 1924, Queeny regained his position as chairman of the company and effectively paid down his debts. It helped that Republicans finally passed a tariff in 1922 that provided strong duty protections for American manufacturers. But the other good news was that financial markets were strong. In the years ahead, Monsanto financed its growth not through bank loans but through bond sales as well as stock issuances. By the 1940s, it was clear that Monsanto's power came less from key inventions it made than through strategic acquisitions it financed through securities trading.[41]

These were good times, though Queeny would only have a few short years left to enjoy them. Now in his mid-sixties, Queeny was living a lavish life. Gray haired and approaching a hefty 220 pounds, this once frugal man allegedly enjoyed "meat and potatoes and three or four martinis every day for lunch." Intemperance extended to tobacco smoking—a Maradona Corona cigar never far from Queeny's bloodstone-bedecked fingers. All this portended poor health for this portly man, who suffered a stroke that left him temporarily speechless. But that was just part of the problem. In 1928 he was diagnosed with tongue cancer and knew he only had a few more years to live.[42]

It was time to hand over the company to his son, Edgar, a scion who would sell chemicals that still course through our veins.

CHAPTER 3

"A Die-Hard Admirer of the Tooth-and-Claw"

IT DIDN'T LOOK GOOD FROM THE COCKPIT. PILOT RALPH PIPER was about to barrel into billowing black clouds, a massive storm that blanketed western Missouri and eastern Nebraska, and the worst part was he had no radar. There were not many good options at this point. The Monsanto executives aboard had to get to a meeting out west meaning Piper was going to have to punch through the cold front blind, something he'd done many times before, but there were no guarantees they would make it out the other side. So he reduced his prop speed, hoped for the best, and entered into the dark beyond.[1]

At precisely that moment, Edgar Queeny, now at the helm of his father's company, barged into the forward cabin. There was a serious problem that needed to be addressed—right away. "Bob," Queeny queried, "where's the ice?" Bob Hinds, Piper's copilot, blushed, knowing he had flubbed badly. Standing over him was the most important man at Monsanto holding a martini shaker sans its necessary chilling cubes. Bob had forgotten to fill the ice machine before they had taken off.[2]

Edgar Queeny liked his gin martinis shaken, and he liked them cold. Ralph Piper, Monsanto's star corporate pilot, knew this well. For decades he flew the Monsanto scion on a plane called the *Prairie Wings* hither and yon as the firm took advantage of the aviation age to spread its commercial roots faster and farther than ever before. On these many journeys, the rattle of ice cubes often filled the fuselage.[3]

Fellow travelers recalled one flight where the firm's boss downed chilled gin blends all the way from Texas to the California coast, getting so inebriated that when he changed planes in Los Angeles, he got on the wrong aircraft. Redirected to Piper's *Prairie Wings*, he forgot to bring appropriate clothes with him, forcing him to meet and greet reporters in Santa Clara wearing his bedroom slippers. The experience disturbed Queeny, who confessed to Piper in the cockpit that he was "scared." "I don't remember a damn thing since we left Texas," he told his trusted pilot, promising to lay off the booze for a few weeks.[4]

"He scared the rest of us several times," admitted Piper. Public relations executive James McKee recalled one corporate party where he and others shuffled Queeny onto a freight elevator because they did not want the drunken company leader to be seen by party guests in a nearby passenger elevator. They had a brand image to protect. Queeny seemed to know he had to work at keeping his habit under wraps. "Next time I pose for photography," he told a close friend, "I'll be careful to hide my customary highball!"[5]

So when Queeny entered the cockpit that morning looking for ice, Piper knew he had to find a fix. "Just a minute, Mr. Queeny," piped Piper, "You go back and sit down and I'll call you in a little bit and I'll have some ice for you."[6]

A few moments later, turbulence rocked the plane up and down, and sleet and rain pelted the windshield. This was a bad time to have a martini problem, but Piper had a plan. He decided to do something unorthodox, choosing not to dispense a deicing solution from a tube that would have allowed him to keep his windshield clear. In a matter of minutes, ice began piling up on the windshield and also on the tube on the front of the plane.[7]

A half hour transpired before Piper saw the welcomed sign of blue skies beyond black clouds. The storm behind him, Piper opened his window and reached out to grab the ice on the alcohol tube. Sloughing it off, Piper was happy to find that the ice dislodged in a perfect cylindrical shape suited for a martini shaker.[8]

"Boss, come on up," a contented Piper called back to Queeny, who brought his mixer forward once again. Piper plopped the ice in the shaker, finding it fit the jar perfectly, and looked up at a grinning Queeny. "I've had martinis made in Alaska from the oldest ice—one of the glaciers up there," he happily exclaimed, adding, "Now, I've had martinis made with the newest ice ever made."[9]

✳   ✳   ✳   ✳

THIS MANIC MARTINI MOMENT says a lot about Edgar Queeny's high-flying, high-society lifestyle—one that was not necessarily predestined at birth. Like his father, Edgar Queeny came into the world right before an inferno sunk family savings. Cash-strapped and penny-wise, John Queeny sent his son to public school for his early education, unable or unwilling to invest in private schooling. But by his senior year, Monsanto sales booming during the Great War, Edgar transferred to the prestigious Pawling School, a boarding academy in the hilly countryside of the Hudson River valley. From there, he went off to Cornell University, where he became a dues-paying fraternity brother and a so-so student majoring in chemical engineering. (He once admitted that he often "ducked classes" in Ithaca and "passed calculus without an exam.") This young man, described by some as "strikingly handsome," drove a flashy car—a Stutz Bearcat—and played the piano, having been schooled by his talented mother, who many credit with helping Edgar achieve a certain patrician polish his father lacked. Two years into his Ivy League education, the United States entered World War I, and Queeny left to join the Navy.[10]

Victory secured, Edgar returned home, becoming Monsanto's first advertising manager in 1919. Like his father, he was moved by sales and marketing rather than the lab work that yielded company chemicals. Over the course of the next decade he took on various managerial positions, none of them related to scientific research, before his father tapped him, in 1928, to become president of the firm. He was just thirty years of age.[11]

Monsanto was now valued at more than $12 million with three chemical plants producing more than 100 chemicals. This young scion had what his father had only dreamed of so many years ago: money and power. And he relished being wealthy.[12]

Tall with dark hair slicked back revealing bushy black eyebrows, Edgar Queeny was the kind of guy who wore expensive socks, shaved twice a day, and enjoyed donning formal wear. In 1935 a "tailor's committee in New York" named Queeny "among the eleven best dressed men in the country," an honor he shared with dancer and Hollywood star Fred Astaire and Henry Ford's son Edsel. The accolade brought jests from friends such as Robert Woodruff, the boss of Coca-Cola, who told Queeny privately, "It must be hell to be both rich and handsome." In reply, Queeny feigned modesty, "I have always tried to go around without patches."[13]

Queeny was, in his own words, a "cold, granitic believer of the law of the jungle," as well as a "crusading idealist." He saw his success as a product of natural selection. He was a devout disciple of Adam Smith, a pen pal of libertarian and Ayn Rand associate Rose Wilder Lane, and a committed social Darwinist, listing the works of Herbert Spencer and William Graham Sumner among his best reads. He despised the New Deal and believed that what made America great was "individualism." As he put it, "The only difference that can be found between the American economic and spiritual climate and that of other nations of the world was the inalienable right of Americans of freedom to pursue happiness without interference from the state." For him, the key to America's greatness was unfettered competition made possible through the liberation of the individual so that he—not she—can "pursue his happiness in his own manner." "Nation's [sic] decay when the people go soft," he warned, "when the individuals lose their competitive spirit and self-reliance and take shelter in what Buckle calls 'the protective spirit' of government—'that mischievous spirit which weakens whatever it touches.'" It was classic Smithian logic: "If the architects of this new and better world could literally transform into reality freedom from want, it would be a tragedy. It would mean the end of ambition. We would be like a nation of contented cows grazing peacefully in an evergreen meadow."[14]

Xenophobia and racism peppered his writings regarding liberty and economy. Queeny did not believe that immigrants could "ever completely grasp the American concept of freedom," though when he referred to immigrants, he did not mean his Irish grandparents who came to America in the early nineteenth century. He was referring to "the jumbled mass of illiterate humanity and a babble of tongues that were strange to our shores—Polish, Italian, Russian, Greek, Hungarian, Yiddish, and Slovakian." He feared an America "with forty or more different nationalities, containing an even greater number of languages, religions, ideologies, and political institutions, and whose standards of living run the entire scale." Though immigrants literally built the business he ran, Queeny would cast them as an insidious force breeding a Communist insurgency within his capitalist homeland.[15]

This man, who some deemed a "die-hard admirer of the tooth-and-claw," was determined to turn Monsanto into a mega-corporation that could change the world. Legend has it that John Queeny feared his son's ambitions, warning others that Edgar wanted to "change everything," which would "ruin Monsanto." He was right about that first part.[16]

*　*　*　*

EDGAR QUEENY'S INITIAL PLAN was to get big by acquisition not by in-house innovation, and in 1929 he initiated a series of mergers, financed through exchange of capital stock, that changed the face of Monsanto forever. That year Monsanto bought the Rubber Services Laboratories based in Akron, Ohio, and Nitro, West Virginia, and then purchased the Merrimac Chemical Company of Boston, Massachusetts. The Akron and Nitro plants allowed the firm to become a major player in the production of accelerators, antioxidants, softeners, and other chemicals needed to treat rubber products. Naturally, this was a booming industry given the rise of mass-manufactured

automobiles. Henry Ford's popular Model T, once an expensive lux-
ury reserved for elites, was now, thanks to assembly-line production,
affordable for many middle-class Americans. And Alfred P. Sloan
Jr., president of General Motors, was churning out new designs for
his cars each year in order to entice consumers to sell old cars and
buy new fashions. In this commercial climate, selling rubber chemicals was big business.[17]

Merrimac, on the other hand, gave Monsanto new capabilities to
serve textile, paper, and leather-making industries in need of tan-
ning compounds, dyestuffs, and other chemicals. Monsanto had
come a long way from being primarily a fine chemicals firm focused
on food additives and sweeteners.[18]

Right after these acquisitions, the economy sank into the Great
Depression, but rather than retreat and recoil, Edgar Queeny became
more aggressive. Some chemical concerns were now deeply in debt,
and Monsanto moved to buy them.

Naturally, there was some debate between the old guard and
young blood about where the firm was headed. Gaston DuBois, now
in his fifties, sparred with Edgar Queeny, saying that chemical con-
solidations were a mere "fad" and not where Monsanto should spend
its energies. "It is but natural that I, as a technical man, should vision
the possibilities of growth from within more clearly than anyone not
so closely connected with technical development," DuBois wrote
Queeny in 1932. This was a not-so-subtle jab at Queeny's lack of lab
experience, but DuBois hoped his appeal could move his new boss to
sink money into internal research rather than outside acquisition.[19]

The relationship between them was frayed. Those who witnessed
debates between Queeny and DuBois in these years described the
battles as incredibly heated, even requiring, at times, John Queeny's
intervention. In the end, Edgar Queeny, headstrong and determined,
got his way. It turned out that many of the firm's board members,
including Charles Belknap of recently acquired Merrimac, had a go-
go attitude about acquisition. Given the green light, and in spite of
DuBois's reservations, Queeny set his sights on the next big takeover.[20]

At this time, Monsanto still lagged considerably behind many of America's leading chemical concerns. In 1932, DuPont, now having diversified into a variety of commodity chemicals, was on top, with more than $560 million in assets (including a substantial financial interest in General Motors). Union Carbide took the second spot with $304 million in assets, followed by Allied Chemical with $285 million in assets. Monsanto was way down the list, holding the eighth spot, behind American Cyanamide, Air Reduction Company, Dow, and Mathieson Chemical. Queeny's $17.6 million in assets was nothing compared to DuPont and Union Carbide, but it was still more than double the size of rival Westvaco and more than five times the size of New Jersey–based Heyden Chemicals. These were the types of firms Queeny wanted to hunt down.[21]

In 1933, Queeny approached deeply indebted Theodore Swann, proprietor of the Swann Chemical Company of Anniston, Alabama, about acquiring an ownership stake in the southern firm. Among other things, Swann specialized in the production of chemicals derived from phosphate rock mined in Florida and Tennessee. Here, millions of years ago, aquatic creatures had once roamed an inland sea. Now their phosphate-rich bones would seed a new chemical industry. Swann tapped these fossil reserves to make phosphoric acid used by a range of food and beverage companies, and there were dozens of other phosphate-based product lines that looked quite promising—including phosphate-based detergents, rust-proofing formulas for steel, and fire retardants. Monsanto completed the merger in 1935.[22]

The decision to get into phosphates had far-reaching implications for Monsanto. Some of the firm's later best-selling and most profitable products, among them Roundup and all® laundry detergent, ultimately came from elemental phosphorus purified from phosphate rock. And the immediate returns were thrilling. In the year of the Swann acquisition, Monsanto boasted that it produced more food-grade phosphoric acid than any other company in the country. Phosphoric acid was used in a variety of products. It provided the tangy flavor for soft drinks, helped preserve jams, and was used as a leavening agent in baked goods. By 1936, phosphate rock had become such a critical resource for the firm that Queeny spent more than $3 million to vertically integrate phosphate mining and processing operations just south of Nashville in a town called Columbia, Tennessee (later named Monsanto, Tennessee, because of the chemical company's large investment there, but then renamed Columbia in the 1960s, just a few years before the EPA became concerned about pollution caused by phosphate mining).[23]

Making elemental phosphorus from phosphate rock was hard, hot work. The company built several massive electric furnaces in Columbia capable of heating phosphate rock to temperatures over 2,700 degrees Fahrenheit, which enabled the separation of pure, elemental phosphorus from other impurities found in the mined ore. All that heat required tremendous amounts of electricity, estimated in 1958 to be enough to power the city of Memphis, a metropolis home to almost 500,000 people.[24]

The energy costs of producing elemental phosphorus from Tennessee mines would have been prohibitively expensive had it not been for President Roosevelt's New Deal investments in the region via the Tennessee Valley Authority (TVA). But this did not stop Queeny from complaining about the government agency. In 1935, he blamed the nation's "slow recovery" on New Deal programs like the TVA, which he believed "caused grave anxiety on the part of many earnest men." This was just the kind of big government program, he argued, that thwarted the competitive spirit of enterprising Americans. In the decades to come, Queeny issued an unrelenting attack against Roosevelt's recovery programs, even though the New Deal fired the furnaces that made Monsanto profits.[25]

Through its Phosphate Division, Monsanto had, once again, become dependent on a finite resource mined from the earth, and Tennessee soon proved inadequate for company demands. Within two decades, the firm went in search of new phosphate deposits in Idaho to keep its profitable product lines humming.[26]

All the while, the company was also tapping other underground resources to seed its future as it began experimenting with new petroleum-based product lines that emerged during the Great Depression. By 1934, the company created the Monsanto Petroleum Chemicals subsidiary and gloated about hundreds of petroleum-sourced chemicals "formerly derived from coal tar or from vegetable sources." Like its chemical cousins, Monsanto was beginning to tap into a new reserve of fossil remains to make its products. The next stage of scavenger capitalism had arrived.[27]

* * * *

PETROLEUM-BASED CHEMICALS CHANGED everything for American companies such as Dow, DuPont, Union Carbide, and Monsanto. For decades, they had been playing catch-up to their European counterparts, who had been pioneers in coal-tar chemistry. With easy access to petroleum gushing from new oil strikes in East Texas (1930), Oklahoma (1927), and California (1928, 1929), companies such as Dow, DuPont, and Monsanto became first movers in petrochemistry.[28]

DuPont for example became a pioneer in polymers and plastics, introducing Nylon, a synthetic fiber derived from oil, in 1938. This product—chemically consisting of a chain of carbon-based compounds—made the company billions and kickstarted a revolution in the apparel industry, which began replacing cotton and wool clothing with new synthetic alternatives. Nylon remained the firm's blockbuster product for decades to come, with synthetic fibers earning DuPont roughly 51 percent of its operating income by 1957. DuPont's success later pushed Monsanto to develop its own line of synthetic fibers through a subsidiary called Chemstrand. DuPont also became a major producer of polyethylene, a key compound used in polyethylene terephthalate, or PET, now used to make most plastic beverage bottles.[29]

Michigan-based Dow followed a similar path. The company began investing heavily in styrene monomers, another carbon-based compound sourced from fossil fuels, to make plastic polymers. In 1937, the firm made a major breakthrough in its styrene labs, creating a versatile insulating material it branded Styrofoam. The company also created dozens of chemicals for the oil and gas industries that aided in unclogging production wells.[30]

Union Carbide, a company formed from a merger of several successful chemical firms in 1917 and a pioneer in polyvinyl chloride, or PVC, plastics, went straight to the source, constructing a petrochemical factory in Texas City, Texas, so it could gain access to petrochemical feedstocks. From these materials, the company manufactured its own synthetic fibers, branded Dynal, as well as Glad Wrap plastic packaging, among a host of other products.[31]

Like Dow, DuPont, and Union Carbide, Monsanto also saw a future in petrochemical polymers. In 1938, the company purchased the Fiberloid Company, a pioneer in plastics manufacturing, and took on a half-interest in Shawinigan Resins, a firm known for producing exceptional safety glass. Both of these businesses were located in Springfield, Massachusetts, and in time they became the foundation of Monsanto's Plastics Division based in New England. By 1939, plastics had become the fourth most important sector for the firm, behind its top-yielding food, pharmaceutical, and glass products. Had John Queeny, who died of cancer in 1933, lived to see the pace of change under his son's leadership, he would have been astounded.[32]

Even though unemployment stood at 19 percent in 1938 as the country entered the "Roosevelt Recession," for wealthy men such as Edgar Queeny it was an optimistic time. "There are unlimited human wants yet to be served . . . [and] unlimited resources with which to meet these wants," Edgar Queeny exclaimed at the end of the decade. "Mention must be made," he said, "of the millions of years behind us in which the materials were evolved—the age-old vegetation, the minerals buried in the earth, and the chemicals that are made from them—out of which today our chemists are producing Plastics." Chemistry was a kind of conquest in which scientists "wrested from Nature's hidden stores, the secrets, yet concealed, which will, only a few years hence, dim today's miracles." "Man cannot create something from nothing," he concluded, "but nature's final products are for the most part useless to man in their native state." Company chemists would unlock a future Americans could not fully fathom.[33]

As Monsanto entered this new frontier, Queeny decided to yield to the advice of DuBois. He purchased the Thomas and Hochwalt Laboratories of Dayton, Ohio, and made it Monsanto's Central Research Department. Run by Charles A. Thomas and Carroll A. Hochwalt — two chemists who had worked with later-infamous General Motors scientist Thomas Midgley Jr. to create leaded gasoline in the 1920s — the Central Research Department became a kind of experimental station where Thomas and Hochwalt could play with chemicals, plastics, and resins in search of new discoveries that could yield profits.[34]

On top of all this, Queeny also added investments in foreign subsidiaries. He purchased the Southern Cross Chemical Company in Melbourne, Australia, set up a joint venture with Mallinckrodt—known as Monsanto Canada Limited in Montreal—and bought a coal-tar distillery in Sunderland, England, to secure raw materials for Monsanto's Ruabon, Wales, plant. By 1938, the firm valued its total assets at roughly $59 million.[35]

The riches made Queeny cocky—so cocky in fact that he felt he could reach out to big brother DuPont—whose assets totaled more than a hundred times that of Monsanto—to see if he could piggyback on the company's catchy slogan, "Better Things for Better Living Through Chemistry." "I would like to have your frank reaction," Queeny wrote president Lammot du Pont II, "to the suggestion that the chemical industry adopt your slogan in all their national advertising." DuPont's president was naturally taken aback, responding a few days later in a letter marked confidential: "We have spent a very large amount of money publicizing this slogan, and feel that, in a way, it is an investment on behalf of stockholders, which should not be given up." "Perhaps I am being too brutally frank," he said in conclusion, but "in a small way it is the old story of individual enterprise vs. Socialism."[36]

Queeny was aghast. This struck at the core of who he was: "I am certain you know me well enough that I need not assure you that any suggestion I make is not instigated by the thought of a 'free ride.'" He hoped Lammot would reconsider.[37]

He didn't.

Though DuPont and Monsanto never saw eye to eye on this point, the truth was both firms had a lot in common. Yes, DuPont dwarfed Monsanto in terms of sales, assets, plants, and employees, but at base, both companies' profits ran back to the same source: fossil remains buried in the ground. Many of DuPont's most profitable products—which included a robust plastics portfolio that included rayon fibers and cellophane wrapping, and various insecticides—ultimately came from coal, oil, and natural gas. The same was true of Dow, Union Carbide, and the other big players.[38]

Some in the chemical industry hinted at the dangers of this deep dependence on fossil fuels. For example, in 1939, Monsanto development director, Francis J. Curtis, laid out the facts: "Our civilization is very largely built on fossil remains, laid down for us in the earth by geological processes in past ages at extremely slow rates over long periods of time." The problem was these mined resources were "being used up in amounts that seem somewhat astronomical." Years from now, there was bound to be a reckoning. "What is going to happen to us and our civilization when the finite quantities of these materials are gone?" The chemical industry needed to start thinking about how it could be the solution: "One of the largest problems which science will have to solve is the gradual transfer of our civilization from dependence upon fossils to living on annual crops."[39]

But in 1939, breaking this deep dependency on resources buried deep in the ground seemed a distant dream. "Of course," Curtis claimed, "as long as petroleum is as plentiful and low priced as it is today, we will continue to use it." The day "must come" when man would free himself from this fossil-fuel economy, but this was a problem for people living years from now. "If oil does have a limit," *Monsanto Magazine* prophesied, "chemistry is pushing it farther and farther into the future."[40]

✳   ✳   ✳   ✳

THERE WERE BIGGER THINGS to worry about in 1939, as Adolf Hitler marched his forces into Poland in September. As was the case in World War I, critical supplies of natural resources were soon cut off. A new quest to synthesize compounds from American fossil fuels began in earnest.

A war to liberate the world from Nazism also wed the American economy, and Monsanto, more closely to chemical creations derived from creatures and plants long dead. One of the biggest areas of research was rubber, traditionally derived from tapped trees in Southeast Asia. Here was a resource that America's fighting men simply could not do without, described by Franklin Delano Roosevelt as "one of the most critical material shortages which this country must overcome to win the war."[41]

In 1941, the government's newly created Rubber Reserve Company contracted with Monsanto to begin manufacture of styrene, an essential carbon-based monomer needed for the production of synthetic rubber and the same compound Dow used to make Styrofoam. The plant, designed by Monsanto's Central Research Department and conveniently located near oil refineries in Texas City, Texas, became operational in March 1943, producing styrene that was then sent on to Goodyear Tire Company in Akron, Ohio, where it was mixed with butadiene distilled from petrochemical feedstocks processed by Union Carbide in West Virginia. The end product was used on the soles of soldiers boots and in Jeep tires that rolled into battle. After the war, styrene became an essential building block in the plastics industry, and Monsanto soon became one of the world's biggest manufacturers of this monomer.[42]

Relatively new chemicals, such as polychlorinated biphenyls (PCBs) became critical during wartime as well. First manufactured for commercial use in 1929 by Theodore Swann in Anniston, Alabama, PCBs consisted of two benzene rings (extracted from coal tar) surrounded by chlorine atoms. The main attribute of these compounds was their ability to withstand high temperatures, making them ideal fire retardants and insulating materials. When Monsanto bought the Swann Chemical Company in 1935, it became the sole producer of PCBs in the country and soon expanded production to Plant B in Illinois. The company's monopoly control over this chemical made Monsanto a lot of money. By the time World War II broke out, PCBs had already become essential to the electrical industry, especially as transformer insulation. A General Electric official put it well in 1937, saying that without PCBs, his company "might just as well have thrown our business to the four winds and said, 'We'll close up,' because there was no substitute and there is none today in spite of all the efforts we have made through our own research laboratories to find one." During World War II, that dependency grew even deeper. PCBs filled the electrical transformers that powered the Arsenal of Democracy, and the military pumped these chemicals into degaussing cables wrapped around warships and doused other war equipment with this fire-retardant material. In time, PCBs ended up in ink used to write confidential cables and even in mundane products such as shoe polish. It was literally everywhere.[43]

The firm took part in secret operations, most notably the Manhattan Project, when Lt. Gen. Leslie Groves Jr. tapped Charles A. Thomas of Monsanto's Central Research Department to complete studies in Dayton, Ohio, on polonium neutrons needed to trigger the atomic bomb. The top-secret operation caused chaos in the Ohio town, with 24-hour security run by armed guards, and workers flooding into the community with little idea of why the government had asked them to come to this place. The military commandeered public spaces, including the city playhouse, and installed spotlights near research buildings that blinded nearby neighbors, who complained about the nuisances.[44]

The precise details of the health hazards Monsanto employees faced behind closed doors is still unclear, given the classified nature of the work, but what is clear is that Monsanto's Dayton labs successfully produced enough polonium to serve as the initiators for the atomic bombs that ravaged Nagasaki and Hiroshima in 1945. That year J. Robert Oppenheimer, the Manhattan Project's scientific director stationed in Los Alamos, New Mexico, wrote a congratulatory letter to Charles A. Thomas just days after US planes dropped the atomic bombs on Japan. "I am perhaps in a position to know more intimately how decisive were the contributions that the workers of your company made and how helpless we should have been without their skill and devotion," he said. "We could not have made the bombs without their help." After the war, Monsanto scientists continued work on nuclear reactors for the newly created Atomic Energy Commission, establishing a new research facility, Mound Laboratory, in Miamisburg, Ohio (atop a historic Native American mound site), which continued in operation through the 1980s, at which point the Environmental Protection Agency declared the facility a toxic Superfund cleanup site.[45]

Cryptic correspondence remains in Monsanto corporate archives regarding the company's work for the US Chemical Warfare Service (CWS), an agency whose roots extended back to World War I. In typed notes about Monsanto's World War II involvement, put together in preparation for a corporate history of the firm, a researcher mentioned "an important research activity of the Company" that "was carried on under a secret contract with the Chemical Warfare Service, under which six of Monsanto's research staff worked on confidential problems." But, he noted, "This cannot be discussed in the book." Perhaps this was a reference to a CWS request, mentioned elsewhere in company records, that Monsanto "explore the possibilities of utilizing carbon monoxide" that was "liberated by the elemental phosphorus furnaces" in the "manufacture of phosgene war gas." Available archival documents reveal that Monsanto ultimately did produce phosgene for the US government, yielding nearly 50 tons a day of the war gas for the CWS by 1945.[46]

Old Monsanto products also found new life during World War II. Saccharin, the firm's first product, became popular once again as sugar rationing pushed consumers toward artificial sweeteners. By the end of the war, the company reported that saccharin sales had increased by 75 percent, with substantial revenue coming from overseas markets.[47]

Monsanto also cashed in on caffeine, in large part because the company's chief buyer, Coca-Cola, now had huge government contracts to service troops overseas (contracts, it should be noted, a very frustrated and enraged Pepsi-Cola did not get). By wars end, Coca-Cola sold nearly 10 billion soft drink servings to American troops, and in those bottles was a lot of Monsanto caffeine. Estimates at the end of the war suggest Coca-Cola's total caffeine demand approached nearly a million pounds per year.[48]

In 1945, Monsanto was doing big business. It now ranked fifth in terms of total net sales among all chemical concerns in the country. DuPont ($611 million) and Union Carbide ($549 million) still had a sizable lead over competitors, but Monsanto's $105 million posting fell right behind Dow Chemical's $124 million. Edgar Queeny despised those who claimed the firm was getting rich off military contracts—"Monsanto Chemical Company is not profiting from this war!" he cried in the 1943 shareholders report—but the truth was government investments had ballooned business operations.[49]

And Monsanto's foreign competitors had lost time. The fact that the petrochemical revolution took off just as the Axis blitzkrieg sent Europe into pandemonium meant that German firms such as Bayer, BASF, and Hoechst—now combined, temporarily, as a cartel called I. G. Farben—fell behind in petrochemical innovation. Farben produced numerous chemicals during the war, including Zyklon B used at Auschwitz and other concentration camps in the ghastly horror of the Holocaust. They also produced synthetic rubber and other chemicals that kept Hitler's war machines humming. But many of these chemical creations drew on coal-tar chemistry and older methodologies. World War II, in other words, was a game-changer for

American chemical companies with close access to large oil reserves, who became the world leaders in petrochemicals manufacture, the future of the global chemistry industry.[50]

* * * *

EDGAR, NOW IN HIS mid-forties and already claiming that he was "too old-fashioned for this age," had decided to take a step back, accepting the board chairmanship and ceding the presidency to Merrimac Division head and former naval commander Charles Belknap in 1943. That same year, Queeny finished writing a book, *The Spirit of Enterprise*, which offered a full-throated defense of unregulated American capitalism and made it for a short time on the *New York Tribune's* national best-sellers list. His longtime buddy, former president Herbert Hoover, heaped praise on the publication and soon requested that Edgar join him in writing for a new conservative paper that would defend the free enterprise system against the disciples of Karl Marx. "Most of my life," Hoover wrote Queeny, "I have been fighting the plague he spread to all mankind."[51]

But if Queeny was penning prose in defense of the virtues of capitalist competition, he was at other times showing signs of getting tired of the fight. He was ready, as he put it, for "a more human life," which included more time away from the madness of the city.[52]

After all, Queeny had long found wilderness restorative. He was an avid hunter, fisherman, and photographer, and he spent hours filming wildlife, even using an underwater camera—designed by his friend Walt Disney—to capture brown bears in pursuit of salmon in the cold river waters of Alaska (a place where he enjoyed those glacier-cooled martinis). He often retreated to Wingmead, his country estate in the rural prairie lands of Roe, Arkansas, an off-the-grid location that only had one "country telephone" shared among thirteen other nearby residents. In 1946, he worked on a book published by Ducks Unlimited that became a film called *Prairie Wings* documenting the rich beauty of fowl flight in his beloved Arkansas wetlands.[53]

Hunting was just part of the allure of Roe. Down south in Arkansas, Queeny liked to farm, planting rice fields and tending cattle. "I am a farmer," Queeny proclaimed to Senator John W. Bricker of Ohio, describing his agricultural operations in Roe, which he often waxed on about in voluminous correspondence to businessmen.[54]

Queeny was an environmental conservationist in the same way that President Teddy Roosevelt was a conservationist. He knew nature as a man on the hunt, taking part in African safaris, as Roosevelt did, in search of big game. His personal papers are filled with letters about big catches and bullet-riddled and dog-bitten birds.

Yet, like Roosevelt, there was something more here than a kind of passion for gut-and-gore blood sports. Queeny clearly appreciated the beauty of the natural world and, at times, lent his voice to fights for preservation. In the mid-1950s, for example, when the Bureau of Reclamation moved to dam much of the Colorado River, Queeny wrote to oppose these projects. "Really," he said, "it would be a great pity to destroy any of our national monuments," and he lent his support to congressional efforts to thwart such undertakings. Of course, part of his concern with the initiative was that it was a costly government enterprise, just the kind of thing he loathed. But his petition indicated his interest in protecting certain ecosystems from the march of a modern world—a world that his chemical firm was helping to usher in.[55]

One way Monsanto was remaking rural worlds was through the sale of pesticides and agricultural chemicals. Queeny kept a close eye on the firm's budding pesticides business that federal dollars had seeded during the war. By 1944, the firm was manufacturing DDT, the powerful insecticide that years later became one of the chief targets of environmentalist Rachel Carson. The military needed lots of this stuff, to kill both typhus-bearing lice and ticks in soldiers' barracks as well as malaria-infected mosquitos in the Pacific theater. Unlike with PCBs, Monsanto was only a minor player in the DDT field, but it created its own branded version, Santobane, in 1944, which the firm continued to produce until 1956, when it got out of the business—much to the chagrin of Edgar Queeny, who believed in the chemical and saw it work effectively on his own property in Arkansas.[56]

Though Monsanto only had a relatively brief flirtation with DDT, executives, including Queeny, learned that pesticides offered big growth opportunities. For Queeny, part of this passion for pesticides stemmed from personal use. He tested DDT on his farm, and in 1945 he told a fellow rice farmer in Arkansas, "If this stuff will increase rice yields several bushels per acre and kill mosquitos as well, it looks to me as though you rice farmers should let us chemists hunt free."[57]

But there were bigger reasons beyond Queeny's personal hobbies that explain why Monsanto began investing heavily in agriculture. During the New Deal, the federal government had pumped capital into rural America. The Agricultural Adjustment Act (AAA) offered farmers loans and price supports, and the Soil Conservation Service delivered money to growers willing to deploy techniques that would control erosion and preserve soil fertility. The Rural Electrification Administration also brought electricity to most American farms by the end of World War II. These federal programs, among others, enabled large landowners to purchase new tractors and combines— which had become technologically more sophisticated in the 1930s and 1940s. Farmers also used government money to mechanize their operations in other ways, which in turn reduced their demand for farmhands. Because postwar federal aid was often contingent upon growers implementing acreage controls, large landowners looked to new technologies and chemicals that could increase their per acreage productivity without increasing their labor demands.[58]

In the 1940s, farmers bought new high-yielding seed varieties developed by scientific breeders. To find the best seeds and machines, farmers turned to the land-grant universities, which after the passage of the Smith–Lever Act of 1914, had developed a far-reaching Cooperative Extension Service that sent agricultural scientists into the countryside to modernize farms. These extension agents looked at farms like factories, promoting high-input agriculture that focused on the use of chemical pesticides, heavy machinery, and synthetic fertilizers (which only became widely used after World War II when munitions factories using ammonia converted their operations to fertilizer production). Federal policy, in other words, was helping to spur the development of "agribusiness," a term that was not yet coined but which aptly captures the type of large-scale industrial farming promoted by the USDA and the extension service in the postwar period.[59]

This move toward factory farming was not just happening in the United States. During World War II, American agronomists were exporting high-input, capital-intensive agricultural techniques to other parts of the world. In 1943, Mexican president Manuel Ávila Camacho worked with the United States government and the Rockefeller Foundation to launch the Mexican Agricultural Program (MAP), an international research project focused on identifying high-yielding food crop varieties. Though some foundation officials and government scientists came to the MAP program with an interest in addressing rural poverty in Mexico, US national security concerns about global political instability brought on by an imbalance between food and population growth quickly became a central motivation driving program objectives. Over the next two decades, MAP would become the center of an international effort to spread high-yielding crops—and the chemicals and synthetic fertilizers they depended on—to developing nations around the world. This Green Revolution, as some would dub it years later, was still in its infancy in the 1940s, but signs pointed to exceptional growth opportunities on the horizon in overseas markets.[60]

It was in this context that Monsanto was expanding its agricultural products portfolio. In 1946, Monsanto hired its first entomologist and built a greenhouse in Webster Groves, Missouri, to begin screening other potential chemicals that could be used, as *Monsanto Magazine* put it, in a "war against harmful insects and other pests." By the end of the 1940s, the firm began selling a powerful (and highly toxic) insecticide, the organophosphate parathion, and branched into two herbicides known as 2,4-D and 2,4,5-T.[61]

Monsanto claimed that pesticides offered a "means of conserving many of our dwindling resources." "In the uneasy dawn after the most exhaustive war in history," warned *Monsanto Magazine* in 1948, "the United States is taking an inventory of its natural resources" and "finding that here and there its cupboard is alarmingly bare." "Must we continue to burn our forests, our houses and give our crops to rodents and insects?" Monsanto promised that "chemistry is pursuing tasks that are constantly extending our material horizons."[62]

Chemistry was conservation, although it was still unclear whether Monsanto's chemistry would have lasting effects on ecosystems so conserved. As late as the end of the 1930s, Monsanto chose not to fund an in-house toxicology lab. "The Phosphate Division has suggested that we consider setting up facilities to study toxicology and nutrition," read the meeting minutes from a research director's meeting in 1940. Creating such a lab had been "discussed several years ago by the Development Committee and turned down as impractical" because it could not be "justified on the basis of utility to the company as a whole." The rub was that paying for toxicology specialists, lab animals, and equipment would require a "huge investment," and such an expensive undertaking simply did not seem warranted given "the added prestige attached to data furnished by outside consultants who are well known in the field." In the end, the research directors felt "the arguments against this far out-weigh any benefits that may be gained."[63]

And yet, there was increasing evidence, produced in those very outside laboratories, that many of Monsanto's compounds were extremely dangerous. In the years immediately after World War I, industrial hygiene had become an established discipline at leading research centers across the country, with Harvard University leading the way. There, industrial hygienist Cecil Drinker had pioneered new animal laboratory experiments designed to test the toxicity of new chemical compounds, and in the mid-1930s he agreed to look into potential adverse health effects associated with Monsanto's popular PCBs. The request for this study came from the Halowax Corporation of New York, one of Monsanto's clients that specialized in production of PCB-based insulating materials. Halowax had become concerned about workers that had come down with debilitating skin lesions and other ailments and wanted to know if Monsanto's chlorinated compounds had anything to do with the outbreaks. Drinker said yes, reporting that his experiments "leave no doubt as to the possibility of systemic effects" associated with exposure to these chemicals.[64]

Monsanto protested the findings, arguing that it had established a medical department under the direction of physician R. Emmet Kelly and had seen nothing that particularly alarmed them in firm factories. Kelly took charge of the company's medical services in 1936, and shortly thereafter the firm began "periodic physical examinations" of its employees. Nevertheless, the company did not do its own toxicological laboratory work—it did not even have a toxicological lab—but Kelly insisted that there had never been "any systemic reactions at all in our men." He questioned Drinker's science, saying, "I don't believe that we can transpose the laboratory results into the actual humans without paying considerable attention to the volatility of different substances and the way they are being used."[65]

Kelly felt that most health problems he saw in Monsanto factories were made up by employees he called the "the disgruntled tenth." These were the "uncooperative and worthless workers, the troublemakers and the agitators." Kelly claimed this terrible tenth took up about 75 percent of all medical visits. It included "the man who does not get along with his foreman, or who is unreliable because of laziness or lack of aptitude." "Ill health may . . . result from conditions of our employment," Kelly admitted, but he said this was "exceptional."[66]

Given the way Kelly looked at Monsanto workers, it is not surprising that this medical director, who served the firm well into the 1970s, failed to see systemic health problems happening right in front of him. Deemed the deceitful cries of disaffected laborers, pleas of harm often went unheeded.

Laborers might have found liberation by quitting their jobs, but for many who worked Monsanto's chemical production lines, this was never really an option. After all, by locating its plants near the remote phosphate mines of Idaho or in West Virginia coal country, Monsanto found laborers who were willing to risk their lives to make company chemicals. Living in small towns known for dangerous work, these laborers often had few choices but to keep at their jobs. They depended on Monsanto to feed their families.

These workers' bodies had stories to tell—if only people would listen.

Monsanto's elemental phosphorus plant in Soda Springs, Idaho,
in 2016. The elemental phosphorus produced at this plant goes into
Monsanto's blockbuster herbicide, Roundup. The EPA designated
this facility a Superfund site in 1990. The EPA allowed Monsanto to
operate its facility, despite the fact that the plant retains its Superfund
listing and even though contaminants of concern still flowed from the
facility into the environment, according to the most recent EPA review
of this site that was available when this book went to press.

PART III

PLANTS

CHAPTER 4

"Wonderful Stuff, This 2,4,5-T!"

IT WAS A FRIGHTENING DAY AT THE NITRO, WEST VIRGINIA, plant on March 8, 1949. James Ray Boggess, a high school dropout then in his early twenties and just a few years into his tenure as a Monsanto truck driver, was in the worker locker room when he heard a loud boom outside. He rushed out into the plant yard and looked up into the sky and saw a gray, odorous "mushroom" cloud billowing up some 150 feet from Building 41. In just a few minutes, "dark powder" began to rain down on Boggess and the other 120-plus workers on site that day. Within hours, Ivan McClanahan, a thirty-two-year-old pipefitter, felt sick to his stomach and started complaining of headaches. Other laborers were similarly laid low. In the days that followed, Boggess soon started noticing bumps that covered his face, back, and legs.[1]

This was just the beginning. For months, Boggess' body broke out with pimples and pustules. He went to Monsanto doctors, but they told him it was just regular acne and would clear up over time. The breakouts did not subside. Ultimately, things got so bad that Monsanto paid to have Boggess' face "peeled" off.[2]

The procedure involved treating Boggess' skin with different solvents that killed the top layer of his epidermis, leaving behind a raw and sensitive new layer. Boggess was often left to finish the job, stripping his skin from his face with a washcloth. "Oh, God, you looked like a mummy," he lamented years later. Day after day, layer after layer was peeled off. "I believe it was approximately five layers of skin," Boggess recalled: "I was young and thought I was tough," but "I told the doctor then, hey, I don't know how much more of this I can stand." After treatment, when Boggess came close to chemicals with his exposed face, he said the pain was "unbearable."[3]

Boggess was not alone. Chester A. Jeffers was thirty-one years old and working as a pipefitter at the Nitro plant when the explosion occurred in 1949. "I seen it blow up," he testified in a deposition years later. Jeffers graduated from Nitro High School in 1935, but, like Boggess, did not go to college. Instead, he went to work for Monsanto. After the 1949 explosion, Jeffers candidly admitted that he was "scared" to go to work, and he had good reason. Jeffers's father had died in the Nitro plant after he was struck by a dislodged piece of lumber while he was working on a chemical processing tank. Jeffers also suffered from what the Nitro workers called "weed bumps," the mysterious skin condition that had so riddled Boggess' body. Monsanto physicians treated Jeffers by popping the painful pustules with methods akin to those deployed by a pubescent teenager—"he would squeeze them out." Jeffers said the doctors told him that his "weed bumps" were "nothing but blackheads" and would go away with time, though this did not come to pass. For years, Jeffers woke up in the morning to find his sheets covered in yellow stains, a common occurrence among Nitro workers leaching fluid from their acne-riddled bodies. All told, more than 200 Nitro employees came down with this debilitating dermatological disease over the course of two decades.[4]

Like many Monsanto men that witnessed the explosion that day, Boggess did not know that the bumps on his body were signs of exposure to a molecule that the Dow Chemical Company later claimed was the "most toxic compound they have ever experienced." Boggess and his fellow workers also did not know that this compound could be found in every batch of herbicide they were sending out of their company factory. Neither did gardeners across America that were spraying this product on their plants.[5]

* * * *

NITRO SITS ALONG THE muddy Kanawha River just fifteen miles down Interstate 64 from Charleston, the capital of West Virginia. A drive past the town today offers a jarring blend of industrial edifices and natural wonders: smokestacks and freight yards sit buried below the shadows of lush and verdant hills that often emit "smoke" of their own, water vapor escaping to the atmosphere. The town is part of what has become known as Chemical Valley, a stretch along the Kanawha that extends just beyond Charleston to the east. It got this moniker in the mid-twentieth century, after Monsanto, Union Carbide, and over a dozen other chemical firms set up shop in the area. Back then, you could smell the industry's scent in the stagnant air of the valley floor.[6]

Long before America's chemical giants came to town, the federal government established a facility for munitions production during World War I that ultimately gave Nitro its explosive name. The government plant produced nitrocellulose, an essential raw material needed in the manufacture of "smokeless" gunpowder. Nitro seemed like nature's perfect arsenal because the proposed site had access to an extensive rail network, and it also sat along a waterway that connected with the grand Ohio River less than fifty miles westward. In addition, coal deposits and natural gas reserves were nearby, ensuring necessary raw material demands could be easily met. Considering these natural advantages, Nitro, officially founded in December of 1917, became one of the top sites for munitions manufacture during World War I.[7]

After the war, the Rubber Services Laboratories of Akron, Ohio, built a facility in Nitro, which is what ultimately brought Monsanto to town. In the spring of 1929, Monsanto acquired Rubber Services, part of that ambitious series of acquisitions spearheaded by company president Edgar Queeny starting at the beginning of the Great Depression. Throughout the 1930s and into the 1940s, Nitro became a central hub for Monsanto's rubber business and a site for agricultural chemical manufacture.[8]

At the time, Monsanto was getting bigger through major mergers, but its strength increasingly came from small towns just like this, where it could experiment with new products that it hoped would bring in profits. One of those chemicals was a powerful herbicide, 2,4,5-trichlorophenoxyacetic acid, or 2,4,5-T for short.

2,4,5-T was one of a new class of chlorinated herbicides that garnered the attention of the US military during World War II. Beginning in 1942, army officers had set up a chemical warfare experimental station at Fort Detrick, Maryland, and by 1944, researchers were looking into the possible use of 2,4,5-T (and another compound, 2,4-D) as defoliants. As the war entered its final years, American soldiers were having to trudge through dense tropical forests as they hopped through the jungle-covered islands of the Pacific en route to the Japanese mainland. A herbicide that could help soldiers cut through the tangle of underbrush could be a major asset to the military as it sought to liberate the world from the grip of the Axis powers.[9]

2,4,5-T and 2,4-D were lab-created plant growth hormones known as synthetic auxins. In the early 1940s, researchers in Great Britain and the United States simultaneously discovered how to synthesize these chemicals that had the ability to kill plants. We now know this is because large concentrations of 2,4,5-T and 2,4-D essentially stimulate cancerous growth of plant cells. But at the time of discovery, the fine details of how this all worked were still being worked out. As late as 1961, the *Washington Post* reported: "How does [2,4-D] kill plants? The answer to this question has not been found, although many theories have been proposed."[10]

The US Army never got the chance to deploy these herbicides in combat during World War II, but news of these powerful compounds soon spread to the private sector, which mobilized to supply postwar civilian demand. The US Department of Agriculture approved the use of 2,4-D and 2,4,5-T in the domestic market in 1945. The agency swiftly transitioned from regulator to promoter, demonstrating the potency of 2,4-D for the public by spraying it on dandelions on the National Mall in Washington, DC, that same year.[11]

The chemicals worked with almost supernatural effectiveness. American Chemical Paint Company, a Philadelphia firm that sold 2,4-D and 2,4,5-T mixtures in the 1950s under the brand name "Weedone," described its new herbicide "as a modern miracle weed-killer." The company said that its product killed "over 100 weeds and woody plants," while having no effect on grass. This was the key selling point. Suburban homeowners could wipe out dandelions, poison ivy, and other plant pests while preserving their green lawns. And the best part, American Chemical boasted, was it was "harmless to people and pets."[12]

Emboldened by the USDA's approval, the Dow Chemical Company entered the synthetic auxins business in 1945. Monsanto also invested in these chemicals, starting manufacture of 2,4-D in St. Louis around the same time as Dow. In 1948, Monsanto began 2,4,5-T production at its Nitro facility. Potential markets for these chemicals seemed endless. In its 1949 annual report, Monsanto noted that "in addition to their applications on lawns and farms, these products are being used on the sugar and pineapple plantations of Hawaii and Puerto Rico." This note came under the heading, "Agricultural Chemicals," the first such use of this product-line label in Monsanto's annual reports. Monsanto's long journey to becoming a major player in the farm business had just begun.[13]

In 1948, the US Forest Service began experimenting with 2,4-D and 2,4,5-T in Oregon's Siuslaw National Forest in an effort to effectively destroy unwanted underbrush. Huge forest acreage now became a potential market for Monsanto's new herbicides. So too did terrain under powerlines and along railroads. Public officials doused these environments with synthetic auxins in the 1950s. And of course, there were the Levittowns of America and other burgeoning postwar suburban havens, complete with gardens and lawns in need of weed-killing chemicals. By the early 1950s, garden geeks published testimonials extolling the virtues of synthetic auxins. One in the *Daily Boston Globe* read, "Wonderful Stuff, This 2,4,5-T!" The *Los Angeles Times* featured "The Garden Doctor," who urged people to use 2,4,5-T and 2,4-D to eliminate weeds. Tellingly, the doctor noted that people should wear gloves when ripping up poison ivy but offered no such precautionary statement preceding mention of the synthetic herbicides.[14]

Forests, gardens, lawns, and fields. These chemicals could go everywhere, and they did. Long before Vietnam became the target of Agent Orange spraying, America used these herbicides to fight a war against weeds at home. As a result, Monsanto was primed to make big profits with its new "agricultural services" business.

That's when Building 41 exploded.

* * * *

A SAFETY VALVE BURST, unable to contain the pressure building up in pipes used to distill sodium trichlorophenol, a key ingredient needed to make 2,4,5-T. Chemicals spewed upward in the atmosphere. James Ray Boggess, the truck driver who would eventually have his face peeled, remembered it vividly. The pollution cloud lingered overhead for several minutes before it came down on him. "It was kind of slow" and had "a terrible odor."[15]

Slow, he said, which was how some of the side effects came on. For some there were headaches, others nausea, but for many the tell-tale sign of trouble was chloracne, the medical term for the nasty dermatological condition Boggess had been dealing with since the explosion. Thirty-six-year-old Harold Young, lead engineer in Building 41, got it bad too, his skin turning a "grayish-brown" just weeks after the incident. Boggess' efforts to get rid of his chloracne by peeling off layers of skin ultimately came to an end when a doctor noted that if he continued, his nerve endings were going to be exposed. "Forget that part," exclaimed Boggess, "I don't want to do that." He was tired of looking like a "zombie."[16]

Amazingly, Boggess went back to work in January of 1950, just a few months after he started getting treatment, but troubles persisted. The "weed bumps" on his legs and back simply would not go away, and his face remained "blood red" and burned. He went to see the plant nurse, who used "a needle that had a big hole in it" to "squeeze those little worm-like" globs of pus out of his bulging bumps. Sometimes she would "stick a knife in" his pimples, Boggess said, "and if they was infected they'd pop out and you could smell it in the next room."[17]

Monsanto's management knew it had a big problem. In the immediate wake of the 1949 explosion, Boggess was one of more than a hundred men whose bodies were riddled with such signs of toxic pollution. Unlike chemical poisonings that worked quietly beneath workers' skin, the costs of contamination were highly visible.[18]

Months after the 1949 blast, the company called in physicians to examine workers with chloracne, first turning to the retired chief of the Division of Dermatology at the US Public Health Service, Dr. Louis Schwartz. Two years into retirement, Schwartz was a legend in his field and a respected authority on dermatological diseases. After examining several employees, he reported that Nitro's workers might be facing serious health issues beyond chloracne. He said the Nitro facility needed to be thoroughly cleaned, and new worker-protection programs should be put in place. According to the testimony of many Monsanto employees years later, this exchange was never made public to workers in Nitro.[19]

In December 1949, Monsanto reached out to physicians Dr. Raymond Suskind and Dr. William F. Ashe of the Kettering Laboratory at the University of Cincinnati in Ohio. The company sent four chloracne-plagued men for examination: Ival McClanahan, Paul Willard, Jesse Steele, and Jonathan Hurley. Suskind and Ashe admitted that they were not sure what chemical was causing workers to break out, but they knew it had to be something associated with the residue of the 1949 2,4,5-T explosion.[20]

Suskind and Ashe's summary assessment of each worker's health was disturbing. McClanahan, thirty-two years old at the time, seemed all out of sorts: "At present the patient complains of a weak feeling, fatigue upon mild to moderate exertion, insomnia, and 'nervousness.'" Paul Willard, thirty-six, was not much better. He too complained of "insomnia and nervousness" as well as debilitating muscle pain that got so bad he couldn't walk for several days. Jesse Steele and Jonathan Hurley both complained of fatigue, insomnia, and nervousness. Hurley, fifty-six, had, like Boggess, used "peeling lotions" to get rid of his chloracne. He said his anxiety worsened to the point that he "found himself unwilling to talk to people and cried frequently." The smell of his pustules almost certainly exacerbated his newly acquired agoraphobia. Ashe and Suskind described it thus: "When these men are in a closed room together, there is a strong odor which suggests a phenolic compound. . . . While we have been unable to prove it, we believe these men are excreting a foreign chemical through their skins."[21]

Ashe and Suskind concluded in their 1949 study—a study Nitro laborers later said they never saw—that McClanahan, Hurley, Willard, and Steele were "suffering from systemic intoxication from a common agent arising out of their employment." "This intoxication," they added, "is characterized by acneform skin lesions, hepatitis, disturbed lipid metabolism, peripheral neuropathy and probably mild central nervous system involvement." They "hoped that" these men "would be able to resume work in the near future in an atmosphere free . . . of phenol chlorinated and nitrated aromatic compounds and other primary irritants." The best solution, they concluded, would be to assign sick workers to "clean outdoor work" where they "will probably have the least trouble."[22]

By this time, Nitro employees were already deeply suspicious about what was going on at the plant, and many turned for support to a well-known ally of West Virginia laborers: the United Mine Workers (UMW). This was common practice in a state with a long history of union organizing. In 1890, coal miners founded the UMW in Columbus, Ohio, just 150 miles west of Nitro, and by the 1940s it had a strong presence in West Virginia. At that time, the UMW had roughly 600,000 dues-paying members nationwide and had gone beyond the coal mines, recruiting workers from a host of other industries, including chemical manufacturing.[23]

Unionized Nitro workers appealed to the West Virginia Workmen's Compensation Commission to hear testimony regarding their health problems. They utilized West Virginia's Workmen's Compensation Act of 1913, which allowed injured laborers to seek compensation for injuries sustained while on the job. Many states passed these workers' compensation laws in the 1910s, urged on by both businesses and labor unions that saw various benefits in the new codes. Industry believed that workers' compensation would forestall growing labor unrest as well as limit corporate costs associated with tort litigation. Labor advocates also supported this legislation because many workers prior to these compensation plans simply were not getting justice in the courts. With labor and big business backing, not to mention the support of former president Teddy Roosevelt, workers' compensation bills passed nationwide. At the beginning of April in 1911, there were no such laws on the books in any state, but by 1925, only five southern states lacked such legislation.[24]

While the workers' compensation case was pending, the UMW reached out to the West Virginia Public Health Department to look into the chloracne problem. State health officials met with Monsanto's Nitro plant manager and walked away from the meeting with sound assurances that the facility was decontaminated. Monsanto's stance at this time was that workers' illnesses came from an isolated event, the 1949 explosion. It maintained that company cleanup efforts would ensure that problems would not persist. Nevertheless, the state public health officials called for toxicological experiments on animals using chemicals found at the Nitro plant. These studies, completed in 1950, determined that chemicals associated with 2,4,5-T manufacture were indeed chloracnegenic in laboratory animals.[25]

In April of 1950, the company asked Ashe and Suskind to complete a follow-up examination of the Nitro workers. In their report, Ashe and Suskind concluded that the original four men exposed to the 1949 blast had "markedly improved" in the six months since the accident. McClanahan's leg pains had "gradually subsided," and some of his chloracne had cleared up. The story was reportedly much the same for Willard, Steele, and Hurley, all showing improved health (though some ailments persisted). Monsanto used this as further evidence to support its claim that the chloracne outbreak resulted from a one-time contamination incident in 1949.[26]

And yet, the company knew it had a problem. In 1953, Monsanto's safety director, Ed Volz, wrote to the US Public Health Service acknowledging that Monsanto "scientists have discovered that it is not as simple to control the toxic agents" at the plant "as one might think." Lawyers that would later bring suit against Monsanto for its practices in Nitro described this as a "startling admission." In 1953, a Raymond Suskind study that was designed for company use and never published in a scientific journal found ninety-five chloracne cases "believed to have originated from the regular operations in the production of 2,4,5-T." In other words, Monsanto health professionals had evidence in 1953 that the breakout that occurred right after the 1949 explosion was not a one-time occurrence. The problem was not contained.[27]

And no one was safe. Suskind revealed in another internal company document marked "protected material" that "medical personnel such as nurses and assistants who treated the workmen . . . developed symptoms of chloracne." Even civilians beyond the confines of Monsanto's facility found themselves in harm's way: "Several wives who were never in the plant area developed acne at the same time as their employee husbands." In 1953, Monsanto assured health officials, saying that the company made adjustments to production and implemented additional safety protections (including giving workers an additional change of clothes), but plant personnel continued to come down with chloracne, even as Monsanto expanded 2,4,5-T manufacture.[28]

By this point, some young men tried to make an escape. In his 1953 report, Dr. Raymond Suskind revealed that seventeen workers afflicted with chloracne had left their positions at the chemical plant in Nitro.[29]

James Ray Boggess was among those who decided to leave. In 1950, he turned to one of the only other sources of employment poor West Virginians like him had: the US Marine Reserves. He served a short stint during the Korean War, participating in training operations down in Cuba, but in 1952, he returned home and once again was looking for work. He went back to Monsanto, the company he first drove trucks for prior to war.[30]

* * * *

BOGGESS WAS LIKE MANY Nitro natives. He was born into a poor family right before the Great Depression and grew up in the years when Monsanto was producing rubber chemicals for the military. His father was a truck driver and was, as his son put it, "down and crippled," which made it difficult for him to secure a high-paying job. Boggess dropped out of high school at the age of seventeen and went to work to try and help out the family. He went in with his father's trucking service—Boggess Transfer—and from 1945 to 1949 he and his dad hauled supplies and waste to and from the Nitro facility.[31]

Monsanto employee Gene Thomas followed a similar path. Reflecting on the reasons he stayed at the Nitro plant so long despite the persistent health problems he faced, Thomas told reporters in the 1980s that the paycheck was a big draw. When he came to the firm, he was eighteen years old and looking for work. Year after year, he suffered chloracne and his health problems got "worse and worse," but he said he "just had to hang on, because" he "just couldn't walk out of there and find a job." "I had a wife and two kids," he added. This was the labor pool from which Monsanto drew its Nitro workers.[32]

Monsanto could hire young people because the work Boggess and others were asked to do was quite rudimentary. Reports from the 1950s spoke of workers bending their bodies into chemical centrifuges and using handheld chisels to remove residue from big vats.

Scraping out chemicals from containers, shuffling barrels across the industrial grounds, and shoveling waste into dumpsters: this was how Monsanto's workers made 2,4,5-T.[33]

Nitro management used high wages to entice men such as Boggess to take on dirty work at the chemical plant. It is unclear how much Boggess was paid in the 1950s, but union contract records reveal that other workers in unskilled positions enjoyed healthy wages. For example, Paul Willard, the chloracne-riddled subject in Suskind and Ashe's 1950 study, earned $2.18 an hour as a "Yard Labor Leader" in 1956. This was more than double the federal minimum wage of $1.00 that year.[34]

✳ ✳ ✳ ✳

MONSANTO COULD AFFORD TO pay these salaries because business was booming. Though its $23 million in net income in the early 1950s was less than a sixth of DuPont's posting, the company saw its profits increase by 346 percent between 1943 and 1952. Over that same period, sales jumped from nearly $83 million to more than $266 million. The firm employed 19,367 people by 1952 (roughly 1,000 less than competitor Dow), up from 10,359 a decade earlier.[35]

Petroleum fueled this explosive growth. The majority of Monsanto's products—Chemstrand synthetic fibers, polyethylene plastics, and new chemical pesticides—came from fossil fuels. Major oil companies had long supplied Monsanto with the feedstocks it needed, but now many of the petroleum firms, Shell and Standard Oil Company of New Jersey among them, were ramping up investment in their own commodity chemical divisions, forcing Monsanto, which had scavenged on by-products produced by the fossil fuel industry for so long, to begin strategizing about ways to gain access to raw materials. In 1955, the firm purchased Lion Oil Company, entering into the petroleum and natural gas extraction business directly.[36]

At the same time, agricultural chemicals were becoming a central part of Monsanto's mission. Having watched the rapid sales growth for herbicides such as 2,4,5-T and 2,4-D, the company was now anxious for more. In the early 1950s, the company conducted a survey and determined that food, pharmaceuticals, and agricultural chemicals were the three major areas where Monsanto could make money. Digesting the data, Edgar Queeny chose not to pursue drugs, with one person close to the decision noting that Queeny "did not like the low cash flow" in pharmaceuticals. Queeny was instead drawn to pesticides and herbicides, in part because of his own personal interest in farming. Company executives made initial investments in fertilizer production, especially after the Lion acquisition, which gave them ample supplies of ammonia, but ultimately determined that the market was "plagued with over-capacity" and not a profitable path forward. Company managers also believed that if the firm invested early in the "non-fertilizer" portion of the ag market it could become "dominant." In 1951, Monsanto hired a plant pathologist to begin intensive exploration for new herbicides. It would take nine years for Monsanto to officially create the firm's Agricultural Division, but the seeds of that organizational transformation had been planted.[37]

Charles A. Thomas, famed atomic scientist who led the Manhattan Project's top-secret polonium program in Dayton, Ohio, now ran Monsanto as company president. Bald and in his early fifties, Thomas had first come to Monsanto when Queeny purchased Thomas and Hochwalt Laboratories back in 1936. He had grown up on a farm in Kentucky and now owned, like Queeny, a country estate outside St. Louis, which some deemed a "proving ground for many of his ideas on agricultural chemistry." He was a big proponent of investing research dollars into pesticides and by the mid-1950s oversaw the development of two new Monsanto chemicals, Vegadex and Randox, both herbicides capable of killing annual grasses and some broadleaf weeds. Randox was used before planting and after harvest on fields devoted to corn, soybeans, and other food crops while Vegadex, as its name implies, was used to clear farms before planting vegetables. Though both of these products were later determined to be "toxic to fish and wildlife" and bore labels warning farmers of health risks, American growers used these two chemicals for decades, though they were later replaced by more powerful (and supposedly more safe) chemicals in the latter quarter of the twentieth century.[38]

Many farmers wanted products like Randox in the 1950s because these herbicides had gotten rid of laborers that once would have managed weeds. A Monsanto scientist stated this explicitly when speaking of company herbicides in the Eisenhower years: "Farm labor was shrinking," he said, "A chemical which could be applied at planting time killing the weeds . . . but allowing the crop to grow normally, was a fabulous new tool for the farmer." The statistics backed up what this scientist was saying. Farmhands were leaving the countryside in large numbers. Between 1950 and 1970 the farm-labor pool in the United States declined by 50 percent. This massive demographic shift meant that growers that stayed on farms became more reliant on chemical inputs to manage their fields.[39]

In the 1950s, much of Monsanto's agricultural research took place at greenhouses near the firm's new headquarters in the St. Louis sub-urb of Creve Coeur, Missouri. Monsanto executives had pushed the relocation in 1951 because they felt that the firm had outgrown its central offices at the John F. Queeny Plant on South Second Street in downtown St. Louis. Queeny, Thomas, and others wanted to cre-ate a kind of collaborative beehive outside the city core where Mon-santo's top minds could work together to solve problems. The office park he envisioned included covered walkways connecting different laboratories and offices to one another.[40]

Monsanto moved to the countryside just as thousands of other white citizens fled St. Louis for the suburbs. The year 1950 was the first year that the Gateway City's population declined, and it would continue to do so for the rest of the century. White middle-class employees—the vast majority of Monsanto office workers, chemists, and engineers—sought to escape a downtown corridor whose Black population had steadily increased in the 1940s, in part because of the draw of wartime employment opportunities. In the 1950s, Mon-santo's middle-class employees utilized new interstate highways and headed westward to St. Louis County, where the Black population stood at just 1.9 percent in 1950. Creve Couer was in St. Louis County, and there homeowners approved stringent restrictive covenants designed to keep Black residents, even wealthy doctors, from buying homes in their community. In an automobile age, Monsanto's suburban employees could certainly have driven downtown for work, but as a Monsanto study pointed out in 1950, "offices . . . have been requiring use of a disproportionate amount of land for parking space" that simply was not available in the middle of the city. The geographic realities of a new hyper-segregated society divided along racial lines shaped the future of Monsanto.[41]

At the Creve Coeur campus, completed in 1957, landscapers remade nature to match an emerging 1950s suburban environmental aesthetic. Company reports on building progress consistently noted ongoing "beautification" programs designed to create a "park-like" atmosphere. Construction updates also highlighted the sowing of grass seeds to create lawns, green emblems of American middle-class affluence. On this corporate campus, nature was as Monsanto would have it: controlled, structured, and designed by man.[42]

A profound corporate transformation was under way: Monsanto was beginning to segregate the dirtier parts of its business from a verdant corporate core. Those who visited its Creve Coeur head-quarters saw an ultramodern company embedded in a pristine environment, the suburban sublime.

* * * *

BUT BACK IN NITRO, Boggess kept at Monsanto's dirty work. In the fall of 1954, after several attempts to avoid jobs in the buildings on site associated with 2,4,5-T manufacture, he was called into the main office and given an ultimatum. According to Boggess, plant superintendent Charlie Smith said, "You can either work in 2,4,5-T or you can take the gate." At the time, Monsanto was facing a serious labor problem. Workers simply did not want to go into a place that seemed to leave people riddled with skin problems.[43]

Boggess made the plant superintendent a proposition. "Charlie," he said, "will you guarantee me that I won't break out? Look, I don't want no more technical skin treatments. I don't want them to take any more skin off my face and get down to the nerve. That's terrible treatment." "No," Charlie said. He must have known others would fill the spot. As Boggess explained, "They went on down the line until they got two people" who "decided to stay."[44]

Monsanto plant managers could draw on a pool of young coal country workers who were willing to sacrifice their bodies for employment. Yes, more than a dozen employees left the plant after the chloracne outbreaks began, but many more stayed on to work. Others, like Boggess, were boomerang laborers, abandoning their post, only to return a few months later to the only work they had ever really known.

For men such as Boggess, the Nitro gig was more than just a job. Boggess and many of his fellow workers nurtured a sense of community at the Monsanto plant. After retirement, Boggess reminisced in a company magazine called the *Nitrometer* about his many years of work at the West Virginia facility, saying, "I'll remember the good old days when everybody worked together salary and hourly both. I'd like to see it return that way."[45]

And Monsanto administrators made clear that if workers challenged the status quo, there would be dire consequences. In January of 1954, a Nitro plant official sent a notice to workers explaining that management "was now faced with the problem of staffing" the 2,4,5-T department. He could only get thirteen out of the needed twenty-six people to work in the 2,4,5-T buildings. So he put pressure on his entire workforce: "A layoff will occur on or after Friday, January 15, 1954, unless an additional 13 employees are obtained to operate the 2,4,5-T Department." He tried to calm workers' fears. "I would like to point out," he told Nitro employees, "that our experience, as far as new chloracne cases are concerned, have been excellent since the new plant went into operation in January of 1953. We feel that

reasonable precautions on the part of the employees in this department from the point of view of cleanliness . . . will eliminate chances
of chloracne." "Cleanliness" was of course a fuzzy term. The company recommended things like frequent overall changes for people
suffering from chloracne and urged workers to wash regularly to limit
potential exposure to toxic compounds. Management messaging that
these measures would prevent chloracne may well have calmed some
workers' fears because, in the end, the plant manager got his thirteen
men. Boggess was not one of them, but even he would return to Monsanto just a few short months after being laid off. He needed a job.[46]

Despite Monsanto's claims of "cleanliness," workers were returning to sites of contamination. In 1954, Dr. Suskind and his Kettering Laboratory issued another report—again, funded by Monsanto
to be used at the discretion of the company—stating quite clearly
"that new cases arise currently in 2,4,5-T synthesis despite careful
reorganization of the process and institution of rigid hygienic precautions," indicating that "acnegens are still being involved and that
potential hazard is still present." A year later, in a confidential corporate memorandum, a Monsanto plant manager echoed Suskind's
finding, warning, "Chloracne remains as the outstanding problems
of the Nitro plant in dealing with hourly personnel."[47]

From 1954 to 1957, scientific evidence continued to show that
2,4,5-T production caused serious health problems. In 1954, Dow
Chemical published a report showing that various animals exposed
to modest concentrations of 2,4,5-T died. Chloracne outbreaks were
happening at 2,4,5-T manufacturing plants around the country
and around the world. In 1956, more than fifty laborers working
in 2,4,5-T manufacture at the Diamond Alkali Company plant in
Newark, New Jersey, came down with skin disorders. That same
year, French researchers examined seventeen men involved in 2,4,5-
T manufacture that had chloracne and noted that they effused an
"*intense odeur chloreé*." In Germany, scientists tested chemicals
found in a BASF plant where dozens of workers had come down

with chloracne. Initial research revealed that whatever toxic chemical was in the BASF plant, it was incredibly potent.[48]

And then, in 1957, the culprit was finally discovered. A Hamburg doctor, following up on earlier research done at the BASF plant, isolated dioxin as the contaminant in 2,4,5-T manufacture that was causing chloracne. The route to discovery was dangerous. The Hamburg doctor experimented on himself, testing various chemicals on his skin until he found the specific molecule capable of inducing pathological change. BASF allegedly reached out to Monsanto to tell the firm about its findings, but company officials later claimed that there were no records indicating receipt of BASF's communication. Dow Chemical executives acknowledged receiving this letter in 1957, but the St. Louis Post-Dispatch reported that firm managers claimed to have "misfiled" the correspondence.[49]

Despite the evidence that was emerging in the 1950s, the West Virginia Workmen's Compensation Commission decided that some 200 Monsanto employees were not eligible for compensation for long-term exposure to toxic chemicals. The commission's decision hinged on experimental evidence offered by Monsanto's paid researcher on the matter, Dr. Suskind. In his 1956 testimony to the commission, Suskind reported that he treated the skin of a dozen Nitro workers with chemicals used in 2,4,5-T manufacture and determined that these individuals did not experience maladies beyond chloracne. As the St. Louis Post-Dispatch revealed, however, Monsanto officials later admitted that these volunteers "were exposed to much lower levels of dioxin than workers at Nitro." Nevertheless, the commission was swayed by Monsanto's argument that dioxin contamination during regular operations had not yielded health problems warranting compensation from the firm. Dozens of men such as Boggess would therefore not find relief from the Workmen's Compensation Commission. This was the first time, but not the last, where men exposed to dioxin would be denied financial reimbursement for pain and suffering due to chemical intoxication.[50]

* * * *

IN OCTOBER OF 1956, the local Nitro labor union signed a deal with Monsanto that provided a four-cent bump to hourly pay for personnel willing to work with 2,4,5-T. Monsanto was now using bonuses to entice laborers to do dirty jobs.[51]

"If sufficient manpower is not obtained," an amended agreement read, "Men in the reverse order of their layoff will be contacted in accordance to plant seniority and given first consideration for the vacancy. If the senior man refuses, the next man will be contacted." This process would be "followed down the entire list of men laid off until the vacancies are filled." In short, if Boggess balked, there would always be someone else to take his place. He had very little bargaining power to push for change.[52]

It seemed there would be no stopping 2,4,5-T production in Nitro. Even the worker's biggest advocate, the local union, was accepting conditions at Monsanto's factory. And state regulators backed down, convinced by Monsanto that there was nothing insidious going on at the company's facility. The problem in 1950s West Virginia was not the lack of government-oversight or worker-advocacy groups. Federal and state bureaucracies were in place to monitor worker health. But Monsanto, armed with some of the earliest health studies on 2,4,5-T exposure, selectively released data to public officials to put them at ease. The company had become an information gatekeeper in the critical debates about this chemical's toxicity in the 1950s. The problem, in other words, was regulatory agencies' substantial reliance on corporately financed studies that proved problematic.

If the Nitro plant had been shut down and its problems exposed to the public in the late 1950s, the toxic hazards of 2,4,5-T might never have been exported to Vietnam. There, half a world away, hundreds of thousands of American soldiers and Vietnamese citizens would soon come to know dioxin's dangers.

CHAPTER 5

"So You See, I Am Prepared to Argue on Either Side"

THE WHIR OF THE PROPELLERS OF THE C-123 "PROVIDER" would have drowned out the chaos of war less than 150 feet below. If it were like most missions at the time, the pilots would have maintained standard delivery altitude for this twin-propeller plane as it approached military targets in the lush, triple-canopy Vietnamese jungle. This was 1962 when C-123s were executing the first assaults of Operation Ranch Hand, the popular name for a US military chemical-spraying operation initially dubbed Operation Hades. The latter name was fitting because these pilots were about to turn tropical forest into hell on earth. Their weapon: Agent Purple, a chemical defoliant consisting of 2,4,5-T mixed with 2,4-D and so-named because of the colored stripe on the herbicide container in which it came.[1]

President John F. Kennedy gave the okay to use this herbicide because he was frustrated. Since the 1950s, American military advisers had been on the ground in Vietnam working to repel Communist insurgencies that threatened to topple South Vietnamese president Ngô Đình Diệm's regime. The Red Tide had been surging. In 1954, the Vietnamese revolutionary leader, Hồ Chí Minh, helped to oust French colonists from North Vietnam, and at the Geneva Convention of that year, Vietnam was divided along the 17th parallel. Over the course of the next eight years, it became clear that Hồ Chí Minh was falling deeper into the Communist sphere of China and the Soviet Union. He became committed to uniting Vietnam under the sickle-and-hammer banner. For "Uncle Hồ," vanquishing Diệm's foreign-backed government was the final chapter in a centuries-long battle to rid Vietnam of alien invaders. For a Cold War warrior like President Kennedy, Hồ Chí Minh's move was a clear and present danger to America's national security that he simply could not abide. Sides had been drawn.[2]

But with Diệm, President Kennedy had made an uncomfortable bedfellow. Known for autocratic governance and brutal repression of Buddhist political opponents, Diệm's regime was hopelessly corrupt, and Kennedy knew it. Nevertheless, the young US president, like his predecessor Dwight D. Eisenhower, felt that he had to do everything he could to keep the North Vietnamese and their guerrilla allies, the Việt Cộng, from taking control in the south. The stakes could not have been higher. If Vietnam fell, so might Laos, Cambodia, and other countries in Southeast Asia. Like dominoes, Eisenhower had argued.[3]

The problem was Kennedy did not want to commit large numbers of ground troops to the region. At the start of 1962, Kennedy was just a year into his first term as president and mired in a Cold War face-off with Soviet premier Nikita Khrushchev. He was trying to build political capital, and to do that he needed to find a way to maximize the US military advantage in Vietnam without having to put too many American soldiers in harm's way.[4]

Agent Purple, and later Agent Orange (which largely replaced Agent Purple in 1965 and consisted of equal parts 2,4,5-T and 2,4-D), seemed a handy solution to this problem. Even with the help of American military advisers, Diệm's Army of the Republic of Vietnam (ARVN) was having a tough time suppressing Việt Cộng forces that seemed to appear and then disappear into the green understory of the lush forests surrounding villages and towns. The US military's powerful herbicides promised to eliminate the Communist enemies' critical natural ally, which would allow US-backed ARVN forces to engage in targeted aerial assaults on guerrilla insurgents. This was the kind of limited war the White House wanted. The Pentagon signed agreements with chemical companies to ramp up herbicide production.[5]

Monsanto played a pivotal role in Operation Ranch Hand. Though a half-dozen chemical companies, including Dow, Diamond Shamrock, and Hercules, produced Agent Orange for the US military, Monsanto was the largest supplier by volume, delivering 29.5 percent of the Agent Orange demanded by the US military over the course of the 1960s.[6]

It seemed like a win–win. Monsanto stood to make millions selling its chemicals to the US military, and the Kennedy administration in turn had a new powerful weapon that could help it root out Communist enemies in Vietnamese jungles. The only problem was that each barrel of Agent Orange rolled onto American air bases contained a dangerous chemical contaminant—the same contaminant that had forced James Ray Boggess to peel off his face a decade earlier.

* * * *

KENNEDY HAD REASON TO be concerned about the environmental and human health costs of Operation Ranch Hand. It was the dawn of the age of ecology. Marine biologist Rachel Carson's bestselling *Silent Spring*, serialized in the *New Yorker* and published in September 1962, had rocked the chemical industry, exposing the widespread environmental and human health costs associated with pesticides such as DDT and other herbicides, including 2,4-D and 2,4,5-T. Monsanto had tried to discredit Carson, publishing a rebuttal in its company magazine that was later rebroadcast by leading national newspapers, claiming that a world without insecticides would be a world ravaged by worms, flies, and fungi. In Monsanto's apocalyptic vision, "Insect and weed raced each other for strawberry patch, garden plot and field of grain." A world without chemicals was a "desolate" world indeed.[7]

But the chemical industry could not silence Carson's clarion call, which reached President Kennedy, who invited the fifty-five-year-old scientist, then dying from cancer, to Washington. Here was a strange situation. Just as Kennedy was giving the go-ahead to drop Agent Purple in Vietnam, he was allying himself with the author of a book that some claimed was becoming the *Uncle Tom's Cabin* of a growing anti-pesticide movement. By the summer of 1962, Kennedy convened a scientific advisory panel to come up with recommendations on how to regulate and manage pesticide use in the United States. In May the following year, that panel concluded that the USDA, the Food and Drug Administration (FDA), and the Department of Health, Education, and Welfare all needed to reassess the potential environmental and human health threats posed by synthetic herbicides and insecticides.[8]

Spurred by *Silent Spring* and this presidential report, Senator Abraham Ribicoff of Connecticut convened a hearing to discuss changes to the Federal Insecticide, Fungicide, and Rodenticide Act of 1947 (FIFRA). Since its enactment, FIFRA tasked the USDA with registering new pesticides and approving appropriate labels that would ensure safe use of those chemicals. For many years, the USDA focused mainly on the efficacy of pesticides—testing whether they truly killed insects and weeds—and conducted only limited human health and ecological analysis. In 1954, USDA began sharing more responsibility for pesticide oversight with the FDA when amendments to the Federal Food, Drug, and Cosmetic Act empowered that agency to set acceptable tolerance levels for pesticide residue found in food. Still, when Ribicoff's hearing met in 1963, it was clear that neither the FDA nor the USDA had adequate funding or infrastructure in place to do the deep analysis needed to make sure pesticides used in the United States were safe. As a result, in 1964, Congress amended FIFRA, increased the USDA budget for pesticide regulation, and broadened agency authority to deny registration for new chemicals.[9]

Yet, none of this halted the use of 2,4,5-T either at home or overseas. By 1965, millions of gallons of 2,4,5-T and 2,4-D mixes rained down on fields and forests in Vietnam and America.

Agent Orange used overseas was not your garden-variety herbicide. The military mixture was spiked with exceptionally high quantities of 2,4,5-T. *New Yorker* reporter Thomas Whiteside estimated that the 2,4,5-T in Agent Orange "averaged *thirteen times* the recommended concentrations used in the United States." The US military was leaving nothing to chance. During the Vietnam conflict, US forces sprayed other weaponized herbicides, including Agent Blue (a mix of cacodylic acid and sodium cacodylate) and Agent White (2,4-D and picloram), but Agent Orange releases far surpassed that of all other herbicides used during the conflict.[10]

Estimates vary on just how much 2,4,5-T ended up in Vietnam, but the best studies suggest Agent Orange releases totaled more than 12 million gallons. The breadth of impact was astounding. Targets included not only jungle hideouts of the Việt Cộng but also cropland suspected of feeding Communist forces. Pilots completed more than 19,900 herbicide sorties, flying as far as Cambodia and Laos. In all, US-led aerial herbicide campaigns between 1962 and 1971 covered more than 10,000 square miles, or nearly the land area of Massachusetts or roughly 8 percent of Vietnam. The majority of these herbicides contained 2,4,5-T.[11]

Nguyễn Thị Hồng remembered this time well. She was just sixteen years old when she and her family joined what she called the "resistance" against American forces in South Vietnam's Đồng Nai province. In 1964, Ms. Hồng and her fellow comrades in arms were deep in the jungle when they noticed what they thought was "mist" enveloping them. Soon, she said, plants all around them started dying. "It was like there was not one leaf left, not one leaf of a tree, every last one had fallen. That's how it was." She later found out that the mist was really an herbicide sprayed by US forces, but at the time she had no knowledge of Agent Orange. "I remember that the animals no longer had any place to stay," Ms. Hồng said. The forest was "gone." It was all still vivid in her mind decades later.[12]

As Hồng's recollections made clear, nature withered in the wake of chemical rains. And the scale of ecological degradation was enormous. In the summer of 1971, former Stanford law student and writer John Lewallen wrote about what he had seen while working in Vietnam in the International Voluntary Services. He started with a world before war. Most Americans simply had no understanding of Vietnam's ecological majesty, Lewallen argued. He cited a Midwest Research Institute (MRI) study conducted in 1967 for the US State Department that calculated "South Vietnam's vegetational biomass [weight of living organisms in an ecosystem] of 600,000 lb/hectare," dwarfing England's and the United States' "average woodland plant biomass of a few thousand lb/hectare." Vietnam was a land of old evergreen forests that stretched into mountainous highlands in the west and north. In the south, these forests bordered biodiverse marshland ecosystems in the mangrove ecology of the tropical Mekong delta that branched out into the South China Sea.[13]

Agent Orange proved particularly devastating to the coastal ecosystem of South Vietnam. In 1970, a study completed by a Harvard-led team called the Herbicide Assessment Commission of the American Association for the Advancement of Science (AAAS) estimated that "50 percent of the coastal mangrove forests of South Vietnam had been sprayed with herbicides." A USDA scientist discovered that just one spraying killed roughly 90 percent of mangrove trees in an affected area, and because it took centuries for this swampland ecosystem to develop its complex structure, he held that what had been wrought here would not be undone for some decades to come.[14]

This prediction turned out to be true. In 1983, Arthur A. Westing, a pioneering scholar of war and nature at the Stockholm International Research Institute, visited the mangrove ecosystem of South Vietnam and reported that the effects were systemic, leaving roughly 40 percent of the delta "utterly devastated." "Lasting inroads into the mangrove habitat" led to "widespread site debilitation via soil erosion and loss of nutrients in solution; decimation of terrestrial wildlife primarily via destruction of their habitat; losses in freshwater fish, largely because of reduced availability of food species; and a possible contribution to declines in the offshore fishery."[15]

Lack of government foresight partially explained why this happened. The fact that the State Department waited until 1967 to commission a study to look into the environmental effects of herbicide use during the Vietnam War is revealing. Five years into a herbicidal assault in a foreign country, the agency was just beginning to ask important questions about what Operation Ranch Hand might portend for Vietnam's natural environment. The 1967 report noted how little was known about potential ecological degradation from widespread 2,4,5-T dispersal: "To our knowledge *no* articles or books have addressed the long-term ecological effects of herbicides, integrated with studies of flora and fauna, rangeland, forests, other non-agricultural lands, waterways, lakes, and reservoirs." Researchers working on the study concluded that they were "confronted with serious knowledge deficiencies." And yet, the Agent Orange spraying continued.[16]

The US military also had a limited understanding of how Operation Ranch Hand would affect soldiers and citizens exposed to chemical sprays. In 1965, senior officials at the US Public Health Service (PHS) denied two scientists' request for just over $16,000 in funding to look into the toxicity of 2,4,5-T, holding that the Centers for Disease Control and Prevention (CDC) was the appropriate agency to conduct such research. The CDC, however, did not fund major studies on the matter until the 1980s, twenty years after the Vietnam conflict had ended. The National Cancer Institute had helped fund a 1966 study by the Bionetics Research Laboratory, a Bethesda, Maryland, private research firm, which found clear evidence of birth defects in rats exposed to 2,4,5-T, but that study was not made known to senior officials in the White House, Defense Department, or USDA until 1969.[17]

All the while, information gatekeeper Monsanto knew it had a problem. In 1965, Dow Chemical called together herbicide producers for a secret meeting to discuss "toxic impurities" in 2,4,5-T formulations. Dow said that dioxin in concentrations as small as 1 part per million could cause substantial health problems. After the conference, Dow's vice president reportedly said, "If the government learns about this the whole industry will suffer." The company fired off a memorandum to Monsanto, which had not attended the meeting, saying that the St. Louis firm's 2,4,5-T batches had dioxin concentrations that exceeded 40 parts per million. This was the same letter in which Dow said dioxin was "the most toxic compound they have ever experienced." Yet, despite these revelations, there is no evidence that Dow or Monsanto shared this information from their 1965 exchange with any federal officials.[18]

In 1966, approximately 5,000 scientists, some Nobel Prize winners, wrote to President Lyndon Baines Johnson, urging him to stop the aerial herbicide campaigns in Vietnam. At the same time, United Nations member states introduced a resolution condemning the United States for its herbicide campaigns, saying they were in violation of the 1925 Geneva Convention banning the use of chemical and biological weapons. By that point, the US government was spraying more than 741,000 acres of Vietnamese vegetation with Agent Orange and other herbicides annually.[19]

Protests continued into 1967 and 1968, but they had little effect on US military policy. In fact, Agent Orange usage increased. In 1969, American forces sprayed nearly 3.25 million gallons of dioxin-laden herbicide in Vietnam, up from roughly 333,000 gallons in 1965. And as they sprayed, US airmen shouted from speakers attached to helicopters, telling all those below that the chemicals raining down on them were safe and harmless.[20]

This was not so. In 1969, the National Cancer Institute–funded Bionetics study finally made its way to White House official Lee DuBridge, head of President Richard Nixon's newly created Environmental Quality Council. It showed that there was a clear link between dioxin exposure and serious birth defects. Given the damning details of the study, DuBridge pushed the president to convene a high-level meeting to discuss the further use of Agent Orange in

Vietnam, and after that meeting, the Department of Defense, citing the Bionetics study, said that it would ban the further military
deployment of Agent Orange in 1970. By then, a broader public
became engaged in protest as journalists across the country started
covering the story. In response, Congress called a hearing about use
of 2,4,5-T in the United States. In that 1970 hearing, the US Surgeon General announced a ban on the further use of 2,4,5-T around
domestic spaces and on most food crops, but to the dismay of many
activists, the government permitted the continued use of the herbicide on rangelands, rice fields, forests, highways, and railroad rights-
of-way for the foreseeable future. Not until the federal government
finally banned all uses of 2,4,5-T in 1985 did the spraying finally
stop—more than three and a half decades after Monsanto's James
Ray Boggess first came down with chloracne.[21]

* * * *

BY THE TIME THE protests against Agent Orange emerged in 1966,
Monsanto was the third largest chemical company in the United
States, earning more than $1.6 billion in annual sales, and employ-
ing about 56,000 people. The firm's empire now reached farther and
farther beyond US borders, with the company opening its European
headquarters in Brussels in 1963 and expanding into Brazil, Argentina, and other parts of Latin America. By the mid-1960s, Monsanto
had forty-three plants in the United States and operated in twenty-
one countries. Many of these factories produced plastics and synthetic fibers, which made up more than 54 percent of the company's
total sales in 1964, but they also churned out everything from detergents to fire-retardant liquids, silicon for computer semiconductors,
and phosphoric acid for sodas. Given its expansive diversification,
the company changed its name in 1964 from the Monsanto Chemical Company to the Monsanto Company, saying it thought "the
word 'chemical' no longer adequately describes the scope of the company's business." The switch also made sense in light of Carson's
book, which had made "chemical" a dirty word.[22]

By then, Edgar Queeny, entering his mid-sixties, was near the end of his days. He had retired as board chairman in 1960, welcoming Charles A. Thomas to the post, and remained chairman of the finance committee for another five years. During that time, he was racked with health problems. "I have an ulcer and must go on a tasteless diet and, worst of all, on the wagon," he told the president of DuPont in November 1960. Then there was a problem with his prostate. "Some quacks who have had me in hand have decided I need slight alterations in my plumbing," he confessed to a friend in 1962. Out of the hospital for just a few months, he went back to the doctors, who diagnosed him with a serious heart condition in 1963.[23]

As his health worsened, Queeny watched as the modern environmental movement gained momentum. It was an interesting time for someone who had always been extremely fond of the outdoors but who also ran one of the biggest chemical companies in the world that was profoundly reshaping ecosystems he loved.

At times, Queeny could not quite square his passion for wilderness conservation with his commercial interests in chemicals. In 1963, just after *Silent Spring* came out, Queeny said that "Rachel Carson's book is a fascinating one but it gives one part of the story," adding, "There is no doubt in my mind about the effect of pesticides on wildlife. On my farm in Arkansas we used to have plenty of quail; now we have none." Here was the conservationist, the idealist. But there was another side to the story: "When we had quail we raised 45 bushels of rice to an acre; now we raise 90 to an acre. So you see, I am prepared to argue on either side."[24]

After 1963, Queeny's outdoor excursions were forestalled by more frequent visits to the hospital. He hated the fact that doctor prescriptions consistently included a reduction in alcohol consumption. "You will not be able to 'trace me from place to place by the empty bottles!'" he jested with a friend after one of his medical visits. "I am allowed up to three drinks—of 'a little Scotch and lots of water.' I tried to convince my 'quack' that it should be the other way around, but the poor guy has no sense of humor—or justice."[25]

Though he continued to jest, he soon was bedridden with a failing heart, and on July 7, 1968, passed away.[26]

* * * *

ST. LOUIS NATIVE CHARLES H. SOMMER now ran Monsanto, a firm he first joined as a sales agent back in 1934. The fifty-seven-year-old chemist, dough-faced and gray-haired, had succeeded Charles A. Thomas as president of the company in 1960, and he immediately made aggressive moves to restructure the firm. Recognizing the growth potential for chemicals like 2,4,5-T, he spearheaded the creation of the firm's Agricultural Division. A proponent of international expansion, he also oversaw the launch of the Brussels office and several other overseas branches of the company.[27]

By the end of the 1960s, Sommer was navigating Monsanto through treacherous straits. John Queeny, so many years ago, had actively sought out government regulation, a Pure Food and Drug Act, hoping that federal oversight would provide legitimacy to a chemical industry many deemed dubious. Now, the prospect of new environmental regulations threatened ruin. Sommer tried to adjust to the times, creating Monsanto Enviro-Chem Systems, a division of the firm focused on developing pollution-remediation technologies, but he grew concerned about the increase in government scrutiny. The new demands of FIFRA were just the first of a series of new regulations on the horizon. By the end of the decade, Sommer fought back against what he saw as weed-like growth of government red tape. Regulatory paperwork, he said, "causes a pollution of its own that brings about a relentless upward pressure on costs," fueling "the fires of inflation."[28]

Sommer knew Monsanto was exposed. The company sold hundreds of chemicals, many of which were now undergoing rigorous reevaluation by federal agencies. Agent Orange was really only one of many serious liabilities. Another flowed through pipes at Monsanto's Krummrich plant, located just across the Mississippi River in a town called Monsanto, Illinois.

"Sell the Hell out of Them as Long as We Can"

IN 1966, SWEDISH SCIENTIST SÖREN JENSEN WAS WORRIED for his family. He had been investigating chlorinated compounds he found bioaccumulating in fish, eagles, and other wildlife in Sweden. The chemicals, which Jensen deemed a poison, were polychlorinated biphenyls (PCBs), and he noted that trace amounts could be found in eagle feathers dating back to World War II and archived in the Swedish National Museum of Natural History. Stunned by the pervasiveness of PCBs, he decided to test his wife and three children and discovered that they too had elevated levels of these chemicals in their bodies. He deduced that his five-month-old daughter probably had become contaminated by drinking her mother's milk. Jensen's professional research into PCBs had become personal; he went with urgency to blow the whistle on a problem that was flowing from one generation into the next.[1]

PCBs were some of the most profitable products Monsanto ever sold. The company was the sole manufacturer of these compounds in the United States. It supplied clients from its facilities in Anniston, Alabama, and Monsanto, Illinois—a small village in East St. Louis now called Sauget that had adopted Monsanto's brand name when the company expanded its operations across the Mississippi in the go-go growth years of the Jazz Age. Monsanto had been marketing PCBs since the 1930s under various trade names—most notably Aroclors—and the scale of production had grown immense three decades later. Electrical companies manufacturing transformers were some of the biggest buyers of this insulating material in the 1960s, but the company had also found new markets. By the time Jensen made his discovery, Monsanto sold PCBs in both liquid and solid forms to clients that mixed these chlorinated compounds into paint used on roadways, boat bottoms, pools, water tanks, and grain silos; other industries used PCBs in ink cartridges, synthetic Christmas trees, and "carbonless" carbon paper; it was added as a fire-retardant chemical to various consumer goods, used as a plasticizer in various plastic products, and added to dishwasher detergents and pesticides.[2]

Jensen's whistleblowing was just the kind of thing Monsanto president Charles H. Sommer had feared. He had watched as environmentalists in the 1960s pushed the federal government to play a more active role in regulating businesses. Rachel Carson, just a few years earlier, had raised the alarm about DDT, and scientists were already making comparisons between PCBs and this now notorious insecticide. Was the roar of Carson's *Silent Spring* about to get louder?

Monsanto medical director R. Emmet Kelly, now in his fourth decade of service to the firm, was concerned by the Swedish news. "The consensus in St. Louis is that while Monsanto would like to keep in the background in this problem, we don't see how we will be able to in the United States," he wrote to a colleague just weeks after Jensen's study became public.[3]

Kelly had to know this tempest was coming. In 1955, as Kelly was debating what to do about 2,4,5-T issues at Monsanto's Nitro plant, he was also already alluding to court battles that might ensue when word got out about PCBs. "We know Aroclors are toxic," he wrote in an internal memorandum to J. W. Barrett at Monsanto Chemicals Limited in London. Kelly's colleague, Elmer Wheeler, the assistant director of Monsanto's Medical Department, knew PCBs were problematic as well. In 1959, he said that exposure to PCBs, like other chlorinated compounds, "can result in chloracne, which I think we must assume could be an indication of more serious systemic injury." (Scientists later dubbed some PCBs "dioxin-like" chemicals because they contributed to chloracne and other similar systemic diseases caused by exposure to dioxin found in 2,4,5-T.)[4]

Wheeler and Kelly did not pass on these sentiments to the press. According to a 1958 internal company memo, firm policy was to not "give any unnecessary information which could very well damage our sales position." Silence was a means of containment.[5]

But if Monsanto officials kept quiet, others raised the alarm. University of California, Berkeley, marine biologist Robert W. Risebrough brought the PCB crisis to the American public's attention in a 1969 *San Francisco Chronicle* article. He offered startling evidence of PCBs' deadly effects on San Francisco Bay wildlife, calling the chemical "a menacing new pollutant" that threatened not only birds and fish, but human health as well. Evidence coming out of Japan seemed to confirm his findings. There more than 1,500 people had become ill after consuming rice oil contaminated with PCBs in February of 1968. Some women sickened by the exposure later had stillbirths that they attributed to PCB contamination.[6]

Wheeler and Kelly's efforts to contain Monsanto's pollution problem was working about as well as the US government's policy of containment in Vietnam. And the news got worse. In October of 1969, a chemist gave a conference paper detailing the ways in which PCBs were adversely affecting bald eagle populations. The symbolism was not lost on Monsanto leadership. "Now the emblem of the heritage of the United States is threatened!" exclaimed company leaders mulling over what to do about the PCB problem. "Lot of screaming" going on over "thin egg shells [sic] in birds," reported a Monsanto research director. "The development of a 'lunatic fringe' post-Rachel Carson," another internal communication warned, "has led to a domination of the media by scare publications in the public and scientific press."[7]

So much was going on in 1969. A few days before the new year began, Americans weary from the Vietnam War gasped in awe at "Earthrise," the first image of our blue planet captured by the *Apollo 8* astronauts—a picture many came to see as a testament to the fragility of humans' home. A month later, a massive oil spill polluted the pristine beaches of Santa Barbara, and in June, the Cuyahoga River, long polluted by Cleveland's industrial giants, caught on fire once again. Lake Erie was also in crisis, phosphate-based fertilizers and chemicals—some manufactured by Monsanto—spurring massive nutrient overloads that fueled explosive algal blooms that created a massive dead zone stretching across the great lake. By the fall, Senator Gaylord Nelson of Wisconsin announced plans for a massive environmental teach-in at college campuses across the country—an effort that would culminate in the first Earth Day on April 22, 1970, the largest one-day mass demonstration in United States history at the time. President Nixon took note, agreeing to sign the National Environmental Policy Act passed by Congress in December of 1969, which called for the executive branch to make permanent a Council on Environmental Quality and required the federal government to complete environmental impact statements before initiating public projects. Months later, Nixon established by executive order the Environmental Protection Agency (EPA) and signed a Clean Air Act that tightened regulations on industrial smokestacks. At the same time, Congress held hearings to discuss changes to the Federal Water Pollution Control Act, originally passed with weak enforcement provisions in 1948. The result was the Clean Water Act of 1972, which gave the federal government broader powers to regulate companies polluting American waterways.[8]

In this political climate, Monsanto began to feel real pressure from state and federal officials. In 1969, Monsanto's medical department notified plant personnel in Anniston, Alabama, that the Federal Water Pollution Control "boys" were going out to investigate waterways near Monsanto's PCB plant. At the same time, the National Air Pollution Control Administration also contacted Monsanto, seeking information on its PCB business. Naturally, top officials at Monsanto were seriously concerned about all this. As one Monsanto employee explained to Elmer Wheeler in 1969, "The Dept. of Interior and/or State authorities could monitor plant outfall and find ppm of chlorinated biphenyls at Krummrich or Anniston anytime they choose to do so. This would shut us down depending on what plants or animals they choose to find harmed." In all official interactions, Monsanto managers urged caution. "Nothing should be volunteered on these type requests unless specifically requested," urged another Monsanto manager: "We can always add but never subtract from something written."[9]

Monsanto's leaders had to be particularly careful in this moment because they were now privy to new in-house research confirming previous findings about PCB toxicity. Beginning in 1963, Monsanto had contracted with Illinois-based Industrial Bio-Test Laboratories (IBT) to look into the toxicology of PCBs. In 1969, one of the scientists collaborating with IBT researchers gave a damning summary of the group's findings: "It seems to the writer that the evidence regarding PCB effects on environmental quality is sufficiently substantial, widespread, and alarming to require immediate corrective action on the part of Monsanto." Other internal communications at the time showed that Monsanto officials were well aware of the broad toxicological consequences of continued PCB production. "We can't defend vs. everything," a company manager wrote in 1969, "Some animals and fish or insects will be harmed." And as for humans, Monsanto research also offered dire predictions. "Data available at present indicate that PCB's may be 'moderately toxic' to man," one company report read. In light of this new evidence, what would Monsanto's leadership do?[10]

* * * *

"SELL THE HELL OUT of them as long as we can," scrawled one Monsanto employee on the meeting minutes notes from an October 1969 gathering of the Aroclor "Ad Hoc" Committee. Elmer Wheeler and top executives in St. Louis had organized this crisis committee a month earlier to brainstorm what to do about the persistent PCB problem. All ideas were fair game, including this one; the goal was to air it all out in hopes of finding a workable solution.[11]

Things seemed to be unraveling at a rapid pace. "The subject is <u>snowballing</u>," an Ad Hoc member emphasized in October. The firm was clearly exposed, and everybody in the room knew it. "The committee has concluded" one 1969 report read, "that the identification of PCB's as an environmental contaminant is certain." Public backlash "and legal pressures . . . to eliminate or prevent global contamination are inevitable and probably cannot be contained successfully."[12]

But halting PCB production would be costly. "There is too much customer/market need and selfishly too much Monsanto profit to go out," one Monsanto executive candidly admitted in 1969. The numbers backed him up. Internal documents noted that the firm brought in roughly $22 million annually from PCB sales by 1970. This meant that the product line generated $10 million in gross profits each year. The firm was getting an exceptional return on the money it had invested in the manufacture of this product line whose sales had grown more than 464 percent over the course of the previous decade. Considering all this, the Ad Hoc Committee was direct in stating its number one objective: "Protect continued sales and profits of Aroclors."[13]

Just how to do that was not entirely clear, but committee members hoped they could delay government regulation by challenging scientific evidence. "Make the Govt., States, and Universities prove their case," argued one Monsanto research director, and whenever possible, "question evidence against us." Just as cigarette company executives would do in their battle to deny links between smoking and cancer, Monsanto officials wanted to sow the seeds of doubt in the minds of the public. Citing a 1970 company-sponsored study that showed "PCB's are exhibiting a greater degree of toxicity . . . than we had anticipated," Wheeler assured company officials that he would make science bend to the demands of the corporation: "We have additional interim data which will perhaps be more discouraging. We are repeating some of the experiments to confirm or deny the earlier findings and are not distributing the early results at this time." One Monsanto researcher was even more direct.

"Some of the studies," he said, "will be repeated to arrive at better conclusions."[14]

The government was only part of the company's problem. Monsanto's doubt-selling would be for naught if corporate clients balked at making future purchases. "Some customers who presently use these materials will be 'scared,'" the Ad Hoc Committee concluded, noting that Monsanto salesmen had to work fast to make sure big buyers, such as General Electric and Westinghouse, stayed loyal to the firm. The biggest area of concern was electrical-insulating fluid, including brands such as Pydrauls and Inertecns, which represented the big dollars in Monsanto's PCB portfolio.[15]

"We want to avoid any situation where a customer wants to return fluid," marketing specialist N. T. Johnson told sales associates in a 1970 internal memorandum. "We would prefer that the customer use up his current inventory and purchase" new formulations of PCB fluids. At this time, Monsanto began offering customers insulating fluid with reduced chlorine concentrations, arguing that these PCB mixtures were safer and more biodegradable. This turned out not to be true, but Monsanto salesman saw this as a key strategy that would help them stay in the PCB trade. "We can't afford to lose one dollar of business," exclaimed N. T. Johnson as he urged his PCB sales force not to be "defensive or apologetic." "Take the offense," he told them.[16]

Monsanto Ad Hoc members were confident that they could expect compliance from its electrical industry customers because their clients were so deeply dependent on their product. General Electric and Westinghouse really had no choice but to remain Monsanto customers. At a St. Louis meeting in January of 1970, General Electric's Edward L. Raab "was most impelling and forceful about the non-replaceability" of PCBs, which he said were "critical or essential" for company transformers. The Ad Hoc Committee concurred: "One of the unique features of PCB is their fire resistance. Here the basic decision whether to risk lives due to fire or risk extinction of some species of birds. In this case the PCB would probably be accepted as a necessary pollutant and tolerated under controlled conditions."[17]

This was the key to success moving forward for Monsanto: convincing the public and government regulators that what Monsanto had to sell was vital to the proper functioning of the American economy. Monsanto and General Electric officials were in "consensus . . . that without availability" of PCBs "large cities like New York would be shut down with no power. Certain industries . . . would go down . . . most of the lights across our country would go out and motors in air conditioners and many industrial applications would not run." Apocalypse beckoned in a post-PCB world.[18]

Monsanto took this message to the public. In a 1970 press release, the company warned that executives had "been advised by one electrical equipment manufacturer that an immediate ban on PCB would result in major power failures throughout the world. This is not the answer." Ironically, Monsanto's press team now believed the best way out of the PCB predicament was in fact to expose just how widespread the company's pollution problem had become. The country simply had no other option but to keep PCBs in the market.[19]

Monsanto had a strong case to make when it came to electrical insulating fluids, which represented about 60 percent of its PCB sales in 1970, but other less-essential product lines proved more problematic. Perhaps most disturbing was the fact that PCBs found their way into food packaging. Federal officials said this was likely because food-packaging companies were using recycled materials that included " 'carbonless' carbon paper" treated with PCBs as well as "printing inks" containing Monsanto's chemicals. In 1971, the FDA discovered that 67 percent of food packaging in a representative US sample contained PCBs. When company officials lamented the "global contaminant" issue they faced, they were not being hyperbolic. PCBs were everywhere.[20]

In the spring of 1970, with evidence becoming increasingly clear that this toxic problem was out of control, Congressman William F. Ryan made a passionate appeal to his colleagues to do something about the issue. Ryan was especially concerned about "open" applications of PCBs, particularly paints. Monsanto officials felt as if they were "walking on thin ice." Momentum was finally beginning to build for major federal legislation regarding this harmful chemical.[21]

News out of Ohio helped propel federal action. There the Ohio State Board of Health discovered PCBs in dairy milk. The problem appeared to be grain silos lined with PCB paint. "All in all, this could be a quite serious problem," R. Emmet Kelly wrote in an internal memorandum. He knew that the company needed to move quickly. "When are we going to tell our customers not to use any Aroclor in any paint formulation that contacts food, feed, or water for animals or humans?" he queried. Pesticides clearly weren't the only worry when it came to Monsanto's effects on the food chain.[22]

If the firm did not act now, it could be held liable for contamination spread by its clients. The problem had to be contained.

* * * *

WILLIAM B. PAPAGEORGE HOPED he could solve the problem. A chemical engineering graduate from Washington University in St. Louis, Papageorge had first come to Monsanto in 1951 and later took over responsibilities for maintaining PCB production at the company's Anniston plant in the 1960s. His intimate understanding of this troublesome product line elevated him to the unenviable position of environmental manager for Monsanto's Organic Chemical Division and the point person for all discussions regarding PCBs. In the years ahead, he would come to be known as "Monsanto's PCB czar."[23]

Papageorge understood the need for bold action. In July 1970, he approved the decision to halt all sales of PCBs for "open" applications, such as paints, plasticizers, and paper products. Notably, however, the company made sure to get rid of any PCB supplies it had in these market categories. Papageorge stated that "all orders of these products which were received up through noon August, 1970 were shipped." By October, Papageorge further applauded the fact that there had been no "extensive returns of materials."[24]

But if Papageorge was willing to execute a quick phaseout of certain sectors of Monsanto's PCB business, he worked hard to retain electrical fluids, the real moneymaker, which still represented the majority of Monsanto's PCB sales. If Papageorge could manage to keep these production lines humming, he would be buying time for the retooling of high-dollar infrastructure while at the same time generating substantial gross profits for the firm.[25]

Here Papageorge's strategy was quite simple: offload liability onto client companies by offering them the choice to reject what Monsanto was selling. As Papageorge explained in a September 1970 status report, Monsanto would "emphasize to all remaining users of PCB's the importance of preventing escape to the environment and . . . [would] ensure that these warnings are fully documented so that they will support the action we have taken in this area should we become involved in legal actions."[26]

But what Monsanto was really offering its customers was the illusion of choice. Company executives said as much in a 1970 presentation to the company's corporate development committee: "Some of our customers have no immediate alternative, some could change only at sacrifices of safety, or cost or various technical factors." By the early 1970s, companies had developed alternatives to PCB-filled transformers and capacitors, but the problem was these new products were expensive. Silicone-filled transformers and capacitors ran roughly five times the going rate of PCB-loaded brands; likewise, "dry transformers"—transformers designed to prevent fires without insulating fluids—sold for 60 percent more than PCB-filled varieties. Considering these financial realities, it made more sense for companies such as General Electric and Westinghouse to risk continued use of a dirty product. An attribute that had first made Monsanto's PCBs so attractive—their cheapness—remained a critical asset in the 1970s.[27]

Monsanto's electrical clients were locked in. They had nowhere else to turn. They too stood to lose a lot of money if they had to abandon PCBs.

By 1971, Monsanto needed the support of its clients more than ever. That year, with news stories continuing to break about PCB contamination in food, President Nixon's Council on Environmental Quality called for passage of a Toxic Substances Control Act (TSCA), a measure that would allow the EPA to restrict the use of certain chemicals and force companies to report and control releases of particularly dangerous compounds. The White House council specifically identified PCBs as a dangerous chemical needing stricter regulation. Perhaps the government would soon move to ban these substances. This was a multimillion-dollar threat for Monsanto and its customer base.[28]

As government scrutiny intensified, Monsanto took a bold step to further protect itself from future liability: in 1972 the firm sent an indemnity agreement to all of its PCB clients. These agreements explained that Monsanto would only continue sales so long as clients agreed to "defend, indemnify, and hold harmless Monsanto" for any damage that might be caused by PCB use. Filled with legal jargon, these contracts covered all sorts of scenarios where Monsanto felt it might be at risk.[29]

*   *   *   *

FOR DECADES, MONSANTO HAD made millions by providing fire insulation to big buyers such as General Electric. Now the company needed these clients to offer it protection as it sought to deal with toxic hazards coming due.

Fighting federal regulators on its own was never a viable option for Monsanto. After all, the company admitted to the EPA that only fifty-five Monsanto employees were directly associated with the production of PCBs at its Sauget plant in East St. Louis, the only remaining PCB production facility in the country at that point. (In 1970, as the PCB problem festered, Monsanto decided under pressure to shut down PCB production at its Anniston plant.) It would be hard to argue that the economic loss of those few jobs justified the ecological consequences continued PCB production would bring. The key was getting the EPA to think about broader economic costs, and that was where electrical company partnerships became critical. The companies had thousands of employees who worked directly with equipment containing PCBs. If the federal government banned these chemicals, their jobs and the jobs of countless other workers that used the electrical equipment they sold would be in jeopardy.[30]

According to a General Electric representative that attended an early agency hearing, EPA officials "ruled that economics are relevant" when weighing regulatory decisions regarding chemicals. Monsanto thus strategized about the best techniques for introducing "relevant economic data" into the regulatory discourse.[31]

In the coming months, Monsanto was sanguine that EPA lobbying efforts coordinated by its big electrical clients, especially General Electric and Westinghouse, would help them stave off stiff regulations. These companies pleaded with the EPA not to take bold action regarding closed system uses of PCBs.[32]

But despite these industry appeals, EPA officials proposed a strict effluent standard under the Clean Water Act in December of 1973 that restricted PCB releases to a level that was barely detectable with contemporary water-testing equipment. "Obviously, the EPA has not listened to any of the written statements by our customers (specifically G. E.)," complained a Monsanto special projects director to Papageorge. In subsequent hearings in 1974, Papageorge continued to try to make the case that banning PCBs would result in "undeterminable incidences of fire" and serious "economic disruptions." PCBs were now confined to "hermetically sealed" applications, such as electrical transformers, he argued. Private industry was controlling and containing the problem. There was no need for stiff regulations.[33]

* * * *

PAPAGEORGE'S CONTAINMENT CLAIMS belied the reality that PCBs continued to seep and slosh into the environment—and into human bodies. In 1975, with the EPA and Congress still debating the final language of PCB regulations, Westinghouse's staff supervisor Dan Albert wrote privately to Papageorge describing a messy situation at its plant where workers making capacitors handled Inerteen, another PCB product. "Employees carry Inerteen home on the soles of their shoes and complain quite a bit about the effect Inerteen has

on wearing out their shoes," explained Albert, adding, "Is this a serious problem?" From Albert's description, it was clear that Westinghouse workers were literally covered in PCBs. "Inerteen penetrates
through [their] leather" shoes, and their hands were also saturated in
this chemical because "employees working in Inerteen are not able to
use gloves since it is an assembly area." "Even if they could," Albert
added, "the Inerteen would destroy the protective glove."[34]

As Albert's statement about worker footwear practices made
clear, the PCB problem was not confined to the factory floor or the
Westinghouse laborers themselves: these workers took their problems home with them. PCBs sloughed off on welcome mats and in
mudrooms, mingling with the footwear and clothing of comers and
goers. Contaminated clothes ended up in laundry machines, there
to mix with a spouse's intimates or children's garments. The final
product, a transformer, might look like a closed container, but the
process of creating it (or repairing it) was hardly the system of containment Monsanto had proposed.[35]

Papageorge spoke in contradictions when responding to Albert's
concerns. In one passage he told Albert that there "should not be any
effect on an employee or his family from home laundering of work
clothing," but in another section he said he could not "overemphasize
the need to properly control the use and handling of Inerteens to prevent their escape into the environment." And there were other mixed
messages. In one paragraph Albert asked a candid question: "Since
Inerteen effects [sic] birds and other animals, how do you explain it
to employees in such a way that they will understand why it can kill
a bird and not a man?" Papageorge pointedly responded, "There is a
potential real effect to humans—including death." Despite this reality, Papageorge wrote: "I strongly recommend that the perspective
gained from 40 years of experience in which no human harm has
resulted be emphasized."[36]

While Papageorge fretted over the Westinghouse situation,
renewed pressure came from the EPA. In June 1975, the EPA wrote
to Papageorge explaining that PCB levels in Lake Michigan were

rising. Robert A. Emmet, a legal chief in the EPA's water enforcement division, broke the news to Papageorge: "As a result of this restriction of PCB use to closed systems, State and Federal pollution control agencies had believed that levels of PCB's found in fish, particularly in Lake Michigan, would soon begin to decline." Sadly, Emmet concluded, "This has not occurred."[37]

EPA now wanted information on all of Monsanto's clients. The goal was to try and figure out how PCBs were getting into the environment from these supposedly "hermetically sealed" containers. In a gesture of goodwill, Emmet wrote to Papageorge, "I am delighted that we are able to make this inquiry in a friendly and constructive manner" considering the "spirit of cooperation which you and Monsanto have displayed in this potentially serious matter."[38]

At the same time as Papageorge was having his discussions with EPA officials, marketing manager Floyd A. Bean was preaching tactics of diversion and obfuscation when training salespeople. "Do not answer any question about Inerteen or PCB's," he told his staff, "even if you know the answer."[39]

∗   ∗   ∗   ∗

BY THE WINTER OF 1975, as Congress completed another round of debates regarding the proposed Toxic Substances Control Act, it now seemed that a federal PCB ban was imminent. EPA administrator Russell Train went on television in December urging Congress to pass TSCA, which would give EPA the authority to prevent dangerous chemicals from entering the environment. "In the past five years," Train said, "an estimated 600 new chemicals a year have been introduced in the US commons . . . without any systematic, advance assessment of their potential impact on human health." To the growing number of environmental activists in the country, this was unconscionable. Train specifically mentioned PCBs among the dangerous compounds that simply had to be eliminated from the marketplace.[40]

In October 1976, Congress finally passed TSCA, which included a ban on the manufacture of PCBs. Sensitive to industry needs,

Congress declared that the ban would go into effect on January 1, 1979, giving companies some time to adjust to the new regulations. As the company's PCB Task Force had urged, Monsanto decided to halt production and stop shipment of PCBs from its facilities in October of 1977, well in advance of the 1979 deadline. Monsanto wanted to show the public that it was choosing, on its own, to take corrective action. In all future public statements on the matter, Monsanto emphasized this point: that it voluntarily got out of the PCB business.[41]

Overseas, Monsanto tried to eke out a few more years of PCB sales. In the United Kingdom, where regulatory momentum lagged, Monsanto's internal memorandum made clear that the PCB pullout would be on a different timetable. There, the firm would try and keep regulators at bay as long as it could. To this end, public relations director Dan Bishop instructed officials at Monsanto's European headquarters to stall if asked probing questions about PCBs. "Avoid any medical questions," he said, especially regarding "PCBs in mothers' milk . . . that have been circulating in the U.S. press." Deflect inquiries and "avoid any comments that suggest liability." The company was going to try and keep its British PCB production facility in Newport, Wales (purchased right after World War II), up and running as long as it could.[42]

* * * *

BACK HOME, TSCA IRONICALLY offered further protection for Monsanto. "With the passage of the Toxic Substances Act," the Monsanto PCB Study Group concluded in an internal memorandum, "the company will have an additional legal defense against . . . litigation." After all, Section 6(e) of TSCA officially sanctioned continued PCB usage so long as the chemicals were "totally enclosed." As the EPA explained, permissable applications included "electrical transformers, railroad transformers, hydraulic systems, mining machinery, heat transfer systems, pigments, electromagnets, natural gas pipeline compressors, small quantities for research and development, microscopy, and carbonless copy paper." These uses were now sanctioned under the law. How could Monsanto possibly be held liable for pollution problems associated with applications that the EPA had officially approved?[43]

The EPA admitted that it considered "the economic impact from restricting these uses" when promulgating its final ruling. Monsanto's lobbying efforts had paid off. Russell Train and the EPA had listened to the dollars-and-cents appeals the firm made about the essential role PCBs played in the proper functioning of the American economy. More than 200 companies had ultimately reached out to the EPA to express similar concerns about PCB regulations.[44]

Here was a strange reality. To defeat a total ban on PCBs, Monsanto worked hard to expose rather than hide how pervasive its toxic products had become in the global economy. Ubiquity was actually an asset in a battle in which economics as well as ecology governed decision making in the early years of TSCA enforcement.

The EPA made clear that it did not have the resources to make sure PCB pollution was properly contained in the years ahead. "The Agency recognizes . . . its inability to regulate some activities, such as disposal of many types of PCB equipment, due to the broad ownership of such equipment at a vast number of sites," an EPA report noted after the passage of the PCB ban. This was a problem too big and too widespread for this federal agency to handle. It would have to rely on private industry to control contamination.[45]

For the Environmental Defense Fund (EDF), one of the nation's leading environmental law organizations, this situation was simply unacceptable. In 1980, the EDF challenged the EPA's decision to allow "totally enclosed" uses of PCBs to continue, urging the federal government to take more aggressive action to eliminate PCB threats. EDF won its initial court battle, but industry pressure resulted in a stay of the federal court's decision to remove exemptions. In the end, the EPA continued to allow closed-system uses of PCBs, in large part because PCB removal was seen as too costly. As a result, PCB-laden electrical equipment remained in place all across the country.[46]

* * * *

IF EPA REGULATORS BELIEVED they had contained the crisis, they were wrong. As transformers and capacitors aged, they became vulnerable to combustion, and when they ignited, PCBs transformed into dioxins and furans—chlorinated compounds considered some of the most toxic chemicals known to man. For the firefighters that rushed to put out electrical blazes, the consequences to health could be dire. Studies showed that emergency personnel often spent more than four hours battling the heat of a transformer fire—a considerable time to stand immersed in deadly chemicals. They took toxic substances back to their stations, where they laundered clothes and washed equipment. Here was yet another way PCBs breached supposedly "enclosed" systems. Monsanto may have been insulated from the toxic consequences of its insulating fluids now set ablaze, but EPA's failure to demand swift and complete PCB removal from the environment meant that American heroes now bore the burden of PCB exposure every time they answered a call to quell a PCB-transformer fire.[47]

And firemen were not the only people in danger. Electrical workers throughout the country faced exposure to PCBs every time they went out to service damaged transformers and capacitors. The public health threat here was real.[48]

Almost every community in the country had a stake in this problem. Millions of PCB-filled transformers and capacitors remained in service in the 1990s, and the pace of replacement was glacially slow. By then, the average annual rate of removal was approximately 2.5 percent. In the 2010s, the EPA was still maintaining a list of PCB-filled electrical equipment in the United States.[49]

The PCB threat did not end when contaminated transformers and capacitors were taken out of service and dumped in local landfills. In 1976, the EPA estimated that 300 million pounds of PCBs were buried in dumps across the country. Monsanto and its electrical industry clients tried to assure citizens that PCBs would not leach from these sites, but some people were not convinced. "That is absurd," exclaimed Hudson River Fishermen's Association president David Seymour in 1975, "Has anybody at G. E. asked what happens when it rains at a garbage dump? If any of the containers holding PCB are broken, the stuff will leach into the ground and may contaminate water for miles around." As Seymour made clear, the PCB problem was still not contained, and it would not be for decades. In the 2000s and 2010s, the EPA finally forced General Electric to spend nearly $1.6 billion to clean up a forty-mile stretch of the Hudson River north of Albany that the firm had used as a dumping ground for PCBs. Safe behind a firewall of indemnity agreements, Monsanto was protected from a contamination problem that might have undone the firm.[50]

*   *   *   *

ODDLY, IT WAS PRECISELY the pervasiveness of PCB pollution that saved the firm from financial ruin in the 1970s. The EPA, still in its infancy during the stagflation of the Ford and Carter years, simply could not justify costly PCB removal programs that it believed threatened the viability of an electrical industry that powered the economy. Pollution had to remain in place because the costs of doing otherwise seemed unconscionable.

The PCB problem would come back to haunt Monsanto in the 1990s, but in the 1970s it had more pressing problems. Two promising products that had brought in profits for decades, 2,4,5-T and PCBs, were being retired. Oil companies and foreign competitors were beginning to cut into the firm's market share in the commodity chemical business. Monsanto needed a new blockbuster product that would help it edge out competition. The firm found it in a new herbicide called Roundup.

CHAPTER 7

"Strategic Exit"

A HUGE TRUCK BEARING MOLTEN RADIOACTIVE MINING WASTE puttered toward the edge of a cliff in the fading dusk. This was the nightshift in June 2016 at a manufacturing plant located in southeast Idaho just north of Soda Springs, population circa 3,000 people. The fiery mess that truck bore was phosphate slag, a by-product of producing elemental phosphorus from phosphate ore mined from nearby mountains. As it approached the edge of the cliff, the truck dumped lava-like sludge down the side of a monstrous charcoal-colored hill of waste. Red-hot orange slag tumbled down the hillside as heat waves rippled out into the air, and farmland just a few feet away lit up in an unnatural glow. Barley, some of which may have been destined for Budweiser beer, hugged the land in the distance.[1]

Monsanto had nowhere else to put its waste. The EPA had decided that radium- and uranium-laced phosphate slag could not be removed from this site. Year after year, this material continued to pile up on the backside of the processing facility.

Monsanto's mega-plant, nestled in the hill country of the Gem State, billowed large amounts of mercury into the air—an estimated 96 percent of Idaho's statewide mercury emissions in 2006. Electrical wires pumped incredible amounts of energy—derived from coal-powered plants—to the furnaces that cooked phosphate rock at temperatures over 2,700 degrees Fahrenheit. Estimates suggest that this one plant used more energy than all the citizens of Salt Lake City, a metropolis just a few hours south.[2]

Monsanto spent millions of dollars to keep this facility in

operation. It had to. After all, it was really the heart of the company's empire, for it was here that Monsanto extracted the elemental phosphorus it needed to make glyphosate, the key ingredient in its world-famous herbicide: Roundup.

*   *   *   *

MONSANTO HAD FIRST COME to this place in 1952. Back then the company's phosphate business was booming as the firm transformed phosphate rock into everything from phosphoric acid used in soft drinks to detergents used to launder clothes. Monsanto's 1952 annual report listed phosphate-based detergents along with soap as the firm's "third most important" product line, just behind plastics/synthetic fibers and rubber, and company chemists were constantly tinkering with all sorts of phosphorous compounds that might bring in more profits for the firm.[3]

One of those chemists was a young man named John Franz, who joined the St. Louis firm in 1955. Like Monsanto's founder, John Queeny, Franz was a midwesterner. He was born to European immigrants of modest means on a dairy farm in Springfield, Illinois, in December of 1929, just two short months after the great stock market crash that sent the nation reeling into the Depression. By his own admission, John Franz was reclusive, someone who preferred tinkering away quietly in a lab. His obsession with chemistry started from a very early age, and as a child he spent most of his time in a makeshift laboratory he constructed in his parents' basement.[4]

Franz's German-born father often fronted the money to buy the chemicals his son needed to conduct his experiments. It was a neighborhood affair. According to John Franz, other kids in his community were setting up their own chemistry labs and often exchanged ideas about molecular meddling. Just as kids in Steve Jobs's generation became masters of computer technology, playing with transistors in 1970s Silicon Valley garages, so too were chemical giants being born in the World War II–era basements of heartland America.[5]

Franz marveled at the tools of modern science. As a ten-year-old boy, he watched movies about Thomas Edison, Louis Pasteur, and Paul Ehrlich. Reflecting on these films later in life, Franz once quipped, "It was very impressive to me that one person could do something that would be so beneficial to mankind." With ambitions of being the next world changer, Franz decided to go to the University of Illinois to study chemistry.[6]

Franz went on to pursue his PhD at the University of Minnesota, and upon graduation, he courted various firms—Standard Oil, DuPont, among others—but ultimately decided on accepting a job at Monsanto, a place that he felt "wanted" him. Besides, Monsanto was smaller; it would be easier for him to stand out there. Franz had dreamed of a professional career that kept him at the lab bench. "I didn't have the qualifications to be a good teacher," he admitted: "I was probably too introverted, and I was also too individualistic; I felt like I just wanted to do research." Monsanto offered him the opportunity to do just that.[7]

When he came to Monsanto's headquarters in St. Louis in 1955, he was assigned to the company's Organic Chemical Division, and he originally worked on chemical processes that generated 2,4-D. But herbicides were not the only product line Franz helped develop. Quickly rising through the ranks, he became a group leader by 1959 for a research team working on various formulations of aspirin, and in the 1960s he worked to develop plasticizers as well as new pharmaceutical products. As Franz put it, "They kept changing my jobs around," but this was something Franz welcomed. Born with an insatiable curiosity for science, he loved his Monsanto gig.[8]

In 1967, Franz's career took a dramatic turn. That year, he left organics for the seven-year-old Agricultural Division, poached by division head Dr. A. John Speziale who saw in Franz a disciplined researcher dedicated to lab work (attributes Speziale himself possessed and admired). At that time, agricultural chemicals had become the "fastest growing product group" within the firm, and Franz was right in the middle of the action. He began trying to develop chemicals that could regulate plant growth but had little success the first few years. He had taken on a "dead area" of research initiated by Dr. Philip Hamm in Monsanto's Agricultural Division in 1960 that involved investigating various phosphorus-based molecules to see if they could be turned into plant growth inhibitors. Hamm worked for almost a decade to find a compound that could be commercialized, but he failed time and time again to make a breakthrough. Many considered him kind of "kooky" for continuing his single-minded pursuit.[9]

So when Franz arrived in the ag division in 1967, Hamm hoped he could turn Franz, the dreamer, onto his project. Though most of the Agricultural Division would give up on this assignment by 1969, Franz, always curious, figured he would continue experimenting.

The company needed some good news. As profitable products such as 2,4,5-T and PCBs came under threat from federal regulators, the company's booming phosphate-based detergent business was also under attack. Monsanto had made major investments in detergents that used sodium tripolyphosphate (STPP) as a primary builder (meaning STPP was the chemical agent that helped remove dirt and grime from clothes). In the 1960s, the company supplied STPP to Lever Brothers, which marketed the detergent "all," a brand created by Monsanto in the 1940s and sold to Lever Brothers in 1957. Over the course of the decade, "all" became such a success that it was included in a 5,000-year time capsule created at the 1964/1965 New York World's Fair, rounding out a collection that included "the Bible, contact lenses and a Beatles record."[10]

But then came a crisis. By the end of the 1960s, scientists, environmentalists, and politicians raised new alarms about the way phosphate detergents were polluting waterways. The problem was eutrophication, an ecological condition in which there is an overabundance of nutrients in a waterway, leading to exponential growth in plant life and algal blooms that in turn exhausts oxygen in the aquatic environment, setting in motion ecosystem collapse. Facing a government crackdown, Monsanto decided in the 1970s to develop a new chemical builder to replace STPP and slowly got out of the phosphate-detergent business. The pressure from environmental groups, scientists, and politicians was too strong.[11]

The question for Monsanto was what to do with all of that phosphate derived from its chemical plant in Idaho that was no longer going into detergents because of these water-pollution concerns. After all, the company reported that almost 50 percent of the elemental phosphorus it produced at its processing facility in Soda Springs had been destined for detergents in 1970.[12]

This was the context in which Franz labored in the ag lab. He brainstormed dozens of phosphorous compounds and started testing them on plants.

He got incredibly lucky. The second chemical he tested in 1970, glyphosate, was a real winner. Later studies revealed that most of the other compounds Franz planned to test were "deader than a doornail."[13]

Shortly after Franz's auspicious finding, Monsanto scientists began testing glyphosate and were delighted with the results. "It's commercial," Phil Hamm blurted out when he saw glyphosate work for the first time in the company greenhouse. But he still had not confirmed that glyphosate was effective in real-world conditions. There was no time to lose. Winter was coming, and if Hamm was going to get confirmation of his greenhouse findings, he had to spray glyphosate on an outdoor test plot as soon as possible. He ordered scientists working under him to skip a round of secondary greenhouse experiments and began field testing in the fall. Soon thereafter a plant pathologist flew over a plot site sprayed with glyphosate, and when he looked down, what he saw was amazing. Weeds withered where glyphosate had been applied. It was exhilarating. So many other chemical compounds had failed when Monsanto finally got to the field-test phase. This product really worked, annihilating a broad spectrum of weeds that plagued farmers. "Eureka!" said the plant pathologist in his summary report. The company branded its new creation Roundup.[14]

Glyphosate offered a "strategic exit," as Monsanto mine specialist

Randy Vranes put it, from an increasingly contentious phosphate detergent business. Rather than abandon its expensive investments in phosphate mining in Idaho, Monsanto could now channel elemental phosphorus into its new herbicide.[15]

John Franz believed he had done good. Years later, Franz explained that it was precisely Roundup's "environmental safety" that made it so appealing to Monsanto executives shell-shocked by the vibrant pro-regulation activism of the modern environmental movement in the 1970s. Unlike herbicides such as 2,4,5-T, now destined to be discontinued because of its toxicity, there were no known human health problems associated with glyphosate. Monsanto noted in product advertisements that glyphosate worked to disrupt a plant enzyme called EPSP synthase, and because this enzyme was "not present in humans or animals," the company argued it was safe for use on crops that humans would consume.[16]

Researchers later challenged this assertion, noting that bacteria in human and animal guts and in the soil do have EPSP enzymes and therefore could be adversely affected by glyphosate. By the 2010s, some speculated that disruption of microflora environments caused by glyphosate might well contribute to maladies in humans. As this book went to press, scientific research in this area was still ongoing, but back in the 1970s, this was not something regulators examined closely.[17]

By the end of 1975, just a few years after Franz discovered the herbicidal properties of Roundup, the EPA approved glyphosate for use on crops in the United States. New legislation in the early 1970s had given the EPA authority to take over pesticide regulation duties previously carried out by the USDA and stipulated that companies had to provide extensive scientific evidence to show that their products were not harmful to humans or the environment before attaining registration. The EPA ultimately gave Roundup approval after reviewing several confidential studies completed by the Monsanto Company. For the moment, Roundup had a clear path to market. A new era in American agriculture had arrived.[18]

* * * *

MONSANTO PRESIDENT JOHN HANLEY, a fifty-something Harvard Business School alum and former Procter & Gamble CEO, must have been elated. He was the first real outsider to run the company. In 1972, a three-man committee of Monsanto's board of directors tapped Hanley to become president of a firm still embroiled in the chaos of the 2,4,5-T and PCB fiascoes. Some of Monsanto's old establishment types wondered what this consumer products guy knew about their business, which reported $2.2 billion in annual revenues in 1972, and Hanley himself even admitted he was a bit green. At a press conference announcing his hire, Monsanto's new president, grinning ear to ear, jested that he came "with little knowledge and great enthusiasm" but promised "to learn pretty darn quickly." Still, he knew more than just the basics. After all, at Procter & Gamble, Hanley had specialized in the detergent and soap business, negotiating directly with Monsanto salesmen. He knew how important phosphates were to Monsanto and must have been happy to find a new product line for those resources mined and processed at great cost.[19]

These were especially tough times for the firm. Big oil companies and foreign firms continued to eat into more of Monsanto's market share in the commodity chemical business, and some products in the firm's pipeline were turning out to be duds. Perhaps the biggest failure was the Cycle-Safe bottle made from Lopac plastic, which Monsanto had hoped to sell to Coca-Cola and other beverage giants. The plan was foiled in 1977 when the FDA moved to ban the bottles, citing studies showing Lopac plastic produced tumors in lab mice. Ultimately, DuPont won the plastic bottle race, patenting the process for making the polyethylene terephthalate (PET) container, which became the beverage industry standard for years to come. It was a major failure for Monsanto, making the company eager for a new success story.[20]

Roundup was just that. Glyphosate sales surged in the immediate years after EPA approval. The broad-spectrum herbicide's ability to destroy dozens of different types of weeds was unmatched. Farmers bought the chemical in large volumes for spraying applicators and told reporters they had never seen anything quite like it. Waterhemp, palmer amaranth, marestail—the nastiest weeds of all—withered when doused with Monsanto's weedkiller. Most corn and soybean farmers sprayed glyphosate before crops emerged from the ground, because there was as yet no way to protect these plants from the weedkiller, but even these applications added up to a lot of Roundup on American farms. Many growers began practicing what was known as "no-till" farming, spraying Roundup to "burn down" all plant life and then seeding fields directly over dead vegetation without disturbing precious soil that could be lost if plowed up. Many growers saw this as the environmentally responsible thing to do. Soon home gardeners had access to the chemical, buying small jugs and applying it to pesky johnsongrass and other weeds. The market was international, with Canadian barley growers and British wheat farmers turning to the new herbicide as early as the mid-1970s.[21]

To sell Roundup in these early years, Monsanto's marketing team deployed feed-the-world language that played on fears of pending food shortages. For the first time in decades, America's agricultural surplus was running low, in part because of a large grain sale to the Soviet Union in 1972. Prices for food were rising in the country, and people were scared. Monsanto fed the fear. "Without chemicals," one 1977 Monsanto advertisement cautioned, "many more millions would go hungry." This rhetoric matched the messaging promoted by proponents of the Green Revolution, which by this point had expanded well beyond the Mexican Agricultural Program (MAP) into Southeast Asia and beyond. In the 1950s and 1960s, foreign-policy makers associated with the Green Revolution had popularized the idea that America's national security depended on promoting agricultural projects around the world that would increase crop yields. By boosting food production, the argument went, farmers would eliminate hunger, thereby preventing the political radicalization of foreign nations. Recent research now makes clear that this logic was flawed. Increasing yields did not eliminate famine or food insecurity in developing nations, in large part because the agricultural reforms of the Green Revolution never addressed fundamental structural inequalities that contributed to poverty in places where MAP-styled programs were executed. But in the 1970s, the idea that insufficient food production around the world was an existential threat to American security was accepted by many. From the 1970s onward, Monsanto advertisers would sell this problem to the American people—and then sell Roundup as part of the solution.[22]

Within eight years of introduction, the company boasted that its annual Roundup sales approached $500 million. The company had a patent on the herbicide for several years to come, giving it monopoly control on this product market. Though other chemical companies invested in new herbicides in hopes of competing with Roundup, nothing really came close to this new wonder chemical.[23]

The Agricultural Division was now the biggest profit generator in the firm, and Roundup was the blockbuster product bringing in much of the cash. The company sold other herbicides, including macho-sounding brands such as Lasso and Machete that invoked images of farmers taming a wild frontier. Still, glyphosate was the real prize, a once in a corporate lifetime discovery.[24]

But back in Idaho, disturbing news threatened to bring it all tumbling down. There federal agents found trouble at the headwaters of the Roundup supply stream.[25]

* * * *

THE EPA NOW HAD new powers to go after corporate polluters. After President Nixon signed into law the National Environmental Policy Act in 1970, businesses seeking to access natural resources on federally protected land had to work with government officials to complete an environmental impact statement (EIS) that would catalog the expected ecological consequences of industry operations. Furthermore, in 1976, Congress passed the Resource Conservation and Recovery Act and the Toxic Substances Control Act (TSCA), giving the EPA new authority to oversee proper disposal of hazardous waste. Thus, though Monsanto had been mining in Idaho for decades, in the mid-1970s the firm faced stricter scrutiny from public officials.

The EIS for the "Development of Phosphate Resources in Southeastern Idaho," completed in 1976, exposed some disturbing facts about Monsanto's operations in the Gem State. The EIS focused on sixteen proposed mining projects directed by Monsanto, the J. R. Simplot company, FMC, and five other firms that covered roughly 15,700 acres of land. The report exposed a litany of environmental problems that would likely attend new phosphate mining in Idaho, noting that mine expansion would result in "accelerated erosion," "reduction in ground water quality," and "severe impacts upon the wildlife" in the area. In addition, the EIS pointed out that planned expansion would mean that water used at phosphate-processing plants would likely quadruple from about 16 million gallons per day (mgd) to 66 mgd (24 billion gallons per year). This was a lot of water for a region getting less than twenty inches of rain a year. And there were other worries. Commenting on the findings of the interagency task force, the EPA highlighted its concern about "radiation impacts associated with by-product and waste utilization of gypsum, phosphate slag, and mine tailings."[26]

Here was a major problem. According to the EPA, "phosphorous ores contain approximately 60 times the levels of natural radioactivity normally found in the Earth's crust." A by-product of producing elemental phosphorus is phosphate slag, which contains elevated levels of radionuclides such as uranium-238 and radium-226. Radioactive materials, including polonium-210, are also sent up into smokestacks when phosphate ore cooks in kilns. In the 1970s, this radioactive material was escaping into the air. To make matters worse, Monsanto was selling its slag waste to the town of Soda Springs as building aggregate for roadways and even home foundations. Regulators were worried about all this radioactive pollution and pushed for investigations into whether it posed serious health risks to the community.[27]

Monsanto was in trouble. Elemental phosphorus mined and processed in Soda Springs was now an essential revenue generator for the firm because it went into the company's new herbicide, Roundup. Expecting sales growth, the company had recently opened new mine pits that funneled millions of tons of ore to the Soda Springs processing plant. Monsanto simply could not abide any disruption to this facility's operations.[28]

And pressure kept coming. In 1980, Congress—pushed by grassroots agitation in Love Canal, a residential community in upstate New York built atop a toxic waste dump—passed the Comprehensive Environmental Response, Compensation, and Liability Act (CERCLA; known as the Superfund Act), which gave EPA new authority to force companies to clean up toxic wastes.[29]

Empowered by the Superfund Act, the EPA focused its sights on Monsanto's 530-acre phosphate-processing plant in Soda Springs. In addition to detailing the troubling spread of radionuclides, investigations produced other disturbing findings: contaminated drinking wells on site and pollution plumes spreading into groundwater that contained dangerous chemicals including cadmium, selenium, and vanadium. Given all this, the EPA ultimately declared the Monsanto facility a Superfund site, placing the plant on the National Priority Listing (NPL) for toxic waste cleanup.[30]

Did Monsanto have another Love Canal on its hands? The prospects were frightening. Federal agencies were shining a spotlight on a pollution problem the company had let get out of hand. If the town rebelled against the firm, the Soda Springs plant might be in jeopardy, and Roundup with it.

The threat of rebellion must have been concerning given what was happening on the other side of the country. There, residents of Nitro, West Virginia, were finally taking Monsanto to court, arguing that another herbicide the company had once sold had caused an ecological disaster in their hometown.

CHAPTER 8

"They Can Have My House; I Just Need Thirty Days to Get Out"

CHARLESTON, WEST VIRGINIA, ATTORNEY STUART CALWELL sits alone on the top floor of a three-story building amidst dozens of cardboard boxes representing more than three and a half decades of litigation against Monsanto and other companies. There are no walls, just one big room with a high ceiling. The building, half a football field in width, takes up a city block, making Calwell's desk positioned in the center of the third floor seem small and completely out of place. Sunshine coming through the floor-to-ceiling windows lights up the room, revealing the telltale signs of an attorney at work. Cigarette butts, Styrofoam coffee cups, briefs, and loose papers clutter Calwell's desk, which is positioned adjacent to a whiteboard featuring scribbled notes about an upcoming legal battle. In 2016, Calwell is still at it, more than thirty-five years after he started his first trial against Monsanto in the 1980s.[1]

This is a historical place, not just because of the litigation history it contains. Calwell sits in a former coal-mining laundering facility. This was where soot-covered miners working in the nearby Kanawha coal field brought their sullied clothes for cleaning back in the early twentieth century. It would be hard to know that history by looking around. Calwell's daughter, an interior designer, has spiced up the building, with vibrant primary colors now covering the exposed beams holding up the place.[2]

On a crisp fall day in 2016, Calwell reflected on his storied battle against the Monsanto Company, which began so many years ago when he first met with Monsanto workers in the town of Nitro, just a few miles away. Calwell spoke with a subtle West Virginia accent and politely asked if he could smoke while he talked. Over the next hour or so, a lit cigarette was always close by, burning on an ashtray or propped between his fingers.[3]

Calwell grew up in an unincorporated town called Cross Lanes just two miles away from Nitro's city center. His father was a drug salesman and his mother a schoolteacher. He attended Nitro High School, and as a teenager in the 1950s, Calwell was well aware of Monsanto's presence. Back then, he remembered seeing a brown sedan "with little gold letters on the door: M-O-N-S-A-N-T-O." This car puttered around town from time to time, visiting the homes of company employees. As a young man, Calwell paid it no mind, but years later he came to learn that this car was picking up "urine samples from the men who lived in those houses for the purposes of seeing if they had developed bladder cancer." The problem was PAB, a chemical produced at the Nitro plant that was used to treat rubber tires so they would not crack in sunlight. Calwell said that Monsanto knew as early as 1934 that PAB was harmful to human health yet allowed its workers to labor in facilities contaminated with this chemical.[4]

If Calwell's biography stopped in his teenage years, it might be easy to cast him as Reba McEntire's "backwoods southern lawyer"—a boy from Appalachia who became an attorney and used his folksy charm to make jurors and judges do his bidding. But this would be inaccurate. Calwell spoke with elocution. He rattled off the names of complex chemical compounds and offered detailed explanations of how they worked in the human body. His memory was capacious. Calwell used his brain more than his charm to get what he wanted in the courtroom.[5]

After high school, he gave college a try, finishing two years at Marshall University, but, as he put, he was "shall we say . . . invited to leave." (By his own admission, his academic performance was less than stellar.) He left for New York City and worked at the 1964/1965 World's Fair before embarking on a cross-country journey that landed him in sunny Santa Barbara, California. There he went diving for sea snails known as red abalones—"otherwise known as *Haliotis rufescens*," he made a point of clarifying. Most days he would head out into the towering crags of the Channel Islands aboard the *Double Ugly*, the *Ándale*, or the *Rat*, diving into the chilly blue waters of the Pacific Ocean seeking treasures below.[6]

Ultimately, Calwell returned to his native West Virginia where he completed his education at Marshall, finishing "in a blaze of glory," before enrolling at West Virginia Law School in Morgantown. He graduated in 1974 and got his first legal job at a law firm in the state's capital city of Charleston, but two years later he saw a golden opportunity to go it on his own when a fellow lawyer took over a judgeship, leaving behind his private practice in Nitro. In 1976, Calwell took over that man's post in the town and began doing personal injury work in the area.[7]

That is when he met Cleo Smith. Then in his early fifties, Smith had worked at the Monsanto plant and for many years complained of cancerous cysts in his body that were "as big as an egg." He had undergone numerous operations to have these growths removed over the years, but he was not getting better.[8]

When he knocked on Calwell's law-office door one day in the late 1970s, he "was in a high state of agitation," Calwell recalled. Smith claimed the "plant had poisoned him," and reportedly indicated to Calwell that he "knew he was dying." At first, Calwell was cautious—"I thought he was crazy." But over the next couple of months, the lawyer saw Smith's health decline. More important, he found medical reports archived by a local labor union that made clear there were serious problems at Monsanto's Nitro facility and that the company had known about these problems for years. Calwell quickly found fifty to sixty Monsanto workers suffering from a host of health problems. He was gathering evidence that could make for a big trial against a corporate giant.[9]

One day, as he got nearer to a decision about filing a case, Calwell decided to stop by to pay a visit to the man who had set all this in motion just a few months earlier. "I'll never forget," Calwell said of his trip to the hospital, "I went down the hall, pushed the door open, and it just had this white, flat, blue light, and the bed was empty." Several seconds of silence passed as Calwell stared off in the distance, his eyes beginning to water. He coughed, gathered himself, and said, "I went back to the office, and filed the suit immediately."[10]

Cleo Smith never got to challenge Monsanto in court, but his family and more than 170 other litigants sought justice through trial in West Virginia's judicial system. It was a long and arduous legal battle that lasted more than four years, and it could have undone the Monsanto Company. In the April 1981 plaintiffs' complaint brought before the US Circuit Court for the Southern District of West Virginia in Charleston, Calwell's clients asked for $2.16 billion in damages. This was a tremendous sum for a company whose net income that year was just over $445 million.[11]

* * * *

AT THE TIME, MONSANTO executives at Creve Coeur headquarters were feeling pressure. Beginning in the 1970s, the company faced extensive international competition in the commodity chemical business, especially from firms in Asia, and by the 1980s, there were real concerns about Monsanto's future prospects for growth.

Another major problem was oil. Roughly 80 percent of the products Monsanto sold came from petrochemical feedstocks, disturbing news as the energy crisis hit. The OPEC embargo of 1973–1974 and the oil-worker strikes associated with the 1979 Iranian revolution sent petroleum prices skyrocketing, driving up basic petrochemical costs. "Big Oil" giants—such as Shell and Exxon—which had been cutting into Monsanto's market share for years, now invested even more money to expand their capacity to produce commodity chemicals. One Monsanto executive described this as a "recipe for disaster." Cheap petrochemicals that oil majors had once "been 'giving away'" at dirt-cheap prices were now being used up at an alarming rate. Monsanto's suppliers had slowly been becoming its competitors over the past several decades. If the firm did not do something drastic in this moment to address this issue, it would be in serious trouble.[12]

Monsanto executives knew their firm had developed a deep dependency on oil that they had to break. For years the company had practiced scavenger capitalism, feeding on the excesses of the booming oil industry, but that strategy for making money now became problematic in a time of scarcity. The commercial ecology had changed, and Monsanto's survival depended on rapid and swift adaptation.[13]

It was in this context that Monsanto's chairman, John Hanley, began channeling money toward life-sciences research. The same year Khomeini loyalists took American diplomats hostage in Iran, Hanley hired Howard Schneiderman, former professor and dean of the University of California, Irvine, to run a new biotechnology program. Schneiderman had become a big name in developmental biology and was well known for his research on genetic codes that regulated cellular growth. In a 1982 shareholders meeting, Schneiderman assured those in attendance that he would be hard at work looking for new product lines that were "less dependent on raw material costs" and which had a "strong proprietary character."[14]

Chairman Hanley laid out the big picture. He told shareholders that the firm's leadership had "candidly debated the proper course for the future" and decided that "raw materials, particularly petroleum-based ones, had become too large a component of too many of our products." Even though the worst of the energy crisis was over and oil prices were starting to come down, the firm was nevertheless "vulnerable to the cyclical fluctuations of the economy."[15]

Hanley had the support of Richard "Dick" Mahoney, who became president of the firm in 1980 and CEO in 1983. Mahoney was a disciplined chemist, having joined the firm as a chemical engineer back in 1962, and for nearly two decades he had worked his way up through various promotions, living by what he dubbed the "FILO" motto—"First In the morning, Last Out at night." (Even in retirement years later, he got up every day for a 6:00 a.m. swim.) Mahoney was a Monsanto insider, someone who felt he understood where the firm had been and therefore knew where the firm should be going. Reflecting on what he saw when he examined the company's financial statements in his early years of leadership, he spoke specifically of the oil problem: "The once cheap hydrocarbons were no longer cheap," he said. "Whereas Monsanto once took 10 cents' worth of oil or gas hydrocarbons and added 30 cents' worth of technology, we were faced with buying 30-cent hydrocarbons and selling a finished product into a marketplace that allowed us to add only 10 cents of our technology. The value we added was dwindling." Mahoney, like Hanley, understood that biotechnology, a budding industry that promised rapid innovation and growth, could be Monsanto's way out of the cul-de-sac it found itself in.[16]

In 1984, Monsanto invested $150 million to create the Chesterfield Life Sciences Research Center just outside St. Louis, Missouri, with Schneiderman at the helm. The company foreshadowed biotechnology breakthroughs just around the corner, noting soon "important traits of plants, including stress-, herbicide-, and pest-resistance . . . may be possible to transfer . . . to important crop species." This was the future of a new, retooled Monsanto. It was the answer to a pesky petrochemical problem and a competitive commodity chemical market. Technologies based on manipulating genetic codes were going to be the big profit generator for the firm in the decades ahead.[17]

Other chemical companies were also diversifying. Dow and DuPont began eyeing lucrative new opportunities in the pharmaceutical industry and investing in biotechnology. Like Monsanto, Dow reported that more than 80 percent of its products came from petrochemical feedstocks, a dangerous dependency in a time of high oil prices that worried executives. In 1981, Dow purchased the Merrell Drug Division of Richardson-Vicks and soon formed a joint venture with Eli Lilly to launch an agricultural biotechnology enterprise. DuPont was also making similar adjustments, making major investments in life sciences as well. But DuPont was in some ways being conservative in its approach, choosing to purchase Conoco oil company in the 1980s in an effort to shore up its supply of critical raw materials for its chemical business. Monsanto was being much more aggressive, closing synthetic fiber plants and shedding many petrochemical product lines as fast as it could.[18]

Considering its vulnerability during this time of costly restructuring, Monsanto feared a damning verdict in the 2,4,5-T case. The firm needed to put an end, finally, to the trouble in Nitro.

* * * *

IT HAD BEEN A long time since James Ray Boggess witnessed the Nitro autoclave explosion in 1949. The company had been trying to keep his and other workers' health problems associated with dioxin poisoning out of the public spotlight ever since. But after the Vietnam War, as scientists and reporters warned about the dangers of Agent Orange, 2,4,5-T, and dioxin, Monsanto executives must have feared retribution.

The company faced threats from all sides. In 1978, Vietnam veterans initiated a suit against manufacturers of Agent Orange in New York federal court. In the Pacific Northwest, Carol Van Strum, a mother from the Five Rivers valley region of Oregon who worried her children were in danger because of herbicide spraying, mobilized an organization called Citizens Against Toxic Sprays (CATS) to fight the US Forest Service's continued use of dioxin-contaminated 2,4,5-T in weed-control programs. Strum's organization pointed to a 1976 explosion at a 2,4,5-T manufacturing plant in Seveso, Italy, that left more than 170 residents and workers with chloracne and thousands of animals dead. Evidence was mounting that dioxin was extremely toxic. Meanwhile, *New Yorker* reporter Thomas Whiteside penned a series of articles, compiled into a book called *The Pendulum and the Toxic Cloud: The Course of Dioxin Contamination*

(1979), which offered a comprehensive survey of dioxin problems plaguing the world. Whiteside reported that dioxin could actually be found in mother's milk in the United States. By now it was clear that other compounds created in Monsanto factories, not just PCBs, were a danger to breast-fed children. Then, in January of 1979, a railroad car transporting dioxin-contaminated wood-treatment chemicals produced at Monsanto's Sauget plant crashed in Sturgeon, Missouri, contaminating the environment there. Frustrated citizens brought suit against Monsanto in what became the longest civil jury trial in US history up to that time and the first major toxic tort case against Monsanto. As concerns about dioxin grew, Jimmy Carter's administration acted: in December of 1979, the president created an Interagency Task Force to look into dioxin contamination and herbicide use.[19]

In light of all of this, Monsanto officials viewed the situation in West Virginia with particular consternation. They knew that the 2,4,5-T they produced at Nitro was problematic because it contained high levels of dioxin—higher than would have been found in herbicides produced by Dow and other competitors. "Ours was a 'dirty' process," admitted a Monsanto official in an internal document dated January 24, 1979. Another piece of company correspondence spoke in blunt terms: "If you'll pardon a little editorial comment . . . the problems at Nitro . . . are 'catching up with us' and are not fundamentally rooted in a hostile press. They will simply be the bellows to kindle the smoldering ashes. Public relations will not be able to completely shield us from criticism; it cannot make past practices (which may have been accepted in the 1950's and 1960's) appear responsible when measured against today's standards."[20]

Company officials believed that West Virginia's political culture offered specific challenges in the dioxin battle. In 1979, a Monsanto official argued that concerns in Nitro were in large part "aggravated by the suspicious and clannish nature of the natives of Appalachia." Years later, as the dioxin trial came to a close, plaintiff James Ray Boggess reacted to this description of the Nitro men, saying,

"Perhaps we are, but we are just fighting to see fair justice is done in this case that means so much to so many people in our valley."[21]

Monsanto's leadership strategized about how to stem the growing mutiny as early as 1977. That year, Nitro management issued a memorandum calling on managers to evade public inquiries for information on dioxin. "Local union representatives will be told time is needed to prepare a response, that it is complicated, or that all parts of the data are not available," the memorandum instructed. Managers made clear that "union requests should be resisted" and that "as limited response as deemed appropriate will be developed" when answering pleas for data.[22]

This was a smart strategy. As journalist Marie-Monique Robin explained, "Monsanto understood by 1978 that it was the only entity that had health data going back to 1949, the date of the Nitro factory accident." The company was the "lord of information," as Calwell's legal team put it. Monsanto had access to a group of workers that had been exposed to dioxin for decades, and through careful management of health studies involving these subjects, Monsanto officials knew they were in a unique position to shape the discourse about dioxin toxicity in the years ahead.[23]

In 1979, Monsanto funded what it hoped would be the definitive research on health problems associated with dioxin exposure. That year, the firm signed a contract with University of Cincinnati Kettering Laboratory professor Raymond Suskind—the same Dr. Suskind that had first examined Monsanto workers after the autoclave explosion in Nitro thirty years earlier—who agreed to carry out a morbidity study involving more than 120 Nitro employees and associated family members affected by that 1949 Nitro plant accident. Suskind accepted $90,000 from Monsanto to complete the $122,526 health study. In 1980, he published his findings in the *Journal of Occupational Medicine* with coauthor and Monsanto epidemiologist Judith Zack. Suskind and Zack concluded that "no apparent excess in total mortality or in deaths from malignant neoplasms or diseases of the circulatory system was observed in a group of workers with a high exposure to [dioxin] who were followed over a period of nearly 30 years." There were only 32 deaths out of an expected 46.41. It seemed dioxin exposure did not correlate with increased mortality rates. Suskind, partnering with Kettering colleague Vicki Hertzberg, later published another study in the prestigious *Journal of the American Medical Association* (JAMA) with broadened scope that considered the health of more than 436 employees at the Nitro plant who may or may not have been exposed to 2,4,5-T production from 1949 to 1969. Beyond chloracne, that 1984 JAMA article also showed no clear evidence of serious health problems associated with dioxin exposure.[24]

Suskind's studies were huge assets for Monsanto, but they did not reveal the whole truth about dioxin. In both the Vietnam veterans case and the Nitro case, Monsanto used Suskind's research to show that dioxin was not carcinogenic and that exposure to this chemical did not result in increased incidence of death. But years later, follow-up studies told another story. In 1991, National Institute of Occupational Health and Safety scientist Marilyn Fingerhut produced a groundbreaking study with several other researchers finding that "mortality from all cancers combined . . . was slightly but significantly elevated" for people exposed to large doses of dioxin. Fingerhut sampled a much larger population than Suskind, investigating the health records of "5,172 workers at 12 plants in the United States that produced chemicals contaminated with" dioxin. Fingerhut confirmed these findings in an even bigger study (21,863 workers) completed in 1997. Fingerhut's work has stood up against subsequent analysis. In 2016, the director of the National Institute of Environmental Health Sciences, Dr. Linda Birnbaum, said in an interview that researchers now know dioxin is linked to "just about every health effect you can think." She ticked off a laundry list of ailments, citing studies showing effects on heart function, vascular systems, and immune response. And when it came to cancer, Birnbaum said there was now "lots of data" indicating dioxin was a "human carcinogen." In 2001, the US Department of Health and Human Services National Toxicology Program listed dioxin on its "known to be [a] human carcinogen" list, and Boston University cancer epidemiologist Richard Clapp declared dioxin "the Darth Vader of toxic chemicals because it effects so many systems of the body."[25]

So how did Suskind and his fellow researchers get it wrong? Stuart Calwell held that Suskind was merely doing the bidding of a company that was funneling thousands of dollars toward his research. But whether such charges of cash corruption were true or not, another problem was sample size, something Suskind admitted in 1980 when he said that his findings "could not be considered conclusive" because "of the small size of the cohort." Other critics took issue with different aspects of the studies. Chemical pathologist and University of Leeds professor Alastair Hays, writing with Environmental Defense Fund senior scientist Ellen Silbergeld, published a scathing criticism of Suskind's work in a 1985 *Nature* article. Hays and Silbergeld pointed out that there were inconsistencies between Suskind's 1980 study and a similar study published in 1983 by Judith Zack and Monsanto's director of epidemiology Bill Gaffey, which also showed no links between dioxin exposure and elevated cancer rates. "In comparing the two papers," Hays and Silbergeld wrote, "some men are listed as exposed to dioxin in one paper, but as not exposed in the other." This was troubling. Moreover, some "19 individuals who died of circulatory disease or cancer" while working at Monsanto "and who meet the criteria for inclusion in the exposed group" were not included in the Gaffey–Zack experiment. Hays and Silbergeld called for completely new assessments given the fact that the "epidemiological picture at Monsanto remains confused."[26]

Despite these critiques from the scientific community, Suskind, Zack, and Gaffey's research became key pieces of evidence as Monsanto went on trial in the 1980s. In the months ahead, Stuart Calwell had to prove that the company knew it had exposed its workers to a deadly chemical, even though published studies funded by Monsanto downplayed the health effects of dioxin exposure. This was going to be a tough hurdle to overcome, but Calwell had a plan.

John F. Queeny founded the Monsanto Chemical Works in St. Louis, Missouri, in 1901, when he was in his early forties. This picture was taken toward the end of his life when he had achieved business success. According to one source, by this point he apparently enjoyed "meat and potatoes and three or four martinis every day for lunch." Hard times no more for a man who had seen his first factory, paid for with his life savings, go up in flames in the 1890s.

Left: Olga Monsanto, featured here with her children, came from a wealthy family. Olga's German mother, Emma Cleeves, was instrumental in the early history of Monsanto, serving as an interpreter for the European chemists that John Queeny hired to make his chemicals.

Below: Edgar Queeny (1897–1968), who took over as president of his father's company in 1928, radically diversified Monsanto's product portfolio, acquiring a number of chemical concerns that made Monsanto the fifth largest chemical company in the United States by the end of World War II. Edgar Queeny loved donning formal attire and once appeared on a list of the best-dressed men in America, alongside Fred Astaire and Henry Ford's son Edsel.

Right: A 1970s photo of Monsanto's chemical plant in Nitro, West Virginia, where the company manufactured the 2,4,5-T that went into Agent Orange, a toxin-laden defoliant used by the US military in the Vietnam War.

Below: In the 1980s, these Nitro workers—John Hein, James Ray Boggess, June Martin, Gene Thomas, and Charles Farley (left to right)—fought alongside more than 170 other plaintiffs, seeking compensation for harm caused by dioxin exposure at the Monsanto-owned plant in Nitro.

L to R: John Hein, James Boggess, June Martin, Gene Thomas, Charles Farley

Molten phosphate slag dumping at Monsanto's Soda Springs, Idaho, facility in 2016. In the 1960s and 1970s, Monsanto sold radioactive slag waste—a by-product of a key ingredient in its blockbuster herbicide, Roundup—to local residents who used the aggregate to build home foundations, sidewalks, and roadways. After the town ultimately banned such reuse, the company began discarding slag at the south side of its plant, creating a mountain of waste.

At the entryway of the Soda Springs facility, an alternative floral representation of what occurs on the backside of the factory grounds.

Subject is Snow Balling

Where do we go from here → 1254 } identified
                              1260
Alternatives!   Possibly 1248 }
                              1242

1.) Go out of Business

2.) Sell ~~the Hell out of them~~ as long as we can and do nothing else

~~What do we tell our customers!~~ The Big Question!

only data — 90 day studies: "no effect in rat & legs"
                              "100/ppm ... c, Phosrisol
                              ... egg shellness in leghorns"

3.) Try to ~~stay~~ in business — controlled applications - Control contamination levels ...

DSW 164934

Left: Scribbled notes from an October 1969 meeting of Monsanto's ad hoc committee set up to consider potential plans for addressing growing concerns that company PCBs were dangerous, including the idea to "sell the hell out of them as long as we can."

Below: David Baker, a resident of Anniston, Alabama, standing over his brother Terry's grave, who died of rare maladies at age 17. Baker fought Monsanto in the *Abernathy v. Monsanto* case in the early 2000s, a suit brought by thousands of residents who held that the St. Louis firm had endangered their lives by polluting their hometown with PCBs, toxic chemicals that today are linked to a host of serious health problems.

An Agent Orange cleanup site at Da Nang airport in Vietnam in 2017. During the Vietnam War, when the airport was a US military base, American forces stored and sprayed large volumes of Agent Orange, a defoliant contaminated with a toxic compound called dioxin that is associated with numerous health maladies. Here, USAID and the Vietnamese Ministry of Defense excavated roughly 90,000 cubic meters of dioxin-contaminated soil. Started in 2012 and completed in 2018, this remediation effort ultimately cost $116 million, with much of that sum spent on electricity to "cook" the soil via the 1,250 heater wells drilled into this concrete container. Though Monsanto was the largest seller of dioxin-contaminated Agent Orange during the Vietnam War, the company did not offer any funding for this undertaking or for similar remediation efforts now taking place elsewhere.

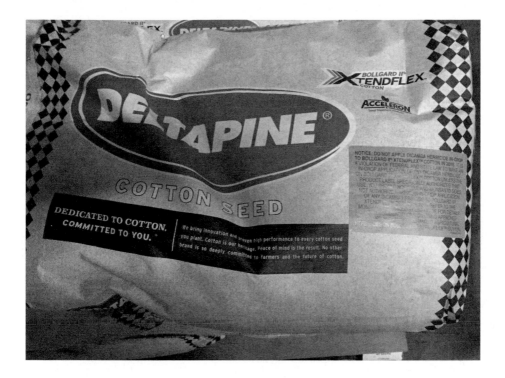

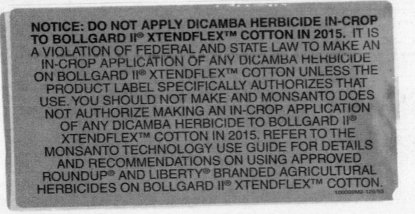

NOTICE: DO NOT APPLY DICAMBA HERBICIDE IN-CROP TO BOLLGARD II® XTENDFLEX™ COTTON IN 2015. IT IS A VIOLATION OF FEDERAL AND STATE LAW TO MAKE AN IN-CROP APPLICATION OF ANY DICAMBA HERBICIDE ON BOLLGARD II® XTENDFLEX™ COTTON UNLESS THE PRODUCT LABEL SPECIFICALLY AUTHORIZES THAT USE. YOU SHOULD NOT MAKE AND MONSANTO DOES NOT AUTHORIZE MAKING AN IN-CROP APPLICATION OF ANY DICAMBA HERBICIDE TO BOLLGARD II® XTENDFLEX™ COTTON IN 2015. REFER TO THE MONSANTO TECHNOLOGY USE GUIDE FOR DETAILS AND RECOMMENDATIONS ON USING APPROVED ROUNDUP® AND LIBERTY® BRANDED AGRICULTURAL HERBICIDES ON BOLLGARD II® XTENDFLEX™ COTTON.

A label Monsanto put on its Bollgard® XtendFlex seed bags in 2015 and 2016. (These "stacked" seeds enable cotton to tolerate dicamba, glyphosate, and another herbicide called glufosinate. It also enables plants to produce a Bt toxin that kills pests.) Though Monsanto had already released dicamba-tolerant seeds into the market in these years, the EPA had not yet approved any dicamba herbicides for use with these seeds during the growing season out of fear that dicamba would spread off-target in hot temperatures. At the time, Monsanto was working on a new brand of dicamba, but the company prevented university scientists from testing the formulation for volatility or drift, arguing it did not have enough in stock. One Monsanto employee laughed in a 2015 email, "Difficulty in producing enough product for field testing. Ha ha ha. Bullshit." Another Xtend team member was worried, telling his colleagues in confidential correspondence that he did not think "One sticker is going to keep us out of jail." As it turned out, there were still dicamba-drift problems even after the new dicamba formulation, XtendiMax, reached the market in 2017.

At Bayer's April 2019 annual shareholders' meeting, CEO Werner Baumann (pictured speaking) faced a humiliating situation when investors, concerned about pending Roundup litigation that could cost the firm billions of dollars, issued a vote of no confidence in the company's leadership. This was the first time this had happened in the history of the DAX, the German stock exchange. In 2018, Bayer acquired Monsanto in a $63 billion deal, just a month before Dewayne "Lee" Johnson won the first case in which jurors determined Roundup could cause cancer. Next to Baumann is a slide showing how far the company's stock price had dropped since Bayer acquired Monsanto.

* * * *

IT WAS NO ACCIDENT that Stuart Calwell became the plaintiff's attorney in the Nitro case. He had written the legal playbook that enabled workers to file suit against employers for "willful, wanton, and reckless" behavior. This precedent became known as the *Mandolidis* decision, and it was the foundation on which Calwell's Nitro case rested.[27]

The *Mandolidis* case involved machinist James Manolidis (his name was erroneously misspelled in court documents), who worked for the West Virginia furniture company Elkins Industries. In April of 1974 he was operating a 10-inch table saw when his right hand slipped, sending two of his fingers through the sharp edge of the rotating blade, severing them from his hand. Two years later, he brought suit against the Elkins company, charging that the firm had failed to install a "safety guard" that would have prevented this from happening.[28]

The plaintiff's attorneys in the case worked to exploit a key provision of the West Virginia workers' compensation law that said a company could be held liable in court for wrongful acts if the firm deliberately intended to harm its employees. This deliberate intent clause had been on the books for years, but the courts had largely offered a narrow construction of this language that prevented successful appeal in civil litigation.

Enter Stuart Calwell, who as a law student at the University of West Virginia composed a brief that sought to expand workers' rights to bring tort cases under this exemption offered under the law. This brief, requested by an attorney Calwell knew, proved crucial in securing the West Virginia Supreme Court opinion in the *Mandolidis* case that acknowledged workers' rights to bring suit against employers for "willful, wanton, and reckless misconduct."[29]

For injured laborers in West Virginia, the *Mandolidis* decision represented a new opportunity to hold big businesses accountable for harm caused to employees. The Nitro workers were going to see if this case could help them win big against one of the most powerful chemical companies in the world.

*   *   *   *

CALWELL THOUGHT HE HAD Monsanto "dead to rights" when the trial began in the 1980s, but this young lawyer soon learned just how hard it was to beat a corporate giant in West Virginia court. For three years preceding trial, Monsanto fought Calwell as he sought to secure access to company files. As late as January of 1984, plaintiff discovery requests were still outstanding, and in a hearing that month, an attorney for the defendant's legal team complained that Monsanto's staff and lawyers had "spent over 40 man-years of people's time" answering discovery requests. Complying with Calwell's outstanding demands for documents would "shut down parts of our company," he grumbled, adding, "We have tens of thousands of stockholders whose money we're talking about here."[30]

The money talk angered Calwell. "We've lost some 14 [plaintiffs] since we filed the suit. They've died from these chemicals." Monsanto pretended "to be confused when they're not" about discovery requests, he said. "That absolutely upsets me to no end," he continued, "particularly when I've lived with 180 of these people for two years." This case was about people's lives—about corporate greed that yielded widows. He demanded Monsanto deliver the documents immediately.[31]

Calwell—sometimes described as "mule-headed" in the Charleston press—was not going away, no matter how long trial was delayed. He could afford to do so in large part because the United Steelworkers union came to his aid, giving him thousands of dollars to litigate the case. He was going to put up a fierce fight.[32]

And so were Monsanto's lawyers. Over the course of the past decade, company executives had watched as legislation, such as the Superfund Act, expanded corporate liability for toxic legacies that had remained buried for years. If Monsanto lost a case like the one in Nitro, it could set a dangerous precedent for other tort litigation involving past wrongdoing. And that could surely ruin Monsanto.

* * * *

ON A COOL MONDAY MORNING, June 25, 1984, Stuart Calwell and his legal team filed into the US Courthouse in Charleston, and prepared delivery of opening statements. To Calwell's left was Charles Love III, another local attorney and Monsanto's counsel. Presiding that day was fifty-five-year-old West Virginia native, Judge John Thomas Copenhaver Jr., who had been appointed to the bench by President Gerald Ford nearly five years earlier. Across the courtroom sat the jurors. They represented a cross section of Charleston: "five housewives, a government worker, a retired bus driver, a miner, a teacher, a clerk, a salesman and a computer operator at Union Carbide Co." In their hands lay the fate of seven Nitro men as well as the future solvency of the Monsanto company.[33]

Recent news from New York would have surely been on the minds of the lawyers gathered in the courtroom. On May 7, 1984, the Hon. Jack B. Weinstein of the US District Court for the Eastern District of New York approved a $180 million settlement in favor of Vietnam veterans who had been exposed to dioxin during the war. Monsanto, the largest producer of Agent Orange by volume, shared responsibility with other manufacturers of the chemical, such as Dow and Hercules. This meant that though Monsanto was the largest producer of Agent Orange by volume, its ultimate contribution to the settlement was minimal, something that pleased Monsanto's lawyers. In fact, when the settlement decision came down, Judge Weinstein, known for his amiable demeanor, popped champagne bottles in his office and shared drinks with the plaintiffs' and defendants' attorneys. Weinstein later explained the aftershocks: "The manufacturers were delighted to get off the hook to the degree they had. Their stock rose on announcement of the settlement, and the government walked away."[34]

But if Monsanto's legal team partied in the judge's Brooklyn quarters after the Vietnam veterans settlement in May, the hangover quickly set in back in West Virginia. The Nitro case was a different scenario. This was Monsanto's facility over which it had direct control. A judgment could mean billions of dollars in liability, not millions. The company was alone in this fight, and it could not afford to lose such a consequential case.

Calwell gave an explosive opening statement. He argued that the company knew it had a toxic problem as early as the 1950s and that it did nothing to address it. As the *St. Louis Post-Dispatch* reported in its coverage of the trial, Calwell "based this allegation on a document drafted in the mid-1950s in which Monsanto computed the medical and legal costs of the health problems caused by 2,4,5-T to be just 4 cents per pound of the herbicide." Calwell argued, "This proves that Monsanto made the conscious decision in the 2,4,5-T process not to change it because it was cheaper to do it the way it was." "We're going to show that this was all done for money," Calwell contended.[35]

He backed up his opening claims with alarming evidence, including a 1965 company memo that clearly showed Monsanto's awareness of dioxin's potential toxicity. Penned by Monsanto's St. Louis–based medical director, Dr. R. Emmet Kelly, this confidential correspondence said that dioxin was "a potent contaminant" and "very conceivably . . . a potent carcinogen." He expressed fear that this compound might cause "another epidemic at Nitro." Kelly's closed-door admission occurred precisely as the US military ramped up requests for Agent Orange. Another four years passed before the State Department terminated its contracts with Monsanto's Nitro facility. In the meantime, the US military exposed thousands of veterans and Vietnamese citizens to dioxin, despite what Kelly and Monsanto officials suspected about the dangerous consequences of such exposure.[36]

During opening statements, Calwell wanted to personalize the plight of the plaintiffs. He singled out his clients, one by one, talking about the number of hospital visits each made over the years (some had gone to the doctor more than 100 times). He wanted the jurors to understand their suffering.[37]

In the face of this assault, Monsanto's legal team quickly tried to quell concern. Love responded to Calwell's claims by first arguing that the firm had indeed made improvements to their facility in the 1950s, that should have reduced workers' exposure to dioxin. He held that, in time, "systemic problems disappeared," evidence that Monsanto was making strides to improve its Nitro facility. And as per Calwell's contention that Monsanto was making sinister calculations about the costs of dealing with workers' health problems, Love made an appeal to capitalist sensibilities: "This company is a company. It is in the business of making money. All businesses are that way. That's our system, that's our country."[38]

On Tuesday, Love concluded his opening remarks. He argued that the plaintiffs' health problems were due to their "lifestyle" and "habits." His evidence came from depositions conducted months prior where he had probed into the Nitro workers' vices. He pointed out that many of the Monsanto employees were smokers and drinkers and that there was no way to determine whether their ailments in old age were caused by these choices or by exposure to toxic chemicals. Furthermore, he lauded the many medical programs and pension plans the company set up to take care of its workers. "What Monsanto has done over the years is something to be proud of."[39]

So began a jury trial that lasted almost a year. In the ensuing months, Calwell worked hard to show that Monsanto knowingly exposed its workers to chemical contamination at its Nitro plant and that it was less than forthcoming about health risks associated with work on-site. In July, for example, Calwell's legal team read from the deposition of medical director R. Emmet Kelly, who admitted that Nitro management did not inform workers of potential "health risks" if such "risks were only suspected." This was something that enraged plaintiffs in the case. In July, Gene Thomas, one of the seven plaintiffs, told reporters he "wasn't bitter until" he "found out they had been withholding information." "They knew in the 1950s it was bad for us," he said, "and they didn't tell us anything, and just kept us working there. They lied to us."[40]

Proving this to the jury, however, was another matter.

* * * *

THE JURY SAT IN deliberations for five days, and when they emerged with a verdict, they requested that a "consensus statement of our feelings . . . be read into the court record after the verdict is announced." This was unusual, and it revealed the moral dilemma the jurors felt they faced. In the statement, the jury admitted that a "preponderance of the evidence showed dioxin causes or contributed to some of the health effects the plaintiffs now exhibit." However, they argued they could not find in favor of the plaintiffs because "Monsanto did not show a willful, wanton, and reckless attitude toward its workers health and safety," the legal bar established by the *Mandolidis* precedent. While chastising Monsanto for not pursuing "a diligent course of action in trying to determine the full impact of dioxin on the health of its workers," they nevertheless concluded that Nitro laborers should have taken personal responsibility for maintaining their well-being: "The workers must exercise reasonable judgement in insuring good health and safety by asking questions when a health problem becomes commonly recognized."[41]

None of the seven plaintiffs received any compensation for the health problems they experienced as a result of dioxin exposure. The jury did award Nitro employee John Hein $200,000, finding that Monsanto had recklessly exposed him to PAB, a rubber chemical that the company knew was carcinogenic, but this was a paltry sum considering the health problems he faced. Hein had contracted bladder cancer while the trial was under way and as a result was going to need months and months of expensive medical treatment. Calwell argued before the court that "$200,000 was clearly inadequate in light of the injury inflicted by the defendant" and filed an appeal immediately on behalf of Hein and the other plaintiffs.[42]

To deter the appeal, Monsanto made an astonishing move. In the judgment handed down on May 17, 1985, Judge Copenhaver approved Monsanto's request for recovery of court costs from the plaintiffs. Calwell's clients were now responsible for more than $500,000 in legal expenses accrued by Monsanto over the past four years (later reduced to roughly $300,000). To ensure payment, the company put liens on the plaintiffs' homes. If they didn't pay the defendant's costs, they were going to have to find another place to live.[43]

The Monsanto liens sparked an uproar in Nitro. Former Monsanto employee Omar Cunningham charged that the company's move was "the biggest darn trick ever played on a bunch of suffering men." Another Nitro plaintiff echoed the same concern: "If the little man can't sue a big company without losing his home, is there any justice?" They understood what this act was meant to signal. "They want to make an example of these men," charged Calwell: "They're saying, 'If you sue Monsanto, it will be the ruination of you.'"[44]

James Ray Boggess explained what the liens meant for him and his family: "They've stopped our lifestyles cold. I was getting ready to refinance my house at lower interest, rent it out to make payments and move south. Now I can't." Boggess' dream of finally escaping a town that had been the site of so much hardship was put on hold. The story was much the same for the other plaintiffs in the case.[45]

But if Monsanto's legal team thought bringing financial pressure to bear on the Nitro men would make them break, they were mistaken. If anything, the move actually emboldened the litigants. Gene Thomas expressed defiance, saying, "It gives you kind of a low feeling to know that somebody's put a lien on your property, something you've worked all your life to get. But still when I get into something like this, I'm going to stick with it." Boggess was of the same mindset: "We believe in what we're fighting for, and we don't believe in backing up to anything less than the truth. It's that simple."[46]

Calwell remembers this moment well. He called his clients into a conference room and said, "You know guys, I think I can go to Monsanto and tell 'em that we won't appeal if you'll just forget these costs." Their response still stuck with him. "To a man," he recalled years later, "they just said, 'You tell them, Calwell, they can have my house; I just need thirty days to get out—do you think they'd give me that?"[47]

Calwell took this message to Judge Copenhaver a few days later who decided to remove the liens on the plaintiffs' homes. That battle was won, but there was more work ahead. Calwell hoped he might be able to get his clients compensation in a higher court, so he filed an appeal with the US Court of Appeals for the Fourth Circuit. On August 29, 1987, the Fourth Circuit upheld the lower court's ruling. The long struggle to show that dioxin had caused harm in West Virginia had come to an end with Monsanto paying no compensation for 2,4,5-T–related health problems.[48]

* * * *

THE NITRO STORY COULD have turned out differently. Immediately after the case concluded, jurors expressed doubts about their verdict. West Virginia resident and juror Nancy Adkins said, "I still don't feel like I made the right decision." Patricia Buford, another juror, said that the real problem was the strictures of West Virginia law: "The judge's final instructions didn't really leave us much choice." In those instructions, the judge made clear that if the jury wanted to find for the plaintiffs, it had to be certain that Monsanto knew about the dioxin problem and engaged in "willful, wanton, and reckless" behavior that perpetuated contamination problems at the plant. But because the only extensive health studies of the Nitro workers came from company-financed labs, how could the jury make such a determination? The company was its own watchdog. There simply was not enough evidence to convince jurors that Monsanto had been reckless in its handling of chemicals.[49]

Except there was. A decade after Monsanto stopped making 2,4,5-T, the natural environment still bore witness to the realities of Monsanto's toxic past. In 1983, the EPA produced a map of the area surrounding the Nitro plant that showed numerous dioxin hot spots. In January 1985, Calwell had tried to submit this map into evidence, but Copenhaver refused to show it to the jury. Three months later, Calwell tried again, but was rebuffed, Copenhaver saying that the map had no bearing on what Monsanto did in the 1950s and 1960s.[50]

The foreman of the jury, Steven Stutter, disagreed. Almost a year after the case ended, Stutter still said he was torn about the decision he had made, especially in light of press reports that detailed dioxin contamination in the Kanawha River. When Stutter learned from *Charleston Gazette* journalist Martin Berg of the 1983 EPA map he had not been allowed to see during trial, he was unnerved. "That would have been relevant," he said, "I wish we'd seen that." Instead, Stutter and his colleagues played cards and Scrabble for hours, locked away in separate chambers out of earshot of court deliberations. "I wish they would have let us see all the evidence," he said, "That's what we were there for."[51]

Stutter's revelation was particularly significant given his background. He had worked at Union Carbide for years, as had his father. He was no enemy of the chemical industry. Nevertheless, he felt the EPA evidence was damning. Speaking of the friends he had made in that Charleston courtroom during the 11-month trial, he said, "I would hope that my fellow jurors would lend their voices in saying that our verdict would have been different." This was all so troubling for him: "I can't get it completely out of my mind."[52]

And neither could the Nitro workers. The Agent Orange saga was not over for them, nor for the thousands of American veterans and Vietnamese citizens whose bodies bore the burdens of the past.

But for Monsanto, the 2,4,5-T cases were finally closed. It was moving on. It had a new herbicide to promote, Roundup, and Monsanto scientists were now working on a breakthrough discovery that was going to make this herbicide a household name and a multibillion-dollar brand.

CHAPTER 9

"Trespassing to Get to Our Own Property"

ERNEST "ERNIE" JAWORSKI HAD BEEN AT IT FOR SOME TIME. The Oregon State University–trained biochemist first came to Monsanto in 1952 and worked in John Speziale's lab alongside John Franz when the firm discovered glyphosate. Herbicides were his specialty, and in the 1970s he began a quest to find a gene that would make plants resistant to the firm's new blockbuster herbicide.[1]

These were heady times in the burgeoning field of genetic engineering made possible by the groundbreaking DNA discoveries of James Watson, Francis Crick, Rosalind Franklin, and Maurice Wilkins roughly two decades earlier. In the early 1970s, Stanford biochemist Paul Berg announced that he and his team had successfully transferred a gene from one type of virus into the DNA of another. This was the first recombinant DNA ever made in a lab. Around the same time, Berg's student, Janet E. Mertz, was developing a method for transferring the genes of a virus known to cause cancer in mammals into *Escherichia coli*, a move that startled one of Mertz's Stanford instructors, considering the fact that this bacterium commonly colonizes human digestive tracts. What would happen if this creation got out of the laboratory? Some researchers, including Berg himself, feared what might come of such experiments, and as a result, the scientific community imposed a voluntary and temporary moratorium on genetic engineering of the kind happening in Berg's lab pending deep discussion at an international meeting held at the Asilomar Conference

Center in the sunny coastal town of Pacific Grove, California, just north of Monterey.[2]

The moratorium was short lived. Those in attendance in Asilomar ultimately decided that the benefits of genetic engineering outweighed the dangers and that scientists could control and contain their lab creations. In the immediate years thereafter, researchers left public institutions for private industry, joining new companies such as Genentech (founded in 1976), Calgene (1980), and Agrigenetics (1981) that stood on the verge of making big-dollar sales.[3]

Jaworski—a dedicated bench scientist and dreamer like John Franz—was one of the biochemists on a mission to make a major discovery in this new scientific frontier. He drew on the research of Belgian scientists Jozef "Jeff" Schell and Marc Van Montagu at the University of Ghent and Washington University professor Mary-Dell Chilton, who, simultaneously in separate labs supported with Monsanto funding, discovered how to use a bacterial vector, *Agrobacterium tumefaciens*, to transfer desired genes into plant cells. By 1979, Howard Schneiderman pulled Jaworski into Monsanto's new life-sciences research, and the fifty-two-year-old researcher began experimenting with *Agrobacterium tumefaciens* with hopes of making a host of commercial products. In 1982, Jaworski along with colleagues Robert Horsch, Steve Rogers, and Robert "Robb" Fraley, finally reported a successful gene transfer using *Agrobacterium*. By then scientists at Cornell were hard at work on a "gene gun," which would allow scientists to shoot small strands of DNA attached to tiny gold and tungsten pellets into plant cells, but Jaworski and Monsanto scientists found that the bacterial vector proved most effective in their labs.[4]

For some time, Monsanto executives had been dreaming about developing genetically engineered crops that could tolerate heavy spraying of glyphosate. If the firm could do that, it stood to make billions. After all, farmers would no longer be restricted to using Roundup before seeds were planted in fields or after harvest. Rather, they could spray Roundup—which was still under patent—over millions of acres of row crops throughout the growing season, killing weeds throughout the spring and summer without hurting their harvest. This was a market opportunity of astounding proportions.[5]

Monsanto's biotechnology team was not just interested in making plants resistant to Roundup. They were also keen on transferring a gene from a bacterium called *Bacillus thuringiensis* (Bt) into crops that would enable these crops to produce a pesticide that the bacterium emitted naturally. The competition was fierce, as new biotechnology start-ups, including Wisconsin-based Agracetus, and older European firms, such as Ciba-Geigy, raced to isolate a Bt gene and get it into plants. The market here was also big, especially for cotton, a commodity crop long plagued by bollworms that the Bt pesticide killed. Whichever firm commercialized this product first was primed to make profits. In the mid-1980s, Monsanto successfully contracted with outside firms to build a Bt gene that would be effective in plants, and in 1988, the company made a successful transfer of the gene into a tomato plant and began experimenting in test plots. The firm was out in front, aggressively spending on R&D to become the leader in Bt technology.[6]

But Roundup resistance remained the big prize. Not only would Monsanto stand to make money by selling its genetic codes for glyphosate tolerance, but also the company would radically boost sales of its most profitable herbicide. The only question was, would the St. Louis firm be the first to patent the technology?[7]

Calgene, a tiny company headquartered in Davis, California, believed it could beat Monsanto to the finish line. Calgene's leading plant biotechnologist, Luca Comai, had scared Jaworski and his team in 1985 when he published a study in *Nature* that showed he had discovered how to make a tobacco plant resistant to glyphosate. The trick was altering the shape of the EPSP synthase enzyme that glyphosate typically blocks in plants. Comai used genetic engineering to change the shape of the actual receptor on the enzyme so that glyphosate could not bind and inhibit the production of amino acids essential to plant growth. Monsanto was alarmed. For months,

Robert Horsch had been tinkering with petunias, introducing a "promoter" gene that encouraged plant cells to make greater amounts of the EPSP synthase enzyme. The idea here was to make plant cells churn out so many tiny amino acid factories that glyphosate molecules would not be able to bind to all of them. But Horsch's petunias, like Comai's tobacco plants, could only handle small doses of Roundup. Neither lab had developed a plant capable of withstanding regular herbicide dousing, though Monsanto was now convinced that Comai's path, if perfected, was the way to success.[8]

Jaworski needed to isolate the gene sequence for a new EPSP synthase that was superior to Comai's engineered enzyme, and he needed to do it fast. Otherwise, the firm might lose out to competitors. But where to look? The hunt was on.

"It was like the Manhattan project," recalled one member of the research team, who talked about the frantic search that lasted for several years. The key insight came from engineers who suggested researchers look at Monsanto's Luling, Louisiana, plant, one of the locations where Monsanto turned its elemental phosphorus into glyphosate used in Roundup. The environment around the plant was heavily contaminated with glyphosate, meaning any microorganisms that could survive in that ecosystem would likely have genes conferring resistance to Roundup. In essence, the firm was hoping to find a profitable innovation by mining its own pollution. And the strategy proved effective. By the early 1990s, Jaworski's team was able to isolate a bacterial EPSP synthase gene sequence that would allow plants to withstand large doses of Monsanto's signature herbicide. They called the system Roundup Ready technology.[9]

If Monsanto could get this product to market, they would completely revolutionize how farmers produce the food we eat.

But there was a problem. Back in Soda Springs, Idaho, the EPA, now roughly a decade and a half into its investigations of Monsanto's phosphate facilities that produced the phosphorus used in Roundup, made a shocking announcement.

* * * *

SODA SPRINGS MAYOR KIRK HANSEN called it a "bombshell." In 1990, the EPA issued a report showing that citizens of Soda Springs and of nearby Pocatello were at risk because of radioactive slag Monsanto and other mining companies had sold to the communities for decades. Radium-226 and other radionuclides contained within that material were emitting gamma radiation throughout these two communities. The 1990s report drew on earlier radiological studies, which showed that more than 30 percent of the sites surveyed in Soda Springs—including homes, churches, hospitals, and schools—demonstrated "elevated gamma radiation levels." In 1977, the Idaho Health and Welfare Department had become concerned about this waste and so banned Monsanto from selling its slag for home construction, but the agency permitted road builders and pavers to use Monsanto's waste for street and sidewalk repairs through the 1980s.[10]

The 1990 EPA report referenced a study conducted by the Department of Energy's Remote Sensing Laboratory, which did a flyover of both Soda Springs and Pocatello and detected radiation in the two towns that was above expected background levels. Maps from this survey revealed hot zones all over the place, with gamma rays emanating from sidewalks, roadways, people's homes, and other buildings. The EPA's Office of Radiation Programs also completed field studies and concluded that if nothing was done to clean up the phosphate materials in these two towns, in forty years the "probability of contracting cancer due to exposure from elemental phosphorous slag" would be "about one chance in 2,500 in Pocatello and one chance in 700 in Soda Springs."[11]

Moved by these findings, the Soda Springs city council voted in June to temporarily ban the further use of phosphate slag for roadway construction. Naturally, Monsanto, now making nearly $1 billion from its Roundup herbicide produced from Idaho ore, did not agree with the decision and pointed to the low cancer rates in Caribou County where Monsanto's phosphate facilities were located—one of

the lowest in the country—as evidence that the slag was not causing a problem. Despite the protest, Hansen and the council favored
precaution.[12]

But even as he acted to forestall further use of slag in Soda Springs,
Hansen tried to downplay the risks his citizens faced. "The very
worst scenario," he argued, "could possibly lead to one more cancer
death in Soda Springs in the next ten years." Hansen was wary of
further EPA intervention and lamented that he and his community
were "struggling for our lives against federal mandates."[13]

Hansen had good reason to support Monsanto and phosphate
companies in the area, for he, like so many other businessmen in
Soda Springs, depended on these firms for his livelihood. His company, Hansen Oil, sold diesel fuel that powered the trucks and
machinery that extracted phosphate from the Idaho hills. He was
willing to stop slag sales, but he did not want to go any further.[14]

Hansen later contracted cancer, and his wife, Debi, died from
cancer in 2006.[15]

In the days after the EPA issued its findings, Idaho politicians and
phosphate businessmen mobilized to attack the federal agency. Senator Steve Symms, a Republican from Idaho and future coauthor of *The
Citizen's Guide to Fighting Government* (1994), called for a congressional hearing to scrutinize the radionuclide report. Symms convened
the hearing in the Soda Springs High School auditorium in August of
1990 and opened the meeting criticizing the EPA, saying the agency
was "considering Superfund listing for the entire community" if the
town did not take more aggressive action to deal with phosphate slag.
He found "such blackmailing . . . to be completely inappropriate."[16]

The same month Symms held his hearing, the EPA officially
announced that Monsanto's Soda Springs plant would be listed as a
Superfund site. But in a rare move, the agency declared that the facility would be allowed to continue operation while managers worked
to clean up pollution problems. Now the question was, would the
entire town become a Superfund site as well?

Idaho state representative Robert Geddes fought back against

EPA's expanded investigation into the township. He said that citizens were upset "about the lifetime sacrifice and investments they have made in their homes, recognizing that property values could be severely depreciated with unfavorable action of a heavy-handed Government agency." For Geddes, the EPA's intervention essentially allowed the agency to "take people's property and devalue it."[17]

Facing intense political pressure, the EPA agreed to submit the 1990 radionuclide report to the agency's Science Advisory Board (SAB)—a body made up of more than 600 scientists selected from various universities and research centers. The review, conducted by the Radiation Advisory Committee within SAB, began in October 1990, and lasted for more than a year.[18]

The SAB's conclusions were revealing. The board said there was no doubt that phosphorous slag contributed to elevated levels of radiation exposure for citizens living in Soda Springs and Pocatello. Nevertheless, it did find problems with the 1990 report. One major concern was that in choosing houses for gamma radiation sampling, EPA agents focused on residences where "homeowners volunteered" to be part of the study. This was a nonrandomized sample. Citing this and other problems, the EPA called for a "fresh start" and created a working group to develop "graded decision guidelines" to help citizens living in slag-filled homes figure out their best options for remediation.[19]

Partnering with the very polluters that created the problem in the first place, the EPA named Monsanto to the newly created slag Technical Working Group (TWG) in November 1992. This group—which included Idaho public officials, EPA agents, representatives from the Shoshone-Bannock Tribes from nearby Fort Hall Reservation, and representatives from FMC and Monsanto—set out to develop guidelines for dealing with exposure to phosphate slag.[20]

The Graded Decision Guidelines produced by the TWG between 1992 and 1995 recommended that Soda Springs and Pocatello residents exposed to less than 100 millirems (mrem) of radiation above background levels annually take "no action" to deal with their pollution problem. (To put this exposure level in perspective, the average citizen experiences 10 mrem when undergoing a chest x-ray and roughly 620 mrem from all sources of radiation—natural and artificial—annually). Despite this recommendation, the Southeastern Idaho Public Health (SIPH) agency, a member of the TWG, warned citizens in 2016 that "current evidence suggests that exposure to radiation at low levels may pose some risk of cancer" and that "scientific opinion differs about how much low-level radiation an individual can be exposed to without harm."[21]

For citizens facing exposure rates above background between 100 mrem and 200 mrem, the guidelines offered some options: residents should consider "spending less time in the basement" or "move primary living areas from [the] basement to upper floors." Essentially the TWG was asking certain citizens to cordon off radioactive quarters of their houses rather than promoting cleanup that would remove the source of the problem and allow citizens to enjoy the liberty to use their homes as they saw fit.[22]

Monsanto was not totally off the hook. The firm had agreed to cover the testing and counseling costs of the program and even agreed to pay for the expenses of "remodeling, shielding, or partial removal" of slag if citizens chose that option. Of course, citizens were urged to consider solutions other than remodeling (which was listed at the end of the guidelines), but it was still an option.[23]

The EPA was keen on offering citizens choices. As one EPA official put it in 1992, the environmental agency intended "to proceed with sensitivity to the concerns of the community" moving forward. This was all part of a larger agency policy. Beginning in the 1980s, the EPA worked to show it was willing to decentralize decision making when it came to Superfund cleanup efforts in order to appease deregulatory Reaganomics disciples controlling the federal purse who might otherwise oppose environmental regulations on the grounds that they were heavy handed. In Soda Springs, the EPA made clear that it empowered homeowners to choose abatement measures that most suited their needs. This was environmental protection implemented from the bottom up.[24]

The EPA held out democratic participation as a powerful selling point for its new cleanup agenda, even though Idahoans' freedom to choose was in reality quite restricted. First, there were economic considerations. Summarizing views expressed in interviews conducted by the EPA in April of 1990, an agency official said that many residents were "mainly" worried "about potential disruption in people's lives, costs, and risks involved in potential" remediation initiatives. Other respondents felt that "property values in the area could drop due to adverse publicity associated with Superfund."[25]

But even if Monsanto paid to rip up foundations and remodel homes, such undertakings would naturally have been burdensome for homeowners, especially those with children. During this process, residents would have to seek shelter elsewhere, perhaps imposing on a neighbor or hunkering down at a local hotel for an indefinite period of time. None of this would have been appealing. Considering the fact that gamma-ray exposures from slag appeared low, removal seemed like a very unattractive choice.

Moreover, the EPA estimated in 1991 that Monsanto employed some 400 people in Soda Springs. In a town of roughly 3,000 people, this was a big deal. In 1988, Monsanto's plant manager Mike McCullough said that his payroll was approximately $21 million. When community members met with EPA in April of 1990 to discuss the pollution problems in the area, agency officials noted that citizens were particularly concerned about the "ripple effect" of job loss caused by remediation measures.[26]

And so, when Soda Springs residents were asked to make "informed decisions" about the best cleanup plan for their town, these were the very real factors they had to weigh in the calculus. The EPA had outsourced the responsibility of making a tough decision about a pesky pollution problem onto a vulnerable population that had lots of reasons to stick with the status quo. So pollution remained in place, and as a result, small amounts of gamma radiation continued to emanate from home foundations. The town of Soda Springs never became a Superfund site, and that was just the way many citizens wanted it.

* * * *

BUT ON THE OUTSKIRTS of town, in the fields surrounding Monsanto's phosphate plant, farmers and ranchers were not happy that the very company that had polluted their land was now being given broad latitude to decide how pollution would be cleaned up. In 1996, six years after the EPA designated Monsanto's plant a Superfund site, the federal government called a public meeting to discuss possible solutions for the contamination immediately adjacent to the facility. With debates about what to do with pollution in the city of Soda Springs largely settled, now the question was how to deal with land abutting Monsanto's fence line that continued to be affected by radioactive polonium-210 particles and chemical plumes drifting and flowing from the company's still active facility. In 1985, the EPA reported that the "areas surrounding" the Monsanto plant were "characterized by high total levels of radiation from a variety of sources." But because the yearly polonium-210 emissions coming from the plant's scrubber stacks remained below EPA air-quality standards, the agency never forced Monsanto to shut down its kilns. To deal with contamination on property adjacent to the plant, EPA proposed ordinances that would restrict land use in contaminated areas, but for the folks working fields around Monsanto's plant, this seemed like a gross injustice.[27]

Sixty-four-year-old civil engineer Robert Gunnell was among the disgruntled property owners. A member of the Church of Latter-Day Saints, Gunnell had grown up in a big family on a barley farm in Soda Springs. He graduated from the local high school in 1950 and left to go to college at Brigham Young University. Later, he began a business in Utah, but by the 1990s, he managed his family's 200-acre farm, which was located immediately north of Monsanto's phosphate plant. His mother, octogenarian Charlotte Gunnell, still lived in Soda Springs and had complained about the pollution problems in her backyard.[28]

Robert Gunnell thought the land-use restrictions proposed by Monsanto and the EPA for his family's estate were discriminatory. In the hearing, he pointed out that failing to clean up the contamination on his land—which included hazardous dust blown from the Monsanto property containing radionuclides—might have real financial repercussions for him and his family. "Is there a possibility," he inquired of EPA project manager Tim Brincefield, "if we were to want to sell that land that a bank would refuse to loan money because of the contamination that's there?" Brincefield candidly replied, "I honestly cannot tell you."[29]

Brincefield's answer naturally left Gunnell unsettled, and in the months ahead, he dogged the EPA. He sent a letter to Brincefield firmly pleading his case. "I am disturbed," Gunnell wrote, "that Monsanto's (and apparently EPA's) preferred solution to this serious problem is to push for adoption of a county ordinance prohibiting certain uses of private property that Monsanto does not even own." He argued that such action was unconstitutional, representing "a regulatory 'taking' of private property in violation of the Fifth Amendment." Gunnell made clear that he was not going to stop fighting: "Monsanto is the polluter. Why should we, the 'Gunnell Family' and the effected [sic] private property owner, suffer the loss of use or be subjected to the potential danger of exposure[?]."[30]

Petitioning the EPA was a family affair. Five days after Robert Gunnell sent his letter to the EPA, his eighty-nine-year-old mother penned a plea to the agency. She explained her family's long history of dealing with corporations that had little respect for their land: "Waste water has been directed to our property, clouds of ore dust have blown across our property and we have been made on occasion to feel that we were trespassing to get to our own property. Now we find that our land has unacceptable levels of contamination. Will it never stop[?]" Like her son, Charlotte was direct in her conclusion, stating emphatically, "I don't want the property zoned for Monsanto's convenience, nor do I want special plants planted on the property to experiment cleanup while taking my livelihood out of production. I just want the property cleaned up and freed of any hazardous waste."[31]

The Gunnells were not the only people frustrated by the public hearing. Other homeowners living next to Monsanto's plant sent petitions to the EPA saying they were frustrated with the way they were being treated.[32]

These protests were effective. The EPA ultimately offered the Gunnell family and others the option of having Monsanto remove polluted soil from their property. The agency reported that the Gunnells approved of the new alternative because they had been given the opportunity "to choose cleanup if they so desired." Tensions cooled and Monsanto was able to begin conversations with the Gunnells about a buyout plan. Months later, Monsanto said they would pay the family nearly half a million dollars for their property. Robert Gunnell accepted the deal.[33]

Other landowners fell in line. By June of 1997, Monsanto could happily report to the EPA that "the Gunnell family . . . had already sold their adjacent property to Monsanto" and noted that other property owners were signing similar deals. The storm was passing.[34]

But nagging concerns within the EPA persisted. Some agency personnel, for example, expressed worries that the solution being worked out in the moment—land-use controls and property buyouts—might be a short-term patch to a massive problem. The big fear was that Monsanto might one day abandon its facility, leaving others to deal with the pollution it left behind. Company officials told the EPA that this would not happen. "The Soda Springs Plant is an integral part of one of Monsanto's major product lines, Roundup™, and will be operating for a very long time," explained a Monsanto employee. In short, Monsanto officials assured the EPA that the polluter would remain atop its pollution for decades to come.[35]

Instead of punishing Monsanto by forcing it to shut down its facility until it fixed its pollution problems, the EPA handed Monsanto a great deal of control over future Superfund cleanup in southeast Idaho. The agency tasked the firm with monitoring effluent plumes, with helping to develop the slag-remediation program, and testing for pollutants around the plant. This was not a story of an overbearing federal agency dictating terms of CERCLA compliance from Washington, DC. The EPA brokered a deal that bore the imprint of local community interests. It was a deal that did not fundamentally disrupt economic life in the region. Soda Springs had its say, and as a result, Monsanto continued to produce elemental phosphorus destined for Roundup at an active Superfund site.[36]

Ultimately, this decentralized system of remediation did not achieve desired results. In 2013, nearly two and a half decades after the Soda Springs plant first earned Superfund status, the EPA issued a disheartening statement: "The remedy for the Monsanto site is currently not protective because concentrations" of "contaminants of concern" remained in groundwater, with toxic plumes expanding toward "domestic wells downgradient" of the facility. In addition, samples taken in off-site soils still showed radionuclide concentrations above "remediation goals." What the EPA dubbed Monitored Natural Attenuation (MNA) of toxic chemicals was simply not happening as predicted. In fact, in some places, concentrations of particular pollutants had increased. The prospects for the future were concerning: "Monitoring trends indicate that the groundwater performance standards will not be met in the foreseeable future." It seemed the policy of keeping pollution in place and promoting polluters to the task of containing hazardous waste was not working.[37]

* * * *

AND MONSANTO'S FACTORY WAS just one part of the problem. As Monsanto was negotiating with the Gunnells, six horses grazing near closed phosphate mines in the hilly country further north of town became mysteriously ill, forcing owners to put them down. Investigations into the cause of the illness revealed that the animals had eaten grass that had accumulated high levels of selenium. Environmental experts pointed to the phosphate mines as the source of this contamination. Overburden piles had leached high concentrations of the chemical into the surrounding ecosystem. Hundreds of animals ultimately died from eating contaminated grass around these mine sites. And studies revealed that selenium was also leaching into pristine cutthroat trout streams in the area. As it had with Monsanto's processing plant in 1990, the EPA ultimately designated Monsanto's old mines as Superfund sites in the mid-1990s to try to contain the problem. But as it did so, the agency nevertheless approved Monsanto's request to expand mining operations into new territory in southeast Idaho.[38]

Members of the Shoshone-Bannock Tribes, many who resided on the Fort Hall Reservation some sixty miles away, decried what was going on in their backyard. Tribal council members held that Monsanto's operations had contributed to the reduction of water resources in the area that the tribes had rights to under long-standing treaties with the federal government. Tribal representatives passionately proclaimed that chemical contamination of the soil and water surrounding phosphate mines was detrimental to Native Americans who hunted and fished on these lands. When one council member was asked what he thought of the EPA after his interactions with the agency in this matter, he said, "They are a joke. They are crap as far as I'm concerned."[39]

Frustration with EPA inaction also ran high among environmental organizations operating in Idaho. The Greater Yellowstone Coalition, an environmental organization with headquarters in Driggs, Idaho, worked tirelessly over the next several decades to expose the troubling selenium problems in southeast Idaho. Yet, the pace of remediation was glacial. Decade after decade, Monsanto opened new mines, even as contamination problems persisted.[40]

EPA clearly sought to encourage democratic participation in the process of remediation in southeast Idaho, but asymmetries of power, which so often emerge in debates about cleaning up toxic pollution, were reflected in policy outcomes that gave Monsanto broad authority to shape remediation. Though Shoshone-Bannock tribal members were invited to participate, they felt the government was not hearing their voices. Environmental organizations continued to raise alarms about troubling issues, but in the end, corporate control of contamination cleanup was never fundamentally challenged.[41]

By the mid-1990s, Monsanto was still managing threats to its Roundup supply-stream in Soda Springs, but it appeared to have the crisis under control. The EPA was going to let the firm continue to operate its facility, even though pollution problems were not fixed. This was good news for the company because Monsanto's first generation of Roundup Ready seeds was about to be approved for commercial release across the country, meaning Roundup herbicide sales were about to explode.

CHAPTER 10

"The Only Weed Control You Need"

ROBERT B. SHAPIRO SHOULD BE A HOUSEHOLD NAME. AS CEO and chairman of Monsanto in the mid-1990s, he oversaw the first commercial launch of genetically engineered (GE) seeds for major commodity crops, sparking an agricultural revolution that transformed the world's food system. The pace of change was astounding. In less than two decades, 89 percent of corn and 90 percent of soybeans grown in the United States came from GE seeds. Today, whether you are a tofu-eating vegan, a corn-fed beef lover, or a high-fructose drinker, it's almost certain that you've downed grain containing the DNA of a Shapiro successor first bred in a Monsanto lab.[1]

For anti-GE advocates looking for a corporate villain "heedlessly profiting from the suffering of victims and the destruction of ecosystems," Bob Shapiro is a bad fit. Shapiro, a Harvard graduate and Columbia-trained lawyer, first came to the firm in 1985 when Monsanto bought Searle, a pharmaceutical company known most notably for such successful products as the aspartame sweetener NutraSweet, the motion-sickness drug Dramamine, and the laxative Metamucil. Shapiro, working under Searle president and CEO Don Rumsfeld, had been the brains behind the pharmaceutical company's masterful rollout of NutraSweet, negotiating contracts with soft drink providers that required major brands, such as Coca-Cola and Pepsi, to feature Searle's NutraSweet swirl on their beverages. The ploy made NutraSweet a premium brand demanded

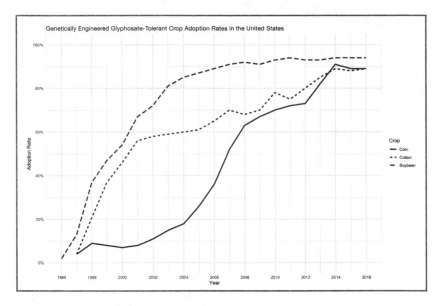

Genetically Engineered Glyphosate-Tolerant Crop Adoption Rates in the United States

FIGURE 1: Genetically engineered (GE) glyphosate-tolerant crop adoption rates in the United States: percentage of planted acres of corn, cotton, and soybean, 1996–2016. *Source: Stephen O. Duke, USDA Agricultural Research Service.*

by other firms. Years later, Shapiro recycled this strategy when he rolled out Monsanto's herbicide-tolerant seeds, requiring seed dealers to include a rainbow logo that said "Roundup Ready" on every bag containing Monsanto genetics.[2]

The 1985 Searle merger came in the middle of Monsanto's restructuring when the firm was transitioning out of its dying petrochemicals businesses and into the life sciences. Shapiro was a big proponent of this transition, believing it would make Monsanto a more environmentally friendly firm. And this fit with his personal politics. Shapiro was a Democrat, someone sympathetic to the political rhetoric of the liberal left. As a young man, the native New Yorker joined his friend Joan Baez on the Washington Mall, plucking chords on his guitar as a means of protesting the Vietnam War—a war Monsanto was simultaneously profiting from through Agent Orange sales.[3]

Shapiro hated pretension, preferring plain clothes to pinstripes, working out of a cubicle like the rest of his employees. He was Bob,

not Robert, around company headquarters, and he had an interest in promoting a collegial corporate culture: "I think it's better
to have an open office than dark-wood paneling and cuff links," he
once said. Reporters often commented on his casual appearance—
"tieless and in a brown Oxford shirt, with black undershirt peeking
out." Shapiro was a nice guy, known for returning emails in a matter
of a few hours after he received them.[4]

Shapiro was Monsanto's "guru," its "image-maker." A fan of early
1990s environmental luminaries, such as Paul Hawken, William
McDonough, Herman Daly, and Amory Lovins, Shapiro believed
that biotechnology could be used to improve the health of the global
environment, a cause he fervently supported. Like early company
leaders John and Edgar Queeny, he too saw salvation through synthetics and preached this message to Monsanto's shareholders and
the firm's biotechnology clients. As he put it in a 1997 *Harvard Business Review* interview, "The conclusion is that new technology is
the only alternative to one of two disasters: not feeding people—
letting the Malthusian process work its magic on the population—
or ecological catastrophe."[5]

To view Shapiro's faith in Monsanto's ability to avert Malthusian
limits as disingenuous would be a cynical read. In the 1997 *Harvard
Business Review*, he spoke of a need to question economic mod-
els that were based on perpetual growth. "A closed system like the
earth's can't withstand a systemic increase of material things," he
told reporters, adding, "If we grow by using more stuff, I'm afraid
we'd better start looking for a new planet." Shapiro held that Monsanto's lifesaving biotechnology could help avert rapacious growth.
He preached this do-good message to those he met, and he could sell
it because he believed it.[6]

Shapiro was just what Monsanto needed in the 1990s. Despite
the firm's best efforts to separate its shiny corporate core from its
pollution-plagued periphery, Monsanto was beginning to bear the
financial burden of its sullied past. The company buried this fact on
page 40 of the company's 1993 annual report, where the firm calmly

admitted that the EPA had identified it as a potentially responsible party (PRP) at eighty-nine Superfund sites around the country. Soda Springs, Nitro—these were just some of the toxic debts coming due in the 1990s. The company outlined liabilities costing hundreds of millions of dollars but assured investors that company profits would "not be materially affected."[7]

The balance sheet did not make it seem that everything was okay. In 1992, the company posted an $88 million loss, and the firm admitted it was going through growing pains. Monsanto had not weathered the early 1990s recession well, reporting a more than 45 percent drop in net income between 1991 and 1992. By 1993, the firm was on the rebound, having unloaded some unprofitable commodity chemical product lines, but investor confidence was anything but assured.[8]

A much slimmer Monsanto became more and more dependent on just a handful of big brands to make its profits. NutraSweet was one of those moneymakers. This product, made from combining amino acids (the basic building blocks of proteins), earned the firm nearly $1 billion in revenues in 1991. Monsanto, a company that said it sold chemicals such as Roundup to produce more food, was also marketing an artificial sweetener as a cure for a gluttonous society downing too many calories. The clash in messaging hinted at the reality that increasing productivity was never really the solution to America's (or the world's) food problems, even if Monsanto was good at selling this idea to the public.[9]

By 1993 aspartame was coming off patent, and sales began to decline, forcing company researchers to seek out new areas for growth. Ambien, a sleep aid introduced in 1993, provided a boost to the firm's pharmaceuticals business as did Daypro, an arthritis-fighting medicine. In the area of animal sciences, the company hoped it had a winner with Posilac, a branded hormone called bovine somatotropin (BST) produced by genetically engineered *Escherichia coli* bacteria and designed to increase milk production in cows. It was the first real commercial biotech product Monsanto brought to market, though the road to commercialization had been tough. Critics, including the popular environmentalist and economist Jeremy Rifkin, had fought Posilac, saying Americans did not need more milk. To prove their point they noted the USDA's ongoing milk-stockpiling program designed to keep prices in a glutted market from collapsing. They tried to block BST, citing evidence that the hormone might cause harm to both humans and cows, but Monsanto fought back, with the company asking public affairs strategists to "ghost-write OpEd pieces, [and] sample editorials" in support of BST. In the end, the FDA sided with Monsanto and approved the product for commercial use in 1993. Monsanto soon boasted that American dairy farmers controlling nearly 30 percent of the US milk market had purchased its new biotechnology.[10]

Beyond these big pharmaceutical brands, nothing was more critical to Monsanto's bottom line than Roundup. The firm's Agricultural Division brought in more than $2.2 billion of the company's $8.2 billion in revenues in 1994, and Roundup was by far the block-buster brand driving growth. As early as 1990, Roundup yearly sales approached $1 billion with the herbicide contributing to approximately 30 percent of the company's overall profits. Monsanto reported that glyphosate sales had increased by 200 percent between 1990 and 1994, and the prospects for growth seemed endless. The company was about to introduce the first Roundup Ready seeds in commercial markets, meaning farmers would soon be able to spray Roundup throughout the growing season on crops genetically engineered to be resistant to the firm's famous herbicide. This was revolutionary. Monsanto had spent in excess of $800 million to create new agricultural biotech products, more than any other global firm. Considering the steep costs of investment, Monsanto desperately needed to make these products profitable.[11]

And it needed to do it fast, because there were startups and new biotechnology firms fighting for supremacy in this new commercial frontier. AgrEvo, the agricultural arm of the German chemical company Hoechst, had recently genetically engineered crops to be

resistant to its herbicide called Liberty (active ingredient, glufosinate). The company was slow to get its product commercialized, but
this was a direct threat to Monsanto's potential herbicide-tolerant
seed business. (AgrEvo's technology ultimately became Liberty-
Link, which BASF bought in 2017.) Pesky Calgene also had a brief
moment of success in 1994 when it introduced Flavr Savr, a genetically engineered tomato, to reach commercial markets. Technically,
this tomato, which Calgene altered to maintain a shelf life longer
than that of non-GE varietals, was the first commercial GE crop
to reach the US market, but there were serious problems. Calgene's
tomato team had not done enough work with tomato breeders to
ensure they got high yields of quality produce. Pests ravaged many
of the firm's tomatoes, and others were damaged during shipment to
stores because they were picked too late in the ripening process. In
the end, Calgene failed because it never mastered the fundamentals
of growing and distributing tomatoes. When the project flopped,
Monsanto bought part of Calgene in 1995, then took a controlling
interest before finally acquiring the company, but other competitors
remained, ready to steal a share of this new GE seed market. Monsanto had to have a successful Roundup Ready launch if it wanted
to be the industry leader.[12]

This was the task before Shapiro in 1995 when he became CEO
of the company. He was the front man orchestrating the rebranding
of Monsanto as it sought to slough off its chemical past and enter
the biotech beyond. The firm had just bet everything on genetically
engineered seeds. Now Shapiro had to convince investors to bet on
these seeds too.

* * * *

A FAVORABLE POLITICAL CLIMATE HELPED. In 1980, the US
Supreme Court decided the landmark case of *Diamond v. Chakrabarty*, which helped to ensure that Monsanto's intellectual property
would be protected in the new biotechnology economy. Microbiologist Ananda Mohan Chakrabarty had filed a federal suit in 1972

after the US Patent and Trademark Office denied his patent application for a bacterium he had genetically engineered for General Electric that could consume oil spilled in oceans. The Supreme Court overturned the US Patent and Trademark Office's decision that "micro-organisms are 'products of nature,' and . . . that as living things they are not patentable." Henceforth, genetically engineered microorganisms could gain patent protection under US law.[13]

But some biotech firms were left with a lingering question: Did the *Chakrabarty* case apply to plants? After all, the Plant Protection Act (PPA) of 1930 and the Plant Variety Protection Act (PVPA) of 1970 specifically outlined plant breeders' exclusive selling rights in the United States. The PPA was quite limited in that it only covered a select number of plants propagated asexually, but the PVPA went much further, offering plant protections for sexually reproduced crops, including commodities such as soybeans, cotton, wheat, and vegetables. Under the PVPA, plant breeders could acquire exclusive rights to sell unique varieties of plants for up to seventeen years. Public agencies fought to limit the reach of this new proprietary regime, successfully inserting clauses in the legislation that allowed farmers to replant seeds produced from protected plants. But despite these provisions, the law was seen as a boon for commercial seed companies. After 1970, major seed firms, such as Dekalb Hybrid Wheat and Delta & Pine Land Company, ramped up their investment in germplasm research. One survey showed that the total plant breeding outlays made by 59 leading seed companies doubled between 1970 and 1979. The *Chakrabarty* case promised to accelerate this pattern of investment, because it meant plant breeders could potentially bypass the PPA and PVPA (which were limited in their protective powers) and seek much stronger patent protections from the US Patent and Trademark Office. But whether this could be done was still not clear. As sociologist Jack R. Kloppenburg Jr. explained, "The potentially overlapping protection provided by these different laws raised substantive and procedural difficulties that could be resolved only by litigation, which pleased no one, except perhaps lawyers."[14]

Ultimately, resolution came in a 1985 case before the United States Board of Patent Appeals and Interferences, *Ex parte Hibberd*. In that case, the board determined that breeders could choose which form of protection they wanted (PPA, PVPA, or a US Patent and Trademark Office patent). Biotech companies were clearly interested in the latter, because it meant that they could claim title to parts of a plant—its genes, cells, and so forth—instead of just the whole organism. It also meant that they would not necessarily be bound by the PVPA rules regarding replanting of saved seeds. If plant breeders thought they could make a lot of money after the passage of the PVPA, the *Hibberd* case made clear that GE seed sales were destined to bring in billions in the years ahead.[15]

Federal courts had offered assistance, but so too did the executive branch. Following the Reagan Revolution of 1981, Monsanto found friends in a pro-corporate White House that made sure Monsanto, Genentech, Calgene, and others did not face unnecessary regulatory burdens when it came to commercializing genetically modified seeds. Berkeley microbiologist David Kingsbury, Reagan's chief adviser on biotechnology, led the charge. Kingsbury was the assistant director of the National Science Foundation and a big believer in the benefits of biotechnology. In 1984, Reagan appointed him to head a special Cabinet Working Group on Biotechnology, which was tasked with developing a pro-industry GE policy that would not "require any additional laws or regulations." Two years later, the group delivered what Reagan wanted: The Coordinated Framework for Regulation of Biotechnology. The new federal framework made clear that regulation of GE products "must be based on the rational and scientific evaluation of products and not on *a priori* assumptions about certain processes." Essentially, the framework held that existing laws already gave the EPA, USDA, and FDA sufficient authority to oversee the commercialization of GE creations. The federal government would give these products no special scrutiny in deciding whether they could be adopted in US markets.[16]

And more good news was on the way. In the latter half of the

1980s, Leonard Guararria, the Monsanto man overseeing regulatory affairs, kept pressing the company's position in DC. In 1986, he set his sights on Vice President George H. W. Bush, the heir apparent to the Reagan dynasty, meeting with him at the White House with the goal of impressing upon him Monsanto's wishes. "We bugged Bush for regulation," Guararria said, "We told him that we have to be regulated." *New York Times* reporter Kurt Eichenwald explained Guararria's objective well: "Government guidelines, the executives reasoned, would reassure a public that was growing skittish about the safety of this radical new science. Without such controls, they feared, consumers might become so wary they could doom the multibillion-dollar gamble that the industry was taking." This strategy of seeking federal regulation of industry to gain legitimacy was the same tactic John Queeny had deployed during the Pure Food and Drug Act debates of the early 1900s.[17]

But when Guararria spoke of regulation, he meant something very specific. What he really desired were regulatory codes that were lax and loose, ones that would not put undue burdens on industry but that would do the trick in convincing the public that federal agencies approved of what was being sold to the public.

Bush seemed to understand what Monsanto wanted. In May 1987, he traveled to a company greenhouse in Creve Coeur, meeting there with scientists who showed him prototypes of the firm's herbicide-tolerant crops. Those gathered expressed concern about potential holdups as the USDA pursued regulatory oversight of the company's GE crop field tests. "They're going through an orderly process," one Monsanto official told Bush. But, he went on, "if we're waitin' until September and we don't have our authorization we may say somethin' different!" Bush responded with a line that roused the crowd: "Call me, I'm in the dereg business."[18]

When Bush became president in 1989, he made good on his word. He created a Council on Competitiveness, headed by Vice President Dan Quayle, which sought to identify government regulations hindering business growth. By February 1990, Quayle wrote to Bush explaining that his team had "decided that biotechnology—already a $1 billion industry with great potential for new drugs, pesticides, environmental, and agricultural products—should be the first target for [the council's] competitiveness review." The president's chief economic advisers were particularly concerned about competition coming out of Japan, where biotechnology companies were growing at a fast clip.[19]

In March of 1990, the council's Biotechnology Working Group consulted with Monsanto's director of plant science technology, Robb Fraley, to hear his thoughts on future policy prescriptions. An internal report penned by Bush's economic adviser Larry Lindsey summarized Monsanto's concerns expressed in the meeting. Lindsey painted a bleak portrait of the current commercial climate for genetically engineered products: "At present, the biotechnology industry faces widespread public resistance, based primarily on ignorance, but fanned by some anti-technology extremists." By the end of the 1980s, numerous environmental organizations in the United States, including the Pesticide Action Network, the National Wildlife Federation, and the Environmental Defense Fund, partnered to create an organization that began lobbying for more aggressive federal oversight of the biotechnology industry. The working group put out a report in 1990 titled *Biotechnology's Bitter Harvest*, which warned that herbicide-tolerant crops would be a "threat to sustainable agriculture." "The direction of agricultural biotechnology is clear," the report read: "The first major products will not be used to end dependence on toxic chemicals in agriculture. Rather, they will further entrench and extend the pesticide era."[20]

In this turbulent climate, Lindsey knew the White House had to act. He said that federal biotechnology "regulatory policy has failed in its mission" because "the public does not believe that approved products are safe and effective." He concluded that the "federal government . . . take positive steps to increase public acceptance of biotechnology."[21]

That is exactly what the Bush administration did. Two years after this Monsanto meeting, Quayle announced the publication of a new FDA policy in the *Federal Register* that doubled down on the position first outlined in the Coordinated Framework for Regulation of Biotechnology issued in 1986. The new federal code declared that the FDA would oversee the regulation of foods produced from GE crops, "utilizing an approach identical to that applied to foods developed by traditional plant breeding." It gave assurances to the public that there was nothing to fear when it came to GE foodstuffs. Scientists had determined that there was "substantial equivalence" between GE crops and conventional varieties, the code read, meaning there was no need for special field investigations of these biotechnological innovations.[22]

Monsanto had influenced policy makers in Washington, DC, and the federal government had given Monsanto approval to bring its GE seeds to market. Now Monsanto needed to make those seeds grow.

* * * *

ENTER BOB SHAPIRO. Although older establishment types initially viewed the man from Searle as an outsider, he embodied the new Silicon Valley vibe the firm wanted to project to investors in the mid-1990s. It also did not hurt that he understood the law, given all the legal trouble the firm was in.[23]

This was the beginning of the dot-com boom, a frenetic financial time where investors threw money at almost anything that sounded like it had to do with bits and bytes. The initial public offering (IPO) for Netscape, a new Web browser, had just blown up the stock market. The brainchild of a twenty-something computer engineer named Marc Andreessen, Netscape was valued at more than $2 billion by the end of trading in the summer of 1995. It was an unprecedented surge for a small start-up. Soon, Yahoo!, Amazon, and other Web-based companies enjoyed stock booms. Brokers made risky investments on these new companies because they were flush with cash. In the 1980s, the emergence of Roth IRAs and 401k plans funneled the retirement savings of millions of middle-class Americans into the stock market. Shapiro wanted to show Monsanto shareholders that his company was a good bet in these high times for high finance.[24]

He spoke the techie language investors wanted to hear. In his first report to shareholders in 1995, Shapiro, then in his late fifties, proclaimed that Monsanto was henceforth in the business of selling "genetic 'software.'" The company was "drastically" reducing its "chemical portfolio," making key investments that would allow the company to become a high-tech, information trading firm. "Nature," Shapiro said in another report, "has already developed amazingly complex and elegant software in the genetic structure of every organism on earth." Through an aggressive acquisition plan, the firm intended to take over seed businesses and biotechnology firms in the years ahead. This was the "Microsofting" of Monsanto.[25]

By 1996, Shapiro could show investors he was not making hollow promises. That year the company successfully launched Bollgard, a cotton seed genetically engineered to produce the Bt insecticide that killed worms and other pests that fed on cotton bolls. Bollgard was a big deal for Monsanto. The company billed it as "the largest and most successful launch of a new product in the history of agriculture." But it was just part of a bigger package. The company simultaneously announced the commercialization of Roundup Ready soybeans and Roundup Ready canola, two seed varieties that had been engineered to be resistant to Monsanto's signature herbicide. These were the first herbicide-tolerant GE crops ever commercialized.[26]

Soybean farmers planting these seeds in the early spring of 1996 were taking a risk because there were still some countries weighing whether or not to approve importation of genetically engineered soybeans. In Europe, Green Party members in Germany and other environmentalists partnered with Greenpeace to try and block GE food shipments to European ports, and in Japan, legislators were still debating whether to allow foods produced from genetically engineered seeds into the country. Archer Daniels Midland (ADM), one of the largest grain processors in the country, even threatened not to handle any GE crops until Japan and Europe resolved the trade issue. But by April, good news came from abroad when the European Commission, the executive body governing the European Union, announced that a majority of member states had approved importation of foods produced from Monsanto's GE seeds. (The commission also permitted cultivation of GE crops in Europe.) A couple of months later, Japan followed suit. With global markets secured, the Roundup Ready revolution had arrived.[27]

* * * *

AT THAT TIME, farmers and scientists were facing a real weed problem. The blockbuster herbicides in the early 1990s were ALS inhibitors, which worked by disrupting a plant enzyme called acetolactate synthase (ALS) that is essential to amino acid production. First introduced in the 1980s, these herbicides allowed farmers to spray weeds that had germinated in fields during the growing season, in part because commodity crops such as corn and soybeans were bred to tolerate ALS inhibitors. Many other herbicides, including Roundup (which acted on a different enzyme in plants than the one targeted by ALS inhibitors), could only be sprayed on fields in large amounts before planting or after harvest. Doing otherwise would kill crops. So farmers used tremendous quantities of ALS inhibitors, including American Cyanamid's popular Pursuit brand.[28]

Flooding farms with these favored chemicals had predictable outcomes. By the early 1990s, weed scientists began noticing waterhemp and other weeds that had developed resistance to ALS inhibitors. Nature was fighting back in the chemical age. Farmers were concerned: How would they clean their fields if these powerful herbicides failed?[29]

The nation's agricultural sector was primed for a Roundup Ready revolution. Monsanto's technology offered a way out of the ALS problem. Basically, instead of killing weeds with Pursuit—which was losing effectiveness as weeds evolved resistance to this overused herbicide—farmers could plant Roundup Ready seeds and use Roundup throughout the growing season. And because Roundup

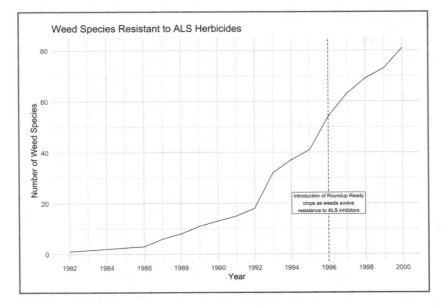

FIGURE 2: Rise in the number of weed species resistant to ALS inhibitors, 1982–2000. *Source: Study by Dr. Ian Heap (2020), International Herbicide-Resistant Weed Database, Weedscience.org.*

was a broad-spectrum chemical that killed an incredible array of weeds, farmers did not have to buy "residual" herbicides typically used on fields before weeds germinated.

Or at least that is what Monsanto's salespeople told farmers. In June 1996, immediately after the commercialization of Roundup Ready seeds, Monsanto technical manager Dr. Eric Johnson told farmers through a company-published outreach magazine, the *Midwest MAGnifier*, "there is **no need for residual herbicides** in the Roundup Ready soybean system." Essentially, Johnson was urging farmers to use Roundup exclusively to keep their fields clean. Roundup was "the only weed control you need," a company advertisement boasted. Why spend money on expensive herbicide complements when the Roundup Ready system was powerful enough to do the job? Nearly a year later, the *MAGnifier* happily reported that roughly 80 percent of Roundup Ready soybean growers used only glyphosate on their fields. And in 1997, a company representative boasted that "field research" confirmed "there is no weed control benefit in using a residual herbicide ahead of Roundup Ultra."[30]

Relying too heavily on particular herbicides had caused the ALS resistance problems of the early 1990s, but Monsanto assured farmers that this would not happen with Roundup. In 1997, Monsanto researcher Laura Bradshaw published a study in *Weed Technology* that laid out the company's logic: "The long history of extensive use of the herbicide has resulted in no verified instances of weeds evolving resistance," Bradshaw and her team explained. Considering the fact that Roundup use dated back to the 1970s, "it is reasonable to expect that the probability of glyphosate-resistant weeds evolving will not increase significantly over that considered with current use."[31]

But in Australian fields halfway around the world, Bradshaw's prediction was already proving false. There, in 1996, weed scientists discovered that rigid ryegrass, a particularly "troublesome" and adaptive weed species, had developed resistance to Roundup. This was before GE seeds had even been commercialized in Australia, but farmers nevertheless had been using Roundup in large quantities, placing heavy selection pressures on weed species. Resistance was already rising at the start of the Roundup Ready revolution.[32]

But farmers would not have known that from the promotional pitches Monsanto made at the time. Roundup Ready technology was "the system that sets you free," explained Monsanto on a company-managed website designed to broadcast messages to farmers in the new digital age. Bob Shapiro, Monsanto's eco-conscious CEO, explained that Roundup Ready seeds were essential tools that would allow the firm to sell "information" not "stuff." In its 1996 Environmental Annual Review—one of the many sustainability publications that appeared in the Shapiro years—the company predicted that Roundup Ready soybeans "had the potential to decrease herbicide use as much as one-third." "Putting information in the gene of a plant enables you to avoid spraying the plant with pesticides," Shapiro told the Society of Environmental Journalists in 1995. GE seeds would also allow farmers to practice "no-till" farming, because instead of disrupting and dehydrating soils through extensive weeding operations, farmers could simply spray throughout the growing season to manage their weeds. In other words, this was good for business and good for the environment. GE seeds would help farmers cut out unnecessary costs associated with weeding their fields, and it would also result in farming practices that would reduce soil erosion. Who could argue with that?[33]

* * * *

IT SEEMED TOO GOOD to be true, and it was, but farmers wanted to believe it. Instead of having to use a bunch of different herbicides to control weeds, they now could just turn to one brand to get the job done. Almost overnight, soybean, corn, and cotton farmers in the United States began switching to Roundup Ready varieties. Overseas markets were next.

All of this was possible because Shapiro made aggressive acquisition and investment moves in the seed business. He worked closely with fiery Robb Fraley, who became vice president of agricultural research in 1992. Described by one executive who knew him well as "the most driven man I've ever known," Fraley, now in his forties and balding, was, in the words of those who saw him negotiate, "merciless." As a scientist in his late twenties, he had joined the Jaworski team that had executed Monsanto's early gene transfers, and he was utterly convinced that the company was sitting on billion-dollar biotechnology products. He was relentless in his quest to get these genes to market, and he knew that to do this effectively, Monsanto had to make some big deals in the seed business. Together with Shapiro, he began a series of commercial negotiations that radically reshaped Monsanto's future.[34]

By the end of 1996, the company had acquired Asgrow seed company (a leading brand in the soybean and corn business), partnered with Delta & Pine Land Company (the premier cotton seed firm based in Mississippi), and made major investments in Dekalb

Genetic Corp. (a major seed company based in Dekalb, Illinois, that invested in genetic engineering). In 1997, Monsanto acquired Calgene and bought Holden's Foundation Seeds Inc., a major corn-seed company headquartered in Iowa. In less than two years, Shapiro and Fraley were turning Monsanto into a GE seed giant.[35]

And in their new seed partnerships, these executives were careful not to repeat missteps the company had made in the past, especially those that led to the disastrous Pioneer Hi-Bred deal of the early 1990s. Pioneer, one of the nation's largest seed companies founded by former New Deal secretary of agriculture and US vice president Henry A. Wallace, pioneered methods in creating hybrid crosses that produced highly prolific corn varieties. By the early 1990s, it controlled more than 40 percent of the entire US corn market and was a leading seller of other commodity crop seeds. In 1992 and 1993, Shapiro eagerly asked Pioneer to carry its Roundup Ready and Bt traits that were soon to be delivered to market. He hastily agreed to let Pioneer use Monsanto's Roundup Ready soybean gene in perpetuity for a meager, one-time payment of $500,000 and also approved Pioneer's use of its Bt gene in corn for just $38 million.[36]

Monsanto executives knew almost immediately how shortsighted the Pioneer arrangement had been. As Roundup Ready soybeans and Bt corn spread across Middle America, Monsanto did not get an additional dime from Pioneer. There were no royalty provisions included in the contract. Monsanto missed out on a lot of money.[37]

In the years ahead, Shapiro and Fraley took this lesson with them to the boardrooms of every other seed company they met with. In discussions with companies like Delta & Pine, Monsanto insisted that farmers pay a "technology fee" that would come back to the St. Louis firm in the form of royalties. Monsanto was not selling the rights to use its genes for a lump sum anymore. Seed companies using Monsanto genetic cassettes would be able to keep some of the technology-fee markup, but Monsanto would get 70 percent of the premium add-on for every bag of seed they sold.[38]

And Shapiro and Fraley did something else that changed the face of American agriculture forever: they required farmers to sign a technology-use agreement (TUA) that prohibited them from replanting saved seeds. For corn growers, this was not all that new. Since the emergence of hybrid corn in the 1920s, many growers stopped saving and replanting seed after harvest, because of the peculiar properties of hybrid offspring that make them less productive than their parents. But soybean growers and cotton farmers still practiced seed saving when Monsanto's TUAs came along. Moving forward, Monsanto sought to end that practice, demanding that its clients purchase their genetically engineered traits every year. Some farmers complained, lamenting the loss of freedom this new regime portended. "Feels like you're living in Russia," one Ohio farmer quipped when describing Monsanto's rules about replanting. But other farmers complied with Monsanto's new demands, entranced by the prospect of pest- and weed-free fields. Four years after Shapiro completed his 1996 seed deals, roughly 54 percent of soybeans grown in the United States and approximately 46 percent of the country's cotton crop were genetically engineered to be resistant to glyphosate. Roundup Ready corn, first introduced commercially in 1998, initially took a smaller share of that commodity market but soon became just as popular.[39]

Shapiro and his seed team had figured out how to make a lot of money from Monsanto's new technology, but if the visionary CEO was going to make his company profitable moving forward, he also had to find a way to offload all the toxic chemical liabilities that were weighing down the firm.

Here Shapiro was crafty. In 1997, he spun off the majority of Monsanto's remaining chemical assets into a new company called Solutia. This new firm was, as the *St. Louis Post-Dispatch* later said, "the solution to many of old Monsanto's problems." Shapiro saddled the spinoff firm with about $1 billion in debt and major Monsanto environmental liabilities. In a few years, Solutia was spending tens of millions of dollars annually just to deal with Monsanto's legacy issues. The Monsanto that was left after the spinoff attracted investors. "Food, Health, Hope," was what this company sold, according to a new company slogan. Three years after Shapiro became Monsanto's CEO and chairman, the company stock price rose an exponential 280 percent.[40]

The "new" Monsanto was actually not all that new. Monsanto remained in the chemical business, holding on to its profitable Roundup herbicide and the industrial plants that had been producing it since the 1970s. It also retained ownership over lucrative pharmaceutical brands, especially Ambien and NutraSweet. In the years ahead, the profit powerhouses of the past would drive Monsanto into its seed future. The company still fed on chemicals, even if it found financial separation from some of the firm's more noxious compounds.[41]

Shapiro had turned environmental liabilities into financial abstractions and buried these toxic assets in Solutia's balance sheet below reports of chemical sales totaling $3 billion annually. In the flush times of financial frenzy, investors were lured in. By 1998, Solutia's share price soared. But gradually, as the stock market binge of the 1990s turned into the financial hangover of the early 2000s, the toxic waste buried within Solutia began to become a real problem.[42]

The reality was that the numbers coolly entered in Solutia's liability ledgers represented the real lives of real people who had been wronged by the Monsanto Company. By the end of the 1990s, some of these people decided they would not become statistics in a shareholders report.

CHAPTER 11

"I Have to Cry for Them"

IN 1995, A YEAR BEFORE COMPANY CEO BOB SHAPIRO announced the first commercial planting of Roundup Ready crops, Monsanto quietly started buying up houses and buildings in the West Side neighborhood of Anniston, Alabama, near one of the company's plants. The firm approached homeowners and church pastors and told them the company was willing to offer big money for their properties and financial assistance to renters wanting to relocate. Many residents Monsanto talked to were African American. Forty-four percent of the people that lived within a mile of Monsanto's plant were Black, and many others were working-class whites. The company told homeowners and renters in its purchase program that "the decision whether or not to participate was yours." Residents had a choice. But Monsanto was making offers that were extremely attractive. The firm sweetened its buyout packages with "bonuses" to entice people to move promptly. Purchase prices sometimes rose 75 percent above market value.[1]

What was going on in Anniston?

David Baker was among the city residents asking questions. He had grown up beside noxious fumes and pollution plumes that billowed and flowed from Monsanto's Anniston plant. But at the time, he did not think much about the chemicals that Monsanto produced at its facility.[2]

He had bigger problems on his mind back then. In 1961, when Baker was not yet high-school age, Anniston was literally ablaze as racist whites, dressed in their Sunday best, set fire to a Greyhound bus filled with Freedom Riders headed for Birmingham. Baker's parents, Imogene and Grover, drove by the blaze that Mother's Day afternoon and watched with horror as young civil rights activists tried to escape from the smoke-filled wreckage. Though Grover wanted to drive on, keep his head down, and avoid confrontation, Imogene forced him to stop the car so she could help those in need. It was a fateful decision. The next morning the phone rang and a young David Baker answered. "We gone bomb your house tonight," said a voice on the other end of the line. Other calls came in with similar threats.[3]

Monsanto benefited from the racial segregation and social separation in Anniston that Baker experienced as a child. It distracted residents and kept them from paying attention to a toxic problem at Monsanto's plant that had gotten out of hand.

For decades, Monsanto treated the West Side of Anniston as a hazardous waste dump, channeling millions of gallons of untreated PCB effluent into Snow Creek, which snaked by homes and churches and playgrounds where children played. Many of the Black and working-class white residents had virtually no knowledge that there was a PCB problem. And even if they had known, most folks were preoccupied with other matters. The fire of the Freedom Rides, racial riots at homecoming parades, and the day-to-day grind of eking out a living in the caste-based society of Alabama—this was what kept people up at night in this part of the world.

David Baker left Anniston in 1971, a year after Monsanto stopped producing PCBs at its Anniston plant, but he would return home in the 1990s to find that the toxic legacies of the past persisted because PCB contamination lingered in the environment. Decade after decade, thousands of Annistonians were exposed to PCBs at levels that far exceeded safety recommendations from the EPA. Many residents on the West Side who had their blood tested in the early 2000s found they had PCB concentrations in their bodies that were thirty, forty, even two hundred times that found in samples taken from average American citizens. And the contamination was not just confined to this one community. Choccolocco Creek, which connected to Snow Creek, and other tributaries coming from the area around the Monsanto plant channeled PCBs toward Lake Logan Martin, a popular recreation attraction, some twenty-five miles southwest. In 1993, a Soil Conservation Service scientist discovered a Choccolocco catfish with PCB levels in its tissues that registered off the charts. In an Alabama culture where fishing and hunting were second nature to most, this was deeply disturbing news.[4]

Which is why Monsanto dealmakers had come to buy homes and buildings. In the 1970s, Monsanto managers may have thought they had solved the firm's PCB problems once and for all, signing indemnity agreements with clients and closing down production facilities in the wake of national environmental protests. But two decades later, men and women in Anniston were raising their voices to say that the firm's historic legacies were still haunting this town. Perhaps Monsanto executives hoped money, much of it earned from their billion-dollar Roundup brand, would keep these folks in Alabama from digging up a past it wanted to keep buried.

* * * *

BUT MANY RESIDENTS WOULD not go quietly. Cassandra Roberts, an Anniston probation officer who was actually baptized in a PCB-contaminated stream near Mars Baptist Church, was moved to action. She and several other women from the Cobbtown and Sweet Valley neighborhoods heavily affected by PCB pollution formed an environmental task force and ultimately filed suit against Monsanto. (This became part of a case involving multiple plaintiffs called *Walter Owens v. Monsanto*.) Andrew Bowie, a deacon at Mars Baptist Church, was another fighter who refused to be bought out. He rejected Monsanto's offer to purchase the building where he and his fellow congregants worshipped and went instead to the office of local attorney Donald Stewart, a former US senator, asking him to file another suit, separate from the *Owens* litigation, against Monsanto. Resistance was rising in Anniston.[5]

Around this time David Baker returned to his hometown. Now in his late forties, Baker found a job with his friends Sylvester Harris and Russell Williams, the proprietors of an undertaker business in West Anniston. Working among the dead, Baker began to note the peculiarities of this place he called home. Baker felt caskets were far too often filled with the bodies of young men and women who had fallen ill under circumstances that could only be described as mysterious. He thought of his brother, Terry, who died of a brain tumor and an enlarged heart at the age of seventeen. Might all this be related?[6]

Then, a few months after starting work with Harris and Williams, Baker found out about a job that would change his life forever. Monsanto was contracting with an environmental-cleanup company to begin soil decontamination and removal in Anniston. Baker decided to take the job, even though the daily uniform made clear this was going to be hazardous work. Donning a hazmat suit and two pairs of protective gloves, Baker began digging up dirt.[7]

While he was working for Monsanto's contracted cleanup company, Baker noted that most of the people that toiled alongside him were not from Anniston but were bused in from other parts of the country, some coming from as far away as Detroit. To his surprise, some of these people knew more about the pollution they were dealing with than locals did. He recalled one situation where a crew member was chastised by higher ups at Monsanto for telling a curious local resident that his property was contaminated. "Anybody out here ask you from now on out, what's going on, y'all direct them to the office," a superior said.[8]

Baker's interest was piqued. In the days and months ahead, he began to organize meetings. He fostered dialogue among city residents, first meeting at Russell Williams's funeral home to talk about the dying and the dead. He took charge of an organization called Community Against Pollution (CAP), originally started by Andrew Bowie a few years earlier, and he, along with his wife, Shirley, began the hard work of mobilizing folks to take ownership of an issue that state regulators and the EPA had let go unchecked for far too long.[9]

These actions cost him his paycheck. In 1998, the environmental outfit Baker worked for told him he was out of a job. The firm gave him no clear indication why he was terminated, but Baker had a pretty good idea what was up. Now, with more time on his hands, he began to devote all of his energies toward CAP. In 1999, he wrote a letter to the EPA, which included a petition signed by many neighbors. The letter charged that local agencies were failing the people of Anniston. The Alabama Department of Environmental Management (ADEM) was not protecting the community, and Annistonians wanted the EPA to do something about it.[10]

Baker was not simply going to wait patiently for ADEM or the EPA to take action. He had seen these agencies' track record when it came to such matters. Instead, he joined Andrew Bowie and roughly 3,500 other Anniston residents in a toxic tort case designed to make Monsanto executives feel the pain the community was enduring because of the firm's past polluting practices. Attorney Donald Stewart, fresh off a $2.5 million settlement victory in a separate Mars Baptist Church case, was eager to take *Abernathy v. Monsanto* all the way to a jury. He relished the rich evidentiary record he found in discovery. What would a jury do when they learned that a Monsanto executive had once suggested the company "sell the hell out of" PCBs despite knowing its toxicity? He was eager to find out.[11]

*   *   *   *

IT WAS IN THIS CONTEXT, with the Anniston trial pending, that Shapiro signed off on the spinoff that brought Solutia into existence. Anniston, like Nitro and other toxic towns, was to be Solutia's responsibility from now on. Monsanto signage was coming down in Anniston, replaced by Solutia branding. As it had done elsewhere in the country, Monsanto was removing its name from a landscape that had become too toxic to touch.

In 1998, right after the Solutia spinoff, Shapiro raked in

approximately $54 million in income and stock options, making him the highest paid CEO in St. Louis, earning more than five times as much as Anheuser-Busch's chairman, and the equivalent of Chicago all-star Michael Jordan's annual corporate endorsements. The *St. Louis Post-Dispatch* billed Shapiro a visionary who was making all the right moves.[12]

But things soon soured in 1999 when a merger with the pharmaceutical firm American Home Products fell through. That same year, DuPont purchased the Pioneer seed company for $9.4 billion. The deal stung. Monsanto had long sought control over Pioneer germplasm, some of the best in the country, but had never been able to successfully court the firm, in part because of clashes between the two companies' executives.[13]

And more bad news came in the form of an "unofficial moratorium" on genetically engineered crops in the European Union, initiated in June 1999, which dampened investor interest in Monsanto's agricultural business.[14]

A series of events led to the European backlash against genetically engineered (GE) foods at the end of the 1990s. First there was a highly publicized campaign led by the Soil Association, a nonprofit organization supporting organic farming. The group worked with an organic farmer named Guy Watson in the small town of Totnes, England, who was trying to block the introduction of Liberty Link seeds that were genetically engineered by German company AgrEvo to tolerate a herbicide called glufosinate. In April 1998, the Soil Association claimed that pollen from Liberty Link crops would contaminate Watson's fields if planted nearby, thereby preventing the organic farmer from being able to label his produce organic. Watson took his case to trial, and it made headline news, rousing environmental organizations, such as Friends of the Earth, which brought attention to the issue.[15]

Simultaneously, news broke about Shapiro's talks with Delta & Pine Land Company, the largest cotton seed company in the United States. What troubled anti-GE activists about the potential deal was the fact that Delta & Pine had recently announced the discovery of a technique for manipulating gene expression that would enable the firm to make the seeds produced from their genetically engineered crops sterile. Critics dubbed this "Terminator technology," a reference to the famous Arnold Schwarzenegger film, and held out the specter that one day, Monsanto would deploy this genetic software to control the world's food supply. (Though Monsanto ultimately acquired Delta & Pine in 2007, there is no evidence that it ever deployed this technology in commercial markets.) Three months later, scientist Arpad Pusztai made public the results of a study he conducted that found rats given a steady diet of genetically engineered potatoes developed problems with their immune system along with other health issues. British Royal Society researchers later challenged the findings in the study, and Pusztai was asked to leave the research institute that had funded his work, but the controversial publication contributed to growing public concern about GE food.[16]

Celebrities, including royalty, were also weighing in. Charles, prince of Wales, added his voice to the growing clamor against Monsanto and GE technology in June 1998, saying he "personally had no wish to eat anything produced by genetic modification." This technology, he added, "takes mankind into the realms that belong to God." By 1999, European companies, including Nestlé and Unilever, announced that they would work to eliminate genetically engineered ingredients in their products. The British retailer Marks & Spencer also said it would do the same.[17]

Finally, in May 1999, Cornell assistant professor John Losey published a paper in *Nature* that concluded pollen from Bt corn killed the larvae of monarch butterflies. Losey's colleagues, and even Losey himself, cautioned that his study was based solely on lab research and did not include extensive field analysis. As such, he could not prove that crops genetically engineered to give off the Bt toxin would kill monarch butterflies in real-world situations. A 2005 USDA study, drawing on fieldwork, subsequently concluded that "Bt corn is not likely to pose a significant risk to the Monarch butterfly population

in North America." However, researchers at Iowa State later hypoth-
esized that the widespread use of glyphosate was eliminating milkweed, monarch butterflies' food source, thereby contributing to
butterfly population decline. But that study came years after Losey's
paper. In 1999, the thought that Bt was the chief culprit of monarch
butterfly deaths was headline news. Environmentalists across the
globe were outraged, and the monarch butterfly quickly became the
symbol of the anti-GE movement.[18]

In the midst of all this, the European Commission reversed its earlier decision to allow GE crops to come into the European continent.
It was a decisive victory for anti-GE activists, with twelve out of the
fifteen member states voting in favor of the ban.[19]

In the end, Monsanto was able to weather the storm. The company
was expanding seed sales to South America, especially Argentina, and
was eyeing major expansion into Brazil. Furthermore, Europe only
imported about 2 percent of all US grain production, so American
exporters just shifted GE-grain shipments that would have gone to
EU countries to other nations willing to accept genetically engineered
crops. And the European Commission did not prohibit US companies
from delivering processed GE corn and soybeans to European buyers
as long as those products went into animal fodder. This remained a
big market for American commodity crop farmers planting genetically
engineered seeds, and it was largely unaffected by the GE food bans.[20]

* * * *

AS MONSANTO FOUGHT OFF backlash in Europe, Baker and his
fellow plaintiffs prepared for a fight of their own in the *Abernathy
v. Monsanto* trial. It was now 2002, six years after attorney Donald Stewart had filed the case in federal court. The plaintiffs had
momentum, in part because the *Owens* case, which involved a separate group of some 1,600 Anniston residents, including Cassandra
Roberts and the women-led Sweet Valley/Cobbtown environmental
task force, had already concluded with a $40 million settlement in the
spring of 2001. In that case, the *Owens* plaintiffs secured damning

Monsanto documents during discovery that were now critical fodder in the *Abernathy* litigation.[21]

For Monsanto, there was a lot at stake, in part because the firm could not rely on Solutia to cover the costs of toxic liabilities in Anniston. This was due to a series of complicated mergers and spinoff negotiations Monsanto took part in around the time of these trials. In 2000, the pharmaceutical firm Pharmacia & Upjohn—known for blockbuster breakouts such as Rogaine and Nicorette gum—acquired Monsanto, partly enticed by the St. Louis firm's arthritis-relief wonder drug, Celebrex. But in 2001, this merged company, known as Pharmacia, spun off all of Monsanto's agricultural assets into what was then called the "new" Monsanto. In that spinoff deal, Monsanto's leadership agreed to assume any environmental liabilities then associated with Solutia, should Solutia go bankrupt. Thus, Monsanto had a vested interest in working with Solutia, which was financially fragile, to find resolution in Anniston.[22]

In the months leading up to the *Abernathy* trial, Monsanto lawyers did what they had done in Nitro, making legal maneuvers that delayed a trial date, hoping to starve out the plaintiffs' attorneys. But Donald Stewart, like Stuart Calwell in the Nitro case, was going nowhere. He was determined to fight alongside his clients. Baker, meanwhile, actively sought out other potential litigants that had been harmed by Monsanto's actions and sent correspondence to California-based attorney Johnnie L. Cochran Jr. to come see what was happening in Anniston. Cochran, now famous after his successful defense of O. J. Simpson in 1995, responded to Baker's appeal, coming to Anniston to meet with him and other community activists. He was moved by what he saw and collaborated with an Alabama law firm in filing another tort case, *Tolbert v. Monsanto*, that involved more than 18,000 litigants.[23]

Thus, when the *Abernathy* case went to trial in 2002, Monsanto was facing attacks from multiple law firms. If the *Abernathy* jury found in favor of Donald Stewart's clients, that would help the cause of other plaintiffs waiting to have their day in court.

The *Abernathy* trial lasted for six weeks. During that time, the jury had read the confidential documents Monsanto had released during the *Owens* litigation detailing the vastness of the contamination problem in Anniston and heard expert testimony about the remarkably high PCB levels that still persisted in the environment more than thirty years after Monsanto stopped PCB manufacture. After deliberations that lasted just five and a half hours, the jury issued a stunning verdict. They had found Monsanto's actions "so outrageous in character and extreme in degree as to go beyond all possible bounds of decency, so as to be regarded as atrocious and utterly intolerable in civilized society." The only thing left to decide was how much Monsanto should have to pay for what it had done. Baker wept.[24]

In this time of plaintiff elation, bad news for them came from Washington. George W. Bush's EPA sought to intervene in the matter, issuing a consent decree in March 2002, just weeks after the *Abernathy* decision, giving what Baker called "a sweetheart deal" to Solutia. According to Baker, the decree once again delegated much of the responsibility for cleaning up toxic waste to the very people that caused the pollution problem in the first place. Under the order, Solutia, not the EPA, would be hiring contractors to conduct contamination assessments in Anniston. The decree also allowed Solutia to forego expensive community blood-testing programs and capped the amount of money the company would need to reimburse the EPA to a modest $6.2 million.[25]

David Baker was angry. With the help of the DC-based Environmental Working Group and Alabama Senator Richard Shelby, Baker traveled to Washington to testify before a US Senate subcommittee in April 2002 and there made a passionate appeal to those gathered. It had been a long journey to get to this point, and the emotions clearly weighed on him, tears welling in his eyes as he spoke of the plight of friends and family members he loved. "If half of the people that were affected by this were here today, you would see tears in their eyes," he explained, "So I have to cry for them."[26]

Baker made clear that it was EPA's decentralized regulatory policy
that had allowed this problem to go on for so long. "The people of
Anniston, Alabama," he said, "have waited for more than 40 years
for the Federal Government to step in and help us clean up the contaminants in our backyards, and in our playgrounds, our rivers, our
creeks and in our bodies. Unfortunately, after 40 years of waiting, I
am here today to report that the Federal Government has failed the
people of Anniston." And if there was any question that delegating
local decision-making to local actors resulted in the empowering of
local people, Baker made clear this was not the case. Children in his
community had to "wash the dogs if they wanted to play with their
own pets in their yard. We cannot plant a garden. . . . No one will
loan us money on our property."[27]

Baker's appeal proved effective. The EPA ultimately abandoned
its proposed consent decree and adopted more rigorous cleanup
measures, but the real news came out of the federal courtroom back
in Alabama. There, as predictions of total damage costs in the *Abernathy* case approached $3 billion, Solutia and Monsanto moved to
stop the situation from getting worse. Lawyers for the companies
met with the judge handling the *Tolbert* case and asked if plaintiffs would be willing to participate in a "global settlement" that
would include the *Abernathy* litigants. Donald Stewart ultimately
agreed to the deal, and after months of negotiation, on August 20,
2003, both sides reached a settlement. The *Tolbert* and *Abernathy*
plaintiffs would collectively receive $700 million, the largest settle-
ment in a toxic tort case involving a particular industrial plant in
US history.[28]

In the Anniston settlement, Monsanto lawyers agreed to cover
$550 million of the $700 million in outstanding costs, with Solutia
picking up $50 million. Monsanto also agreed to cover a substantial
chunk of the $100 million set aside for medical monitoring in Annis-
ton. This might seem like quite a deal for Solutia, but the firm was
in serious financial trouble, in large part because of the millions of
dollars it was spending to clean up Monsanto's problems in other

parts of the country, estimated at $100 million annually by 2003. Because of those financial obligations, Solutia ultimately filed for bankruptcy in December of that year, less than four months after the Anniston settlement.[29]

Monsanto took a financial hit, posting a $23 million loss in August 2003, but the Anniston settlement was really a bargain deal for the firm, in part because corporate insurance covered a substantial portion of Monsanto's financial obligations. By 2004, Monsanto was rebounding, celebrating $267 million in net income that year, in large part because of its Roundup Ready seed system that enabled it to make billions of dollars. Within three years, net income more than tripled. For a company that had just gone through substantial reorganization, this was really good news. Monsanto was going to be just fine.[30]

✳   ✳   ✳   ✳

SHAPIRO, NOW IN HIS early sixties, was rich. As he went into the 2000 Pharmacia merger, he agreed to stay on for eighteen months at the new firm before calling it quits. When he did leave, he took with him a severance package worth $14 million and a pension yielding $1 million annually. In Anniston, the average payout to *Abernathy* plaintiffs in the *Abernathy–Tolbert* settlement was roughly $49,000. Most *Tolbert* litigants received less than $9,000.[31]

Money did not heal Anniston. In the months ahead, plaintiffs complained about disparities in payouts, which were based on PCB levels in people's blood. Baker faced charges from other citizens that CAP, the organization he led for so long, profited from the whole affair. He denied these allegations, but nevertheless received phone calls at his house, with people threatening to kill him and his wife for their perceived greed. Baker, a man who as a child had heard similar threats over the phone in the wake of the Freedom Rides, once again feared for his family.[32]

And in terms of cleanup, there was waiting and more waiting. First Solutia and the EPA had to agree on what cleanup would look like and who would pay for it. Further feasibility studies were done

before PCB removal got under way. In 2013, air sampling by EPA officials confirmed that PCB levels in Anniston still exceeded national
averages. At that time, an estimated 5.5 to 10 million pounds of
PCB-contaminated pollution lay buried in a nearby landfill. As of
the publication of this book, cleanup operations were still ongoing
in Anniston, now nearly half a century after Monsanto stopped PCB
production there.[33]

All this quickly became old news at the "new" Monsanto. The firm
had survived the Anniston scare. The cycle of spinoffs had bought
the firm's GE seeds salesmen time to do their work, while Solutia
took on many of the burdens of Monsanto's past. By 2004, Monsanto was once again making substantial profits and was primed for
more growth.

That's when the firm got tangled up in a mess that threatened to
ruin everything Shapiro had so carefully put in place.

Soybean harvest in the Cerrado, a vast savannah in Mato Grosso,
Brazil. Monsanto spread its genetically engineered (GE) seed
business to mega-farms there beginning in the early 2000s. Because
the Cerrado is in the tropics, there is no winter downtime, meaning
farmers can plant corn crops right after harvesting soybean fields,
with tractors bearing corn seed sometimes following just fifteen
minutes behind combine harvesters.

PART IV

WEEDS

CHAPTER 12

"Oh Shit, the Margins Were Very, Very, Very Good"

HE WANTED TO LEAK HIS STORY TO THE PUBLIC, BUT HE WAS not sure it was a good idea. He spoke via an encrypted phone service and said he was weighing the risks. This was not a game. He had a family to think about; his pension could be put in jeopardy. He had signed a nondisclosure agreement (NDA) that would have opened him up to serious legal liability for breach of contract. Monsanto's lawyers could come after him for a lot of money. And the "scariest" part of the whole thing, he said, was that even if the firm did not seek damages, his NDA stipulated that he would have to cover Monsanto's court costs. The workers in Nitro, West Virginia, whom Monsanto lawyers punished by putting liens on their houses in the final days of the Agent Orange court battle in the 1980s, would have understood his anxiety.[1]

Still, it bothered him. He felt the public needed to know what he knew. He debated whether or not to tell his story for weeks, then months.

He wanted to talk about an old herbicide Monsanto had repackaged and was hoping to turn into a moneymaking brand. Pitched as the next new innovation that would liberate farmers from their weeding woes, this herbicide had a serious defect, and the man on the phone said he had information that could shed light on how this flaw made it past regulators.

The troubled man from Monsanto took his time, pondering his

options. Could his insider knowledge help farmers seek redress? It was an ethical conundrum, concern for family weighed alongside concern for others. And what if his insights had little to no effect on the conversation? It would have all been for naught. So much to consider.

He sought out lawyers, seeking any advice he could get. Was there a way to secure "whistleblower" protection? Any kind of statute of limitations on the NDA? Were there any caps on damages they could seek?

No, no, and no. It didn't look good. If he was going to tell his story to the public, he was going to have to risk it all.

And he just wasn't willing to do that. In the end, he decided not to go on the record.

Farmers would have to learn the hard truth firsthand. Deep in Middle America, as growers walked their fields, they came to know what Monsanto had tried to hide. There was a serious problem with the company's new weed-control system, a problem that was spreading across the countryside and threatening, as one observer put it, to "tear the social fabric of farming communities" asunder.[2]

Few farmers envisioned things playing out this way. Back when the Roundup Ready revolution began, Monsanto's technology seemed to offer salvation to farmers buried in weeds that had become resistant to ALS inhibitors. Monsanto had sent salesmen to tell people the good news.

One of the people who received the message was a man from Scott, Ohio. He ran a business that most people outside rural America knew nothing about. He was a seedsman.

※   ※   ※   ※

FRED POND, FATHER OF two in his sixties, remembers Monsanto's warning well. It was that big a deal, a "9/11-type thing," as he put it. He was at a professional seed-growers conference in Columbus, Ohio, nine years after taking over Pond Seed, a family-owned seed-cleaning business started by Pond's grandfather in 1927. He remembers deciding to attend a presentation by a Monsanto man, which he thought was odd at the time, considering the fact that Monsanto, to that point, had very little to do with selling seeds. What was this guy doing here?[3]

Pond was about to learn some shocking news. According to the Ohio seed dealer, the Monsanto representative "got up and said that Monsanto had invented this technology and the majority of us in this room could either sign on or we'd all be out of business within the next two years, and they were going to dominate the entire seed industry and there was nothing anybody in this room could do about it." The room fell silent, as hundreds of seedsmen took in the news. Pond remembers his "jaw dropping to the floor." What bravado. "It struck me as impossible," he explained in an exasperated tone years later.[4]

But it wasn't impossible. Between 1996 and 2000, Bob Shapiro expanded Monsanto's seed business, gobbling up independent companies, which helped it become the largest seed seller in the world by 2005. Seed money accumulated during Monsanto's chemical days made these seed mergers possible. Big players, such as Seminis and Dekalb, became part of the company portfolio, but so did smaller operations—Midwest Seed Genetics, Crow's Hybrid Corn Company, among dozens of others.[5]

Prior to 1996, soybean growers often saved seed, perhaps paying a small fee to people like Fred Pond to clean their germplasm of impurities before planting the following year. Now, farmers, bound by Monsanto's technology-use agreements (TUAs), had to buy all their seed every year—and at a premium price. Fred Pond recalled that in 1995 farmers might have purchased a bag of soybean seeds for $12, but by 2018, farmers were paying $55 to $60 a unit. Monsanto salespeople said that these higher prices reflected value generated by the company's Roundup Ready technology, which saved farmers money by reducing the need for herbicides, eliminating labor costs associated with weeding, and boosting yield. But some people were not so certain. In the two decades following the introduction of Roundup Ready

technology, seed costs for soybean and corn farmers nearly quadrupled, causing some to question whether farmers had gotten a raw deal.[6]

Pond was one of those people that felt ambivalent about what had
happened, in part because he became an agent, if somewhat reluctant one, of Monsanto's conquest. After the shocking moment at the
Columbus conference, Pond made the two-hour drive back west to
his processing facility in small-town Scott, Ohio, population circa
330 people. Over the next few seasons, he continued to try to operate without a Monsanto license, but by 1999, his business was not
doing well. He remembers being in his office holding an acceptance
letter from Comair, a commercial airline affiliated with Delta, and
weighing the decision of whether or not he should head to Florida
for flight school, a dream he had always had. But in that moment, he
got another offer to start selling Monsanto genetics, and he decided
to give it a go. He put the letter in his desk, where it remains today.
The pilot thing would always be there if things did not work out.[7]

Business boomed. From the very beginning, Monsanto made Pond
offers that were just too tantalizing, giving him attractive and lucrative incentives to sell as much of Monsanto's Roundup Ready technology as he could. If he met Monsanto's sales targets or exceeded
those targets, he would receive substantial financial bonuses from
the company. And cash was only part of the enticement. If a seed
processor was really moving seeds at high volume, Pond said that
Monsanto would offer to pay for a two-week vacation or fly a processor and his family somewhere exotic. He recalled the rush of
being driven in a race car around Indianapolis Motor Speedway, all
at Monsanto's expense. This was how it worked in this new high-
dollar seed business. The Monsanto model soon spread throughout the industry. Bayer, Pioneer (owned by DuPont after 1999), and
other big genetically engineered (GE) seed companies did the same
thing, offering everything from barbecues to tractors, and not just to
seed processors. Fred Pond knew a big farmer that received an all-
expenses-paid trip to Hawaii. In this changing industry such practices became standard fare; other seedsmen around the state spoke

of the extravagant parties and trips Monsanto financed in these years. "There was just so much money in it," Pond said. For a struggling family business like his, it was too good an offer to turn down.[8]

Through Fred Pond, Monsanto spread its seeds to hundreds of farmers. This is how it worked. In the case of Roundup Ready soybeans, Pond sourced the latest GE seeds from Monsanto. These seeds sometimes came from crops Monsanto propagated in Argentina (the idea being that Monsanto could take advantage of an opposite growing season in South America so that brand new seeds—"version 1.2" as Pond put it—could be available at the start of spring in the United States). Pond then planted these seeds on his property and offered them to contract growers, who agreed to reserve a portion of their crop for seed production. After October harvest, trucks coming from contractor farms and from Pond's own fields delivered seeds to Pond's processing facility. There, the seeds traveled through an air screen machine that separated seed based on size (processors look for the mid-cut seeds, not too big, not too small), then on to a gravity separation table that disaggregated seeds based on weight. Next, Pond treated the seeds with fungicide and pesticide chemicals using equipment developed by Bayer. During the treatment process, Pond added a polymer "lipstick" to the beans, a technique that served the marketing objective of making these seeds stand out as well as allowing Pond to comply with federal laws designed to prevent chemically treated beans from entering food markets. The final touch involved drying seeds in preparation for packaging.[9]

In 2018, Pond said he served some 300 farmers, which included folks working small patches of land as well as growers managing thousands of acres, and many of these people he knew well. "Their key to success," Pond said of Monsanto's salesmen, "was to bring people like me on board that would endorse them." Pond's business had been in Scott for almost a century. Fred had earned the trust of the people that lived around him. He said that while farmers would come into his office and want to know fine details about yield data, many clients simply were "relationship buyers"—"Whatever you say is fine with me, Fred."[10]

Monsanto knew that focusing on people like Pond was the key to their biotechnology business. By the time Roundup Ready seeds came online, rural depopulation had radically changed Middle America. Migration from the countryside was already well under way in many parts of the country by the 1920s. But in the years after World War I the mechanization of America agriculture, in part aided by federal policies that encouraged rapid farm consolidation and expansion, set the stage for the acceleration of farmer flight in the decades after World War II as family farms gave way to mega-monocrop fields doused in synthetic fertilizers and pesticides. By 1990, less than 2 percent of the US population still lived and worked on farms in the United States. Monsanto salespeople fretted over the 2 percent, not the 98 percent. And it focused largely on commodity crops, such as soybeans and corn, because that made the most business sense. In 1995, American farmers planted more than 71 million acres of corn and spent on average $23.98 per acre on seed. The soybean market was also big, with growers sowing more than 62 million acres of land (up from 15 million acres in 1950) with seeds that cost $13.32 per acre. If Monsanto could sell to these farmers, they could make a lot of money. Those remaining in rural America held the keys to the machines that could spread Monsanto's products over miles and miles of American farmland. And seedsmen were like super sellers that could push enormous seed inventories out the door. They would make the farm choices that would determine consumer choices in American markets that represented billions of dollars in potential profits.[11]

But trust and a handshake were not the only things that sold seeds. When these farmers took Fred Pond products back to their farms, the important thing was, they worked—"just like magic." In the 1990s, Fred recalled the visual spectacle that compelled compliance with Monsanto's new seed regime: "When the farmer drove down the road, and all his neighbors' fields were perfectly clean and his fields looked like a disaster, and they talked about it at the coffeeshop, it was an overnight success." In the short term, Roundup really did wipe out thistle, waterhemp, marestail, and the other tangled messes that drove farmers mad. The fact that a farmer could now spray this stuff throughout the season without hurting his crop was a game changer. Everybody wanted in.[12]

Pond got rich, but for him it was all bittersweet. "Quite honestly, they made me a lot of money that I probably wouldn't have had had I stayed in the custom cleaning business," he said looking back. But he then added, "I guess I would also say they probably took advantage of the American farmer more than any other company in the history of America." He was referring to the steep fees these farmers paid for their seeds, fees he knew helped pay for the race cars, major league baseball games, and luxurious treatment he received. He remembered the first few years when farmers were actually shown the technology fee that was attached to seed prices and how farmers "went through the roof" when they learned how much they were paying for Monsanto's genes. The company then forced Pond to switch to "seamless pricing" in which the technology fee would no longer be separated out of the total cost on farmers' bills. Henceforward, "the farmer could never know what they were charging" for the technology fee. That was between Pond and Monsanto.[13]

"I don't want to come across as being negative about Monsanto," Pond said, but looking back, it bugged him how things happened. Pond said that Baby Boom seedsmen "felt guilty about going to these lavish parties. . . . We didn't think it was right. . . . It was just so different than anything we'd ever been exposed to." But he went along with it because it seemed to be the only way to stay in business. Still, an uneasiness persisted. He felt bad for the growers who he thought had been exploited by the system. Monsanto "played on a group of people in society that were very honest," Pond said. This was the summary assessment of a man who had watched the Roundup Ready revolution unfold in America's farm country.[14]

Monsanto's marketing department called men like Fred Pond key influential persons (KIPs). Marc Vanacht, Monsanto's manager in charge of the company's Roundup account in Europe beginning in the 1980s, explained exactly the kind of individual the firm was interested in pursuing. A KIP "was a person that over the years had accumulated credibility either because of their technical knowledge as a scientist" or, in the case of farmers, because of their "social presence" and sometimes "size." When it came to Roundup sales in Europe, Vanacht explained that the company did not necessarily seek noblemen—"the best, most expensive" chateaus in Bordeaux, France, for example—but people that were "known," someone who farmers would say was one of them.[15]

KIPs were important, but so too were field demonstrations. Monsanto's marketing in the Old World mirrored its playbook in America's Midwest: get Roundup into fields as fast as possible and let farmers see, with their own eyes, just how well this stuff worked. Vanacht spoke about his efforts to turn communists wedded to the Kremlin into capitalists committed to Monsanto chemicals through marketing campaigns in the fields of East Germany after the fall of the Berlin Wall. "I basically hired a whole lot of students," he said, "and the only condition was that they needed to have [an] agronomic degree and a car." These students drove throughout the countryside, spraying small plots and showing anyone who would watch this chemical wonder. They were happy to do it because Monsanto made it worth their while. Vanacht admitted that they paid these young people a salary that was "equivalent to the Prime Minister of Germany." Monsanto was not just selling chemicals; it was selling a high-dollar lifestyle to young people escaping a Cold War past.[16]

If all this sounds a bit like how a pharmaceutical company operates, the analogy would not be far off the mark. After all, Monsanto's 1985 Searle purchase had made the company a big player in pharmaceuticals. It is not surprising, then, that its marketing arm would resemble that of the big drug firms. When Vanacht spoke of Roundup, he described it as having a "pharmaceutical effect" and said that even if his marketing team was not necessarily in the same "league of the pharmaceutical companies," his firm worked with key farmers the way that drug firms courted doctors.[17]

Monsanto could throw money around because Roundup generated an incredible amount of cash. The magic was in the markup. "Oh shit, the margins were very, very, very good," Vanacht said, referring to the late 1980s and early 1990s, even before Roundup Ready seeds were introduced. Prior to 1991, Monsanto had patents in other countries on its product and could charge a premium price for its chemical because the firm did not yet compete with many companies in the glyphosate market. International protections in Western Europe ended that year and in other countries shortly thereafter, but the company retained an exclusive patent on its Roundup formulation in the United States until 2000. By that point, several overseas companies were producing glyphosate generics, but Monsanto remained dominant internationally, controlling 80 percent of the global market. This was in part due to the fact that many companies were spooked by Monsanto's enormous throughput and capitalization and decided to accept Monsanto's offer to license the selling of its chemicals. At the start of the new millennium, Monsanto earned $2.8 billion just from Roundup, which represented roughly half of the company's net sales that year. "It is the best-selling agricultural chemical product ever," the *New York Times* reported in 2000.[18]

But even in the early 1990s, Vanacht remembers just how flush the company was due to its blockbuster chemical. He said the company plowed Roundup money into "disgustingly, scandalously high marketing" budgets that helped the company make inroads in places where their product had not yet gained popularity.[19]

Part of this involved creating brand awareness via television advertisements. Vanacht explained that in the United States, Monsanto enjoyed affordable rates when paying for Roundup commercials in rural areas because regional television stations in the United States served small markets in Iowa, Nebraska, and other farm states. In Europe, the game was very different. National broadcasters ran the show, so Monsanto had to pay rates that included big metropolitan centers such as Paris, Marseille, and Toulouse. It was "Goddamn expensive, basically unaffordable," Vanacht recalled, but he had a plan. To get around high costs, he decided to pick commercial spots at pre-dawn hours, when rates were cheap. This way he could get Roundup in front of farmers that were having their morning coffee before heading out to the fields.[20]

Pond spoke about what it was like on the receiving end of Monsanto's marketing in the 1990s. "Every piece of literature you picked up," he recalled, "from *Ohio's Country Journal* to every magazine, was 'Buy Roundup, Buy Roundup.'" Driving through Ohio farm country, farmers saw billboards promoting Monsanto's products. These signs along roadways cutting through soybean and corn fields in Middle America would have looked out of place in Los Angeles or New York. But even if big-city citizens never saw Monsanto advertising in the rural countryside, these small-town investments slowly seeded change that shaped a global food system.[21]

Farmers bought what Monsanto was selling. Between 1994 and 2014, glyphosate use in some areas of the country increased by more than 1,600 percent in the United States. The reason farmers could use so much Roundup was because almost all of their crops were resistant to the herbicide. By 2014, 94 percent of all soybeans and 91 percent of all corn seeds sold in the United States were genetically engineered to be glyphosate resistant, up from zero in 1995 for both commodities.[22]

For a while, all seemed well. Because Roundup was so effective, total herbicide use in the United States declined by about 15 percent between 1996 and 2003. Some of these now-abandoned chemicals were pretty toxic, leading some to claim that the Roundup Ready revolution was helping to make farming more environmentally friendly. The grain elevator operator that worked with Fred Pond in Scott, Ohio, noted that his inventory of herbicide chemicals dramatically diminished as Roundup quickly became the miracle chemical everybody wanted. Monsanto annihilated many of its chemical

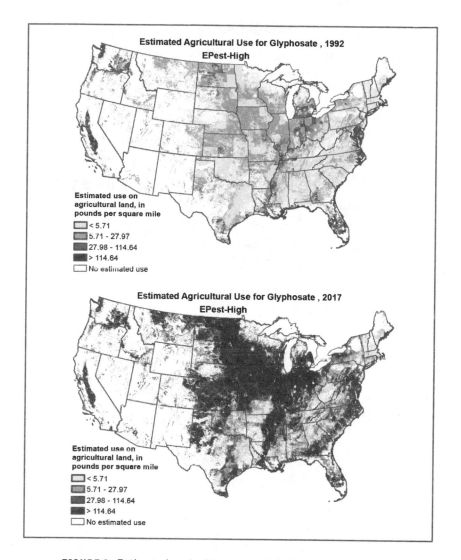

FIGURE 3: Estimated agricultural use of glyphosate in the United States, 1992 and 2017. *Source: USGS Pesticide National Synthesis Project.*

competitors, with the total number of commercial herbicide sellers dropping from 29 in 1980 to just 8 by 2005.[23]

Farmers realized labor savings because they did not have to spend as much time weeding as they had before. One 2009 study found that soybean farmers in the United States reduced household labor by 14.5 percent after adoption of herbicide-tolerant crops. (The same study, however, did not find any statistically significant labor savings associated with herbicide-tolerant corn.) USDA researchers discovered that some soybean farmers, freed from time-intensive weed-control tasks, went on to earn money through nonfarm employment. To be sure, these same researchers also noted that "the empirical evidence on the impact of adopting herbicide-tolerant soybeans on net returns is inconclusive." But just because these crops did not necessarily yield higher on-farm incomes, free time that attended a much more streamlined weeding system offered other benefits that kept farmers wedded to the Roundup Ready system.[24]

"It was a party free-for-all," Pond remembers of those early years. His sales soared as GE seeds rolled through his processing plant and out to hundreds of farms in Ohio. But like many good parties, the binge produced a hangover. By the mid-2000s it became clear that there were serious problems with the Roundup Ready system.[25]

* * * *

IT STILL GETS MARK LOUX fired up when he thinks back on it. Loux is the kind of guy farmers trust. A weed scientist and state extension specialist at The Ohio State University since the late 1980s, he has logged an incredible number of miles in Ford and Chevy trucks—many paid for with money that came from corporate supporters, including Monsanto—visiting soybean and corn fields in the Buckeye State over the past three decades. Salt-and-pepper hair tops his tall build, the signature of a man who has been around and seen some things. Loux speaks in a low, gravelly voice when he delivers no-nonsense advice about the latest science related to weed-control technology. The way farmers in the state express their respect for him, it seems his actions back up his words.[26]

Like the work of most weed scientists across the country, Monsanto offered financial support for Loux's research over the years. Facing dwindling state and federal appropriations for universities and colleges, agricultural science departments welcomed contributions

from corporations with deep pockets. To be sure, chemical companies and farm-equipment manufacturers had long enjoyed a cozy
relationship with land-grant schools, extension service scientists,
and USDA officials. But by the twenty-first century, the boundaries
between public agencies and private institutions became even more
blurred. A 2012 Food and Water Watch report detailed these trends.
The watchdog group found that by the early 1990s, industry financing "surpassed USDA funding of agricultural research at land-grant
universities" for the first time. By 2009, private-sector funds covered
nearly "$822 million in agricultural research at land-grant schools,
compared to only $645 million from the USDA." Between 2006 and
2010, the University of Illinois at Urbana-Champaign's agricultural
school "took 44 percent of its grant funding" from private companies, with Monsanto contributing millions of dollars. And the story
was much the same at The Ohio State University. In the 2010s, some
researchers in Ohio State's College of Food, Agricultural and Environmental Sciences earned grants from Monsanto, with some individual donations totaling nearly a quarter of a million dollars. Mark
Loux was no exception. Between 2008 and 2021, Loux received more
than $330,000 directly from Monsanto for his research.[27]

But even though Loux owed part of his funding to Monsanto, he
was certainly not a Monsanto man. He was on the front lines when
weed-resistance problems started emerging on American farms in
the early 2000s. First it was horseweed (also called marestail), which
University of Delaware crop scientist Mark VanGessel discovered to
be resistant to glyphosate in the early 2000s. VanGessel published a
study in 2001 documenting his finding, but Monsanto would have
none of it. The company did not advise farmers to change the way
they were applying herbicides, and it did not argue for the use of
other herbicides.[28]

Then came waterhemp and giant ragweed. Resistance was
spreading, which was not particularly surprising to Loux. "Never
bet against mother nature," he said, and that was exactly what
Monsanto had done. Through the overuse of glyphosate, farmers

were placing incredible selection pressures on weed species that bred genetic mutations. It was a classic story of Darwinian survival, set in motion by a calculus that demanded Monsanto sell more of its trademark chemical year after year.[29]

Things got ugly. Loux said that Monsanto wanted to be "the gatekeeper" on resistance. The firm began challenging scientists that tried to declare new weed species resistant to Roundup. He remembered a Monsanto representative coming to contest his research team's claim that a weed called lambsquarters was showing some signs of low-level tolerance to glyphosate. According to Loux, the rep told him, "You guys obviously don't know what you're talking about," citing Monsanto studies that clashed with Loux's findings. Loux, his integrity called into question, exploded. He asked the researcher that had done the work on this study to leave the room, and then turned to the Monsanto man and "read him the riot act." "You don't come in here and tell us we don't know what the fuck we're doing," he recalled saying to the rep. "I came really close to just throwing him out." Loux didn't like being controlled by corporate minders.[30]

And neither did farmers. While Monsanto was sending men into greenhouses to scrutinize scientists' work, the firm was also sending Pinkerton detectives and ex-officers of the Canadian Mounted Police onto farms in the United States and Canada to ensure that growers were not saving Monsanto's patented seed. Farmers agreed to this invasion of private property when they signed the technology-use agreement, which stipulated that growers had to "provide the location of all fields" featuring Monsanto genetics and "cooperate fully with any field inspections" the company deemed necessary. Monsanto was being aggressive. The company sent detectives among corn and soybean rows and also hired pilots to fly helicopters over fields to root out what the firm called "seed piracy." In 1999, the *Washington Post* reported that Monsanto had initiated more than 500 investigations on farms in North America, some of which the firm took to trial. By 2004, the Center for Food Safety counted some ninety cases that had been brought by Monsanto against farmers for violation of patent policies. These were the day-to-day operations of the Roundup Ready regime—a "system that sets you free" the company had promised farmers.[31]

The consequences for those having crossed Monsanto were severe. Ray Dawson, an Arkansas farmer, was perhaps one of the hardest hit. He ultimately had to pay a $200,000 to get Monsanto to stop hassling him, despite the fact that he denied ever pirating seeds. In the end the pressure on his family was too intense. At one point, Monsanto lawyers allegedly threatened to incarcerate Dawson's wife, arguing she had lied under oath during a deposition. Dawson, a brave man who often donned a hat that read "Dawson Farms, Monsanto Folds," ultimately decided it was better to pay off Monsanto than watch his family be torn apart. Others would do the same. The average payout per farmer from these cases amounted to around $100,000 by 2004.[32]

Some farmers never settled, instead fighting fierce legal battles against what they called a growing "police state." Perhaps the most famous case was that of Canadian farmer Percy Schmeiser, who lost a battle with Monsanto after nearly five years of litigation. Monsanto investigators had found Roundup Ready canola on Schmeiser's property in Saskatchewan in 1997 despite the fact that Schmeiser said he had never bought GE seeds from Monsanto and denied illegally obtaining germplasm from other sources. Schmeiser claimed that pollen from neighbors' farms or seeds dumped from passing trucks must have carried these GE traits onto his property. Monsanto did not believe him. Company field tests on Schmeiser's farm revealed that a majority of Schmeiser's crops had Roundup Ready technology. Schmeiser later explained that he had tested a small plot of canola on his farm with Roundup and saved seeds from that plot after determining those crops were resistant to glyphosate, but he insisted he had done nothing wrong. He simply saved seeds from plants growing on his property, something farmers had done for centuries.[33]

The Supreme Court of Canada did not agree. Chief Justice Beverley McLachlin made clear that the court was not concerned "with the innocent discovery by farmers of 'blow-by' patented plants on their land." That was not the issue in this case. Rather, the court found that Schmeiser had intentionally saved seeds that he knew included Monsanto's technology. This was a violation of Canadian patent law that protected Monsanto's genetic innovation.[34]

But putting pressure on farmers by taking them to court was just part of Monsanto's playbook; public shaming often worked just as well. In the 1990s and early 2000s, the company worked the airwaves, broadcasting on local radio stations the names of farmers that violated company patents. The firm also launched a toll-free hotline, 1–800-ROUNDUP, and encouraged farmers to call in if they believed their neighbors engaged in seed piracy. "If you have information about the misuse of seed, press 2," announced a friendly female voice on the company switchboard in 2018. Callers selecting "2" were connected to a live operator working for Monsanto's "Seed Stewardship Team." It all took a matter of seconds. For those farmers concerned about being outed for outing their neighbors, the company tried to reassure them up front: "Your call will remain anonymous." By 1999, just a year after the hotline launched, Monsanto reported more than 1,500 calls coming in from around the country. Ordinary farmers, not just Pinkerton detectives, were doing the hard work of rounding up violators of Monsanto's Roundup Ready system.[35]

As neighbors turned on one another in Middle America, resistant weeds continued to spread. By 2006, just ten years after the Roundup Ready revolution began, scientists discovered glyphosate-resistant weeds in more than twenty states. The list of resistant weed species grew to include common and giant ragweed, Palmer amaranth, and ryegrass. By 2008, an herbicide-tolerant Palmer amaranth had swept through the American South, causing major problems for farmers that had become wholly dependent on the Roundup Ready system to clean their fields. For almost a decade, Monsanto had done little to combat this resistance problem,

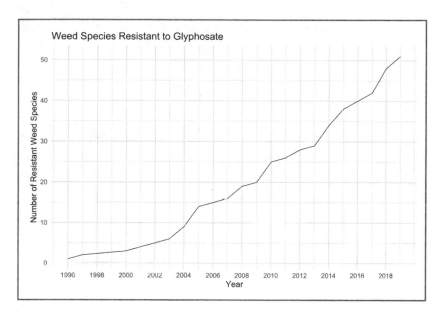

FIGURE 4: Rise in the number of glyphosate-resistant weed species,
1996–2019. *Source: Study by Dr. Ian Heath (2020), International
Herbicide-Resistant Weed Database, Weedscience.org.*

choosing not to radically adjust its label recommendations to combat glyphosate overuse. But the resistance problem simply would
not go away.[36]

Monsanto could no longer deny a problem that farmers could see
quite clearly in their own fields. Loux remembers Monsanto's swift
about-face around 2010. He was at a distributor's meeting, and a
Monsanto representative he knew "basically said, 'I want to apolo-
gize to you because we were told to tell everybody that you were full
of shit.'" Around the same time, Monsanto helped to organize several conferences on resistance. Company officials said they wanted to
be partners with people like Loux whom they had hounded back in
the day. At a dinner during an agricultural conference, Loux remembered a Monsanto representative announcing that his company now
wanted to know what Loux and his colleagues thought about how
best to combat resistance. Loux was flabbergasted. For years, he had
been recommending the use of other herbicides alongside Roundup

to reduce incidence of resistance, but he and many of his colleagues had been ignored. He felt the company was being so "arrogant."[37]

Monsanto's mea culpa could not undo the damage already done. Roundup resistance was a serious problem because there did not seem to be anything in the herbicide pipeline quite like glyphosate. The gravity of the situation could not have been overstated. Leading global glyphosate expert and Australian scientist Stephen B. Powles argued "that glyphosate was up there with penicillin as a once-in-a-hundred-years discovery." To lose this resource would be devastating to the farming community, which had become so dependent on this one chemical. How would farmers get out of this mess?[38]

* * * *

HUGH GRANT, A FORTY-FIVE-YEAR-OLD scientist with a thick Scottish accent owing to his upbringing in the old coal-mining and textile town of Larkhall, Scotland, was at Monsanto's helm when the weed-resistance issue became a big problem. In 2003, the board of directors chose Grant to replace Belgian chairman and CEO Hendrik Verfaillie, who had fallen out of favor when the firm's profits began to sag in the early 2000s. At that time, Monsanto's problems were hardly Verfaillie's making and included bad crop cycles caused by inclement weather and global sales lags stemming from GE seed bans in Europe and slow approval processes in South America. Facing all this, the directors wanted a new leader in the top office.[39]

Grant, like 1990s Monsanto CEO Bob Shapiro, was good for Monsanto's image in part because he shunned braggadocio and bombast. Sure, he had moments of wild spontaneity—like the time he impulsively jumped into the bone-chilling waters of the Arctic Ocean while on a National Geographic expedition to see melting ice caused by climate change. Or the time he decided to take a group of farmers out for a rousing round of karaoke. But, for the most part, Grant was known for his Zen-like calm and levelheadedness. Like Bob Shapiro, he opted for a small cubicle-like office at company headquarters instead of a lavish corner suite filled with mahogany. "Gracious,"

"jovial," "altogether mild," "a quiet consensus-builder"—these were
the labels popular press reporters assigned to Monsanto's middle-
aged manager. For a company suffering from an image problem,
Grant's cool manner was just the kind of thing company directors
wanted to see in the C-suite. There was a recognition among many
top executives, including Grant, that the firm had been a bit arrogant
in its efforts to introduce GE technology to the world, causing the
blowback the company faced in Europe and elsewhere. The firm's
imagemakers now wanted to find ways to exude a bit more humility.[40]

It probably helped that Grant came from humble beginnings. His
first job was as a mail clerk in a men's-suit store owned by his father,
and he was the first in his family to go off to college, matriculating
at the University of Glasgow in the 1970s where he studied molecular biology and agricultural zoology. He had a passion for plants, a
love he first developed while harvesting lettuce and tomatoes during
summer breaks. He nurtured this passion by enrolling in a master's
degree program in agriculture at the University of Edinburgh, and it
was there, at the age of twenty-three, that he saw a Monsanto want-
ad that changed his life.[41]

Monsanto was looking for young professionals that could sell
Roundup to barley farmers in Scotland. The new herbicide had
only been on the market for a few years, and the company needed
local salesmen that could help pitch this product outside the United
States. A Scotsman who could banter in the brogue common to this
place, Grant was a perfect fit for the job, and he soon proved himself to be a talented salesman. He earned promotions quickly, leading a Roundup team in Northern Ireland before admiring managers
in St. Louis called him to company headquarters, appointing him
the new lead strategist for the company's Roundup brand. This was
an incredibly significant position, especially for a young professional
still in his thirties.[42]

As lead Roundup strategist, Grant devised a brilliant plan for
dealing with pending patent expiration for the firm's signature herbicide. He figured that in order to keep competitors from absolutely

devouring Monsanto's glyphosate market-share his firm needed to precipitously drop prices immediately prior to US patent termination in 2000. The strategy mirrored what powerful German firms had done to John Queeny's little chemical business in the early 1900s. The idea was to lower the price point for Roundup so that competitors would not have enough economic incentive to build their own glyphosate plants. Reduced prices would also stimulate larger-volume sales, which would help bring in earnings. Grant drove down prices for Roundup from $44 a gallon in 1997 to $28 a gallon in 2001 and simultaneously reached out to competitors, offering them the opportunity to use Monsanto's glyphosate to produce their own generic formulations. Many companies bought in, enticed by the expanded markets made possible by Roundup Ready seeds. The strategy worked, staving off, at least in the short term, a flood of direct competition in the glyphosate-manufacturing business. In 2001, the company happily reported that Roundup still accounted for the majority of its revenue, bringing in $2.8 billion in sales.[43]

Thus, Grant was a Roundup man, someone who owed his entire career and professional success to one powerful chemical, even though in his public pronouncements he tried to distance himself from this chemical legacy. In the early 2000s, he echoed sentiments first articulated by predecessors Dick Mahoney and Bob Shapiro that Monsanto was no longer a chemical firm. "We're selling information," he told PBS in 2000, arguing that the company's GE seeds would reduce "the amount of herbicide that was applied in the crop, and [reduce] input costs to farmers." According to Grant, Monsanto would be selling less stuff, not more as it continued its transition to becoming a life-sciences company.[44]

But the appointment of the firm's lead Roundup salesman to the highest office in the "new" Monsanto should have made clear that the firm's chemical past was very much tethered to its biotech future. It mattered that Grant had been with the company since 1981. His vision of the future was inherently shaped by his corporate rearing in the days when Roundup sales grew exponentially. In the end, he was a chemical salesman, someone who always wanted to see the brand he knew and loved do well. He would never abandon the herbicide that had gotten him where he was.

* * * *

FARMERS, HOWEVER, WERE GETTING frustrated with Grant's favorite chemical. By the mid-2000s, many growers knew that the only solution for the foreseeable future was to go back to the past—to turn to chemicals that GE seeds were supposed to make obsolete, herbicides other than Roundup.

The data were clear on this issue. Beginning around 2003, total annual herbicide use, which had been on the decline since 1996, began to rise. A decade later, the total amount of herbicides used on American farms increased to 353 million pounds, up from 294 million pounds in 1996—a 20 percent rise since the start of the Roundup Ready revolution. Focusing on just soybeans, the *New York Times* reported that herbicide use on US farms increased by 150 percent in just two decades after Roundup Ready seed introduction, despite the fact that total soybean cultivation only rose by less than 33 percent. Increased glyphosate purchases contributed to increased overall herbicide use during this period, but that rise was also due to old chemicals that started making a comeback in many markets because they were, in many cases, the only defense against glyphosate-resistant weeds.[45]

One chemical that saw a major resurgence because of Roundup-resistant weeds was 2,4-dichlorophenoxyacetic acid (2,4-D). First commercialized by several chemical companies in the 1940s, 2,4-D, which had been mixed with toxic 2,4,5-T in Agent Orange, had long been used as a residual herbicide in weed-control programs. Mark Loux recommended it to farmers because, as he put it, the chemical had been used for decades and there seemed to be no clear signs that it was affecting farmer health. But other researchers were not so sure, including Michigan State agricultural economists Scott Swinton and Braeden Van Deynze, who cautioned that glyphosate was "ten times less toxic than 2,4-D." In 1995, the University of California

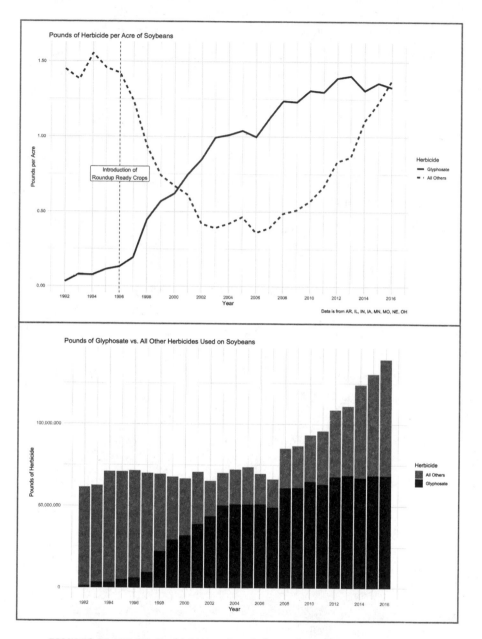

FIGURES 5A AND 5B: Herbicide use trends for soybean, 1992–2016: pounds
of glyphosate per acre compared with pounds of all other herbicides per acre
(top); pounds of glyphosate compared with pounds of all other herbicides
(bottom). Data compiled from Arkansas, Illinois, Indiana, Iowa, Minnesota,
Missouri, Nebraska, and Ohio. *Source: USDA National Agricultural Statistics
Service and USGS Pesticide National Synthesis Project, Estimated Annual
Agricultural Pesticide Use by Major Crop, 1992–2017.*

Environmental Health Program ranked 2,4-D among the top four most hazardous agricultural herbicides sold in California out of 150 chemicals in use at that time. Nevertheless, farmers felt they had to turn to old chemicals such as 2,4-D because in many cases, this was the only stuff that would kill glyphosate-resistant weeds.[46]

Another chemical that Monsanto resurrected at this time to deal with the resistance issue was a compound called dicamba. This herbicide had a long history. The federal government first approved dicamba for use on US farms in the 1960s, and within a few years, it became a popular weapon in farmers' herbicide arsenal. By 1990, growers sprayed roughly 15 percent of all corn acreage with dicamba, and use increased in the 1990s. But dicamba's popularity was hampered by a serious design flaw, one that had worried farmers for decades.[47]

Back in the 1970s, researchers discovered dicamba's tendency to vaporize and spread from one farm to another, particularly during warm weather, something Roundup typically did not do. Subsequent reports of dicamba drift continued in the years that followed, and it remained a topic of concern when Monsanto began pursuing the idea of developing dicamba-resistant crops in the early 2000s. At that time, company representatives were hounding folks like Loux, trying to keep the Roundup resistance problem under wraps, but the firm clearly knew that it needed to change course if it was going to get out of this resistance fiasco.[48]

Monsanto's farm clients were growing frustrated, as Roundup failed to keep weeds at bay like it had back in the 1990s. Monsanto needed a solution for a problem it helped create. The firm turned to dicamba for salvation.

* * * *

IT TOOK ALMOST A decade for Monsanto to develop and get federal approval for a new GE seed that could make cotton and soybean crops resistant to both dicamba and glyphosate. (Monsanto would also seek approval for a new dicamba-tolerant corn crop system, but that review would take longer.) The company branded its new creation the "Roundup Ready Xtend Crop System," overcoming the final regulatory hurdles in 2014 and 2015.

This was some good news, but at company headquarters, everything seemed to be spinning out of control. News had leaked that the World Health Organization (WHO) was going to come down hard on glyphosate, labeling it a "probable human carcinogen." Resistance was a manageable nuisance; this was a bombshell. "The activists [are] well organized," Dan Jenkins, Monsanto's regulatory affairs liaison to federal agencies, said in a text message to colleagues in 2014—"They smell blood" and are "circling." A few months later, Jenkins texted another company official to express his growing concerns: "I think we need to talk about a political level EPA strategy and then try to build a consensus plan . . . on several fronts: . . . Dicamba, glyphosate, resistance mgt . . . we're not in good shape, and we need to make a plan."[49]

Jenkins was right. Monsanto was not in good shape. In 2015, the WHO issued the report Monsanto had feared, causing company officials to begin a frenzied effort to prevent other government regulatory bodies from following the international agency's lead. This is when the company moved to ghost-write articles designed to prove that glyphosate was not dangerous when used appropriately. Good news came from the EPA in October 2016 when the agency issued its determination that glyphosate was "not likely to be carcinogenic to humans," but Monsanto managers were still worried. They feared the CDC's Agency for Toxic Substances and Disease Registry (ATSDR), which was interested in completing a review of glyphosate that many company officials believed would be at odds with the EPA's report. Jenkins and his team got to work.[50]

"Spoke to EPA," Jenkins said in a text exchange, "pushed them to make sure atsdr is aligned, said they would." Another Monsanto official, Eric Sachs, texted Mary Manibusan at EPA with the same request, and Manibusan wrote back, "Sweetheart, I know lots of people so you can count on me." This was good news, Sachs said, because Monsanto officials "were trying to do everything we can to keep from having a domestic IARC occur with this group. May need your help." Dan Jenkins also applauded Jess Rowland, an EPA official who allegedly said in a phone call with Jenkins regarding the ATSDR review, "If I can kill this I should get a medal."[51]

In the end, the CDC study was delayed, but there were still many troubles that lay ahead. In 2016, California schoolyard groundskeeper Lee Johnson filed his Roundup case in a San Francisco court. Thousands of other plaintiffs were awaiting trial right behind him. For the foreseeable future, glyphosate was going to be a major concern.

But so was another herbicide: dicamba. The company knew this herbicide could spread to farms that did not have dicamba-tolerant crops, creating catastrophic losses to farmers that did not buy Roundup Ready Xtend seeds. The company was trying to fix the problem by creating a new dicamba formulation that would not vaporize and drift off-target so much, but this issue was not fully resolved in 2015 or 2016 even as Monsanto rushed its new seeds to market.

As with so many other chemicals Monsanto had sold, the heart of the plan was to control information flow. Monsanto did not want third-party researchers slowing down delivery of a seed system it had spent great time and resources developing. In 2016, after Monsanto had gained federal approval to launch its Xtend seeds but before it had received the go-ahead from the EPA to sell its new dicamba formulation, company officials made a bold move. They prohibited university scientists in Arkansas, Missouri, Illinois, and in other big ag schools from conducting volatility tests on XtendiMax herbicide, Monsanto's new dicamba mix that included "VaporGrip" technology designed to keep the herbicide from spreading off-target. Arkansas professor Jason Norsworthy told Reuters, "This is the first time I'm aware of any herbicide brought to market for which there were strict guidelines on what you could and could not do." Monsanto justified its actions by stressing the urgency of the crisis. As company global strategist Scott Patridge explained, "To get meaningful data takes a long, long time," and there simply was no time for delay. Farmers needed Monsanto's technology now, and the company intended to give it to them.[52]

By this point, the dicamba-resistant seeds were already out of the bag. Farmers, desperate to solve a growing weed-resistance problem, had started planting Xtend cotton seeds in 2015 and then Xtend soybeans in 2016. (A dicamba-tolerant corn crop system that would make corn seeds resistant to five different herbicides, including dicamba, was still under regulatory review in the United States as this book went to press.) And because the EPA had not yet approved Monsanto's XtendiMax herbicide for use on Xtend crops, farmers used cheap dicamba generics in 2015 and 2016 to kill weeds in their fields, even though these products were prone to vaporize in hot temperatures. Immediately there were reports of dicamba drift, but the EPA did not take action to halt the spread of these new GE seeds. Monsanto later claimed that the widespread dicamba-drift problem could have been prevented had farmers not violated the law and waited for the company to get its new, non-vaporizing formulation to its customers. But in the 2017 growing season, when XtendiMax finally entered markets with EPA approval, volatile dicamba spread throughout the American South.[53]

It was a disaster. In Arkansas, one of the hardest-hit states, the Plant Board received more than 1,000 complaints from farmers claiming that dicamba had drifted onto their farms and damaged their crops. Overwhelmed, the board decided to halt further dicamba application until it could figure out how to keep this herbicide from spreading beyond where it was sprayed. The problem stretched across America's farm country, from Georgia to North Dakota, though hot southern weather seemed to trigger some of the worst outbreaks. In all, the EPA reported in 2017 that there were some 2,648 ongoing investigations into damage caused by dicamba.[54]

The problem even drifted onto farms in Monsanto's backyard. That is when Bill Bader, Missouri's largest peach grower, witnessed devastating losses to his crops when soybean growers around his property started spraying dicamba in large quantities. Fruit and orchard growers like Bader were often hard hit during the dicamba debacle because they could not purchase traits that would protect them from drift. After all, there was no such thing as a dicamba-resistant peach, something Bader well knew when he watched his peach trees wilt in 2015 and 2016. With his business in decline, Bader felt he had no choice but to take Monsanto to court.[55]

Mark Loux remembered those days well. He recalled confronting a herbicide representative selling BASF's Engenia, a dicamba mix designed to compete with Monsanto's XtendiMax. He was blunt: "Y'all are pulling a fast one on the world. It's really hard to watch this. It's like corporate immorality. How can you do that, like walk away from this whole thing? You have a defective product."[56]

Weed scientist Jason Norsworthy, now free in 2017 to conduct volatility experiments on Monsanto's XtendiMax, showed that the company's dicamba herbicide could drift more than a football field—390 feet—away from target locations. It seemed Monsanto's "new" technology did not fix the old drift dilemma. Norsworthy's study led some EPA scientists to recommend 443-foot buffers around dicamba-sprayed farms to ensure drift would not affect neighboring land.[57]

Monsanto officials fought back. They did not want to tell farmers to limit herbicide use so dramatically; it went against the company's quest to sell more chemicals. Monsanto's regulatory affairs team therefore tried to bury Norsworthy, challenging his integrity and suggesting his research was tainted because he had in the past "endorsed" a product manufactured by Bayer that competed with XtendiMax—a strange accusation, both because Monsanto funded similar university research all the time (should those studies be called into question?) and because Monsanto was about to merge with Bayer, a company that had called Norsworthy one of the "pre-eminent weed scientists" in the country. After hearing Monsanto's complaint, the EPA—now under the leadership of President Trump's second EPA administrator, Andrew Wheeler—decided to renew registration for Monsanto's herbicide with labels that called for 110-foot buffers on downwind sides of farms and 57-foot buffers on all other farm edges.[58]

Soybean and cotton farmers that had not already made the switch to dicamba were now trapped. Ohio State weed scientists had predicted this scenario seven years earlier. The "risk of off-target movement," they said in a 2011 report, "could drive more rapid adoption of" dicamba-resistant traits "because growers will want to protect their soybean and cotton crops from nearby applications" of dicamba. In other words, drift forced compliance with Monsanto's new seed system. Farmers who chose to reject Monsanto's Roundup Ready Xtend package would open themselves up to potentially serious financial injury. Who would risk it? For many farmers, there really was no choice but to buy the seeds.[59]

Real solutions seemed nowhere in sight. In the end, a GE system that relied so heavily on dicamba was bound to create weed-resistance problems that would have to be fixed sometime in the not-so-distant future. This, perversely, would actually be good for business—at least for Monsanto. In the short term, dicamba drift would force GE-holdouts to turn to Monsanto's seeds. And in the long run, the flood of dicamba would create weed resistance that would then cause farmers to clamor for new GE varieties capable of combating that resistance. It was a vicious cycle.

✳ ✳ ✳ ✳

ALL OF THIS—THE BULLYING, corporate control, resistant weeds, return to old herbicides, and the dicamba-drift problem—might have been worth it if indeed these GE crops had radically enhanced food production. This, after all, was Monsanto's declared mission. "Worrying about starving future generations won't feed them," a company advertisement exclaimed in 1998: "Food biotechnology will." GE seeds "can certainly increase yield and improve productivity, which is a condition of feeding people," Shapiro had told the world in 1995. Here was a noble cause, and one that many must have expected would entail unforeseen costs. But the reality was these seeds never produced the bounty Monsanto or Shapiro had promised.[60]

In 2012, the USDA published the first big agency study comparing yields of herbicide-tolerant (HT) crops and conventional varietals. Agricultural Research Service scientist Stephen Duke, dubbed "Mr. Roundup" by some in his field because of how closely he studied the

Roundup Ready revolution, was the lead author of that 2012 report. Duke concluded that "yield data from the years before introduction of GR [glyphosate-resistant] crops, continuing to the present show that the same yield trends before introduction continued after introduction."[61]

Two years later, USDA scientist Jorge Fernandez-Cornejo concluded that "the effect of HT (herbicide-tolerant) seeds on yields was mixed." "Over the first 15 years of commercial use," he explained, GE crops "have not been shown to increase yield potential," which, as opposed to actual yield, is "the yield of an adapted cultivar when grown with the best management and without natural hazards such as hail, frost, or lodging, and without water, nutrient, or biotic stress limitations (water stress being eliminated by full irrigation or ample rainfall)." Fernandez-Cornejo did cite studies that showed some HT crops were "associated with a statistically significant, but small, increase in yields," and he offered nuanced analysis, pointing to evidence that indicated some growers who used seeds with stacked traits—that is, seeds that conferred resistance to two or more herbicides or included the Bt gene to repel insects—did see increases in crop yields, especially in particularly pest-ridden environments. But he also added that "several researchers found no significant difference between the yields of adopters and nonadopters of" HT crops. In other words, herbicide-tolerance traits, in and of themselves, did not offer significant yield advantages for farmers.[62]

Researchers overseas looked into the question of GE crops and yield in the 2010s. The *Journal of Agricultural Science* published a 2013 paper that was based on field studies in the United States, Canada, Argentina, and Romania and which found "HT technologies do not increase yields significantly." The authors of this study made clear that GE seeds did "facilitate more cost-effective weed control" and noted that Bt crops, engineered to help control insect pests, did produce higher yields on some farms, but the statistical evidence simply did not support the claim that HT crops dramatically increased yields. Similarly, a team studying glyphosate-resistant soybeans in

Brazil reported in 2014 that the use of glyphosate-resistant soybean crops "did not effect . . . plant yield."[63]

In 2016, the *New York Times* investigated United Nations data on commodity crop farms in Europe and the United States. Because some countries in the European Union had banned the use of GE crops in the late 1990s, reporters realized that they had real-world data that could help them assess whether GE crops were in fact more prolific than conventional varietals. The newspaper's investigation echoed what other scientists were finding. After comparing cultivation in Western Europe and the United States, the *Times* concluded "Over three decades, the trend lines between the two barely deviate."[64]

Twenty years after the first Roundup Ready seeds hit markets in the United States, the National Academy of Sciences (NAS) published the first comprehensive study on the biological, ecological, and economic effects of GE crop adoption in the United States and around the world. This landmark study was published as a weighty tome, totaling more than 600 pages of text and appendices, and it included extensive sections comparing yields for GE crops and conventionally bred varietals. NAS admitted there was clearly "disagreement among researchers about how much GE traits can increase yields compared with conventional breeding" and noted that the "sum of experimental evidence indicates that GE traits are contributing to actual yield increases," especially on pest-ridden farms using Bt technology, but concluded that "there is no evidence from USDA data that they [GE traits] have substantially increased the rate at which U.S. agriculture is increasing yields." The authors of the NAS study explained their findings in clear language: "There was little evidence that the introduction of GE crops were resulting in a more rapid yearly increases in on-farm crop yields in the United States than had been seen prior to the use of GE crops." In 1996, Monsanto had said Roundup Ready crops would drive productivity gains critical to feeding a hungry and growing world population, but the historical evidence, two decades later, showed that this was not happening as planned.[65]

"This does not mean that such increases will not be realized in the future," NAS scientists were careful to add—a caveat that reflected associated researchers' general faith in GE technology. Like Stephen Duke and others at the USDA that worked on these groundbreaking reports, NAS researchers were not vehemently opposed to GE seeds. In fact, many believed there was much good that could come from drought-resistant seeds and other GE innovations. The problem, as they saw it, was not that GE technology was inherently bad; rather, they simply pointed out that promises made by GE companies twenty years before fell far short of realities on the ground.[66]

North Carolina State professor Fred Gould, chair of the 2016 NAS study, expressed this sentiment well. "Early on," he said, "people were promising all sorts of things that would happen with these crops in terms of actually increasing yield" by selecting for drought tolerance and other desired traits, but he concluded "that has turned out very difficult to do." As Gould intimated, the vast majority of GE crops cultivated around the world, some 80 percent according to a study in 2008, were genetically engineered to be resistant to herbicides. Only a small fraction of seeds sold globally had drought-resistant GE traits by the 2010s. For the small percentage of GE crops that were engineered to be drought-resistant, major advancements were not necessarily linked to biotech breakthroughs. Speaking about drought-resistant corn, for example, Gould explained that the "biggest leaps of the day" came from "conventional breeding" not from the insertion of GE traits into seeds.[67]

And it is critical to note that the vast majority of the GE traits Monsanto sold went into three main commodity crops: soybeans, corn, and cotton. If this was Monsanto's plan for feeding the world, it was a strange one at that. After all, roughly half the corn produced in the United States by the mid-2010s went into feed for livestock, and another 30 percent into ethanol production. Approximately 70 percent of America's soybean harvest in 2013 became fodder for livestock. The conversion of this feed into food fit for human consumption was incredibly inefficient. Dr. Jonathan Foley, director of the Institute on the Environment at the University of Minnesota, wrote in *National Geographic* that for "every 100 calories of grain we feed animals, we get only about

40 calories of milk, 22 calories of eggs, 12 of chicken, 10 of pork, or 3 of beef." To provide food for billions of people in the future, Foley argued, farmers would need to begin shifting away from a monocrop farming culture that sent more grains into the bellies of farm animals than humans. But that was not something Monsanto was eager to promote given its extensive investments in GE corn and soybeans. Increasing yield remained Monsanto's mantra in public conversations about its contributions to eliminating hunger around the world.[68]

And yet, it seemed old breeding techniques, not new genetic engineering technology, were driving yield gains in the twenty-first century, a reality that clashed with Monsanto's messaging. The logo for Xtend seeds and herbicides featured two dynamic arrows embedded in the front X, both pointing forward—to the future. The company stressed the novelty of its innovations. "A powerful new soybean system is coming," Monsanto prophesied in a 2015 Xtend advertisement. It would be a system that would provide "more options" and "greater flexibility." It was an "advanced" technology that offered "future solutions" to farmers. This was progress.[69]

But was it really? Farmers were now going back to old chemicals to try and keep back weeds born of Monsanto's previous crop system. It seemed the past was being sold to farmers who were having to deal with a present that did not match up with what Monsanto had promised.

And American growers were not alone. From 2000 to 2020, Monsanto had sought to expand its seed sales into global markets. It had to in order to keep growing. After all, by the mid-2010s, roughly 90 percent of all cotton, corn, and soybeans grown in the United States was already genetically engineered to be glyphosate resistant.[70]

Monsanto's financial future lay in foreign lands, but to be successful, the firm had to act fast. After all, the technologies it was selling in overseas markets were the same ones under fire back home. Could Monsanto get growers in Vietnam, Brazil, and beyond to buy its seeds before they fell out of favor?

CHAPTER 13

"They Are Selling Us a Problem We Don't Have"

"IT WAS INCREDIBLE," THE SPECTACLED SEXAGENARIAN JOÃO Paulo Capobianco said over and over again as he reflected on the battle over Monsanto's Roundup Ready seeds that took place in Brazil back in the early 2000s. At the time he had been the national secretary of biodiversity and forests within the Ministry of the Environment, Brazil's equivalent to the EPA, working in the administration of Luiz Inácio Lula da Silva—a political leader whose storied rise from the working class to the nation's highest office had captivated Brazil. Capobianco had headed the key government commission tasked with investigating the pros and cons of legalizing genetically engineered seeds in Brazil.[1]

Back then, farmers in Brazil's southernmost state of Rio Grande do Sul had already been illegally cultivating Roundup Ready crops in violation of a Brazilian court ban on such practice that had been issued in 1998 after Greenpeace and a local consumer advocacy group complained that adequate environmental and human health assessments had not been performed in Brazil. The Rio Grande do Sul farmers were planting what were known as "Maradona" seeds, so-named because they came from arch-rival Argentina, home of the famed soccer player Diego Maradona. In 1996, the Argentinian government had approved the use of Roundup Ready seeds, and Monsanto began selling its products there almost immediately through local seed dealer Nidera. Initially, Monsanto sold these seeds without demanding royalties from growers, but within a few years, the firm began to request technology fees. Some cried foul, arguing that this was a clear bait-and-switch strategy designed to hook farmers into Monsanto's crop system so it could then raise prices and make big profits. Many growers rebelled when the American company tried to extract fees for seeds in the years ahead. Facing pressure from growers' groups, Argentinian regulators resisted Monsanto's efforts to enforce patent protections, upholding laws that allowed farmers to save seeds and use genetically engineered—transgenic—traits without paying fees to Monsanto. All the while, Roundup Ready technology spread throughout Argentina and over the border into Brazil.[2]

At the time, Monsanto really needed Brazilian farmers to adopt its genetically engineered (GE) germplasm. This country, which represented nearly half of South America's landmass, was an unmatched market for expansion. And by the end of the 1990s, Brazil was even more critical given the fact that the company had mixed success expanding its GE seed system elsewhere. As late as 2001, only the United States, Argentina, and Canada had extensive GE seed dispersal, and Argentina and the United States accounted for roughly 90 percent of all GE acreage by 2003. The Indian government approved in 2002 the use of the company's Bollgard cotton seeds, which were genetically engineered to produce toxins that could kill insects, but in other parts of the world, Monsanto ran into resistance. The European Union, for example, blocked the planting of genetically engineered crops in the early 2000s, with consumer health advocates and environmentalists charging that Monsanto seeds would cause ecological harm. Other polities, such as Canada and South Africa, approved the use of Roundup Ready seeds, but these countries had nowhere near the amount of land devoted to commodity crop cultivation as South America's biggest nation. Given all this, Brazil was the lynchpin of Monsanto's envisioned global empire.[3]

Which is why some people believed Monsanto was behind the secret sale of illegal Maradona seeds from Argentina to Brazil. In an interview in Buenos Aires, a government official admitted that

Monsanto had filed for a license to export Roundup Ready seeds from Argentina to Brazil even though GE seeds had not been approved for sale in Brazil. As one scholar put it, "Argentina was the 'door' Monsanto used to enter" Brazil and the rest of South America. "Of course," an exasperated Capobianco said when pressed on whether he knew Monsanto was behind the Maradona seed sales, "everybody knows about that."[4]

It seemed like an "irreversible situation," Capobianco said. What were they to do? Southern farmers were begging President da Silva's office to approve transgenic soybeans. The law as it stood required the government to destroy these farmers' crops, which amounted to thousands of tons of soybeans. "Can you imagine?" exclaimed Capobianco. They had to do something to help the farmers out, so Lula acted. In 2003, he issued a temporary decree allowing the southern farmers to bring their grain to market. But farmers could not sell seed for future use. Lula did not have much choice. It was a patch to buy time so that his administration could figure out what to do next.[5]

In this moment, Capobianco wanted to pause. He urged the federal government not to approve any further introduction of GE crops into the country until the regulators had completed extensive studies on the ecological impact these plants would have on Brazil's ecosystem. Capobianco said these studies might take years, but they were important to protect the health and safety of Brazil's people and its environment. He believed it was unconscionable to approve the use of GE seeds using studies that focused on American farms and fields and did not take into consideration Brazil's unique environment.[6]

But Brazilian legislators did not heed Capobianco's warnings. In March 2005, Brazil's Congress passed a law approving the use of Roundup Ready seeds without demanding any further environmental impact studies.[7]

For Capobianco, there was no other way to look at this than to see it as a total defeat. "For me, it's terrible, it's sad," he said. He had urged politicians for some time—five years—to see if there were

unforeseen ecological problems with the Roundup Ready crop system before the government gave the go-ahead to plant Monsanto's
seeds. If he had gotten what he had wanted, he might have been
able to prevent problems already playing out in the United States
from coming to Brazil. After all, just as the Brazilian Congress was
approving Roundup Ready seeds in Brazil, weed scientists back in
the United States were learning that there were troubling issues with
Monsanto's crop system, and Roundup-resistant weeds were just
becoming the subject of critical scientific analysis. Monsanto was
trying to keep this all hush-hush, even badgering people like Ohio
State weed scientist Mark Loux, who were beginning to investigate
resistance issues, but in time it would all be so clear: Monsanto's GE
system had a critical flaw.[8]

But Capobianco did not get more time. Instead, history was set to
repeat itself in Brazil. Because the Brazilian Congress put no regulatory restrictions on the introduction of Monsanto's GE seeds, the
company's transgenic germplasm spread fast across the country. In
less than a decade, adoption rates for transgenic soybeans reached
roughly 88 percent, and glyphosate sales boomed.[9]

* * * *

MONSANTO'S HUGH GRANT HAD seen this all coming. For years, he
had been plowing fertile ground for his beloved Roundup brand in
Brazil. Prior to becoming CEO, Grant oversaw the construction of a
new glyphosate-manufacturing facility in the small industrial town
of Camaçari, on the northeast coast of Brazil. Completed in 2001,
this facility turned phosphorus received from Soda Springs, Idaho,
into chemical intermediates that could then be shipped to company
factories in São José dos Campos, Brazil, and Zárate, Argentina, for
final formulation. The idea was to Brazilianize the Monsanto brand
by making targeted investments in the country. "We will begin to
produce the raw material of Roundup in Brazil, favoring the reduction of imports by the country and by ourselves," explained a corporate affairs representative to the Brazilian press.[10]

These investments proved a smart move as Monsanto tried to stave off Chinese glyphosate manufacturers that began flooding Brazil's markets with cheap generics in the 2000s. The company made the case that Chinese competitors were threatening local industries by dumping low-priced glyphosate on the Brazilian market. Even though Monsanto employed just a few hundred people at its Camaçari plant and a few hundred more in São José dos Campos, the company used its local plants to appeal for government protections. And the strategy worked. Responding to Monsanto's corporate pressure, the Brazilian government issued a tariff on Chinese glyphosate in 2003 that amounted to a 35.8 percent markup on the product.[11]

The timing could not have been more perfect. Two years after the government-imposed tariff protections, the Brazilian Congress approved the use of Roundup Ready technology, and Grant watched as GE seeds and Roundup spread throughout the countryside. Brazil became the largest exporter of GE soybeans in the world and the second largest global exporter of GE corn. Only the United States had more land devoted to GE crop cultivation a decade and a half later.[12]

The center of Monsanto's Brazilian seed empire was a huge savanna in the country's core known as the Cerrado. This unique biome is a vast expanse, larger than all the land east of the Mississippi River and north of the Florida Panhandle in the United States, and had long been seen as unfit for agricultural development, in part because of the poor soils and the thick brush that sprawled across the landscape (thus, Cerrado, "closed," "thick," or "dense" in Portuguese). But in the 1970s, soybean researchers at Brazil's leading agricultural research institution, Embrapa, made an amazing discovery that turned this seeming wasteland into prime real estate.[13]

What Brazilian farmers needed were soybeans that were adapted to the hot and humid tropical environment of the Cerrado, and there were several hurdles to overcome. First, day length was shorter in the tropics than in the temperate regions where soybean cultivation flourished. It was also hotter in Brazil than in the lands where Chinese farmers first cultivated soybeans millennia ago. If Cerrado

cultivation was going to work, soybeans needed to be bred to thrive
in the higher temperatures and shorter photo periods of the tropics. In addition, the soils of the Cerrado were atrocious for farming,
which meant soybeans suited for this environment would have to be
hearty varieties capable of handling the highly acidic soil that lacked
basic plant macronutrients.[14]

In 1975, Embrapa launched a special soybean research institute
in the city of Londrina, in the southern state of Paraná, and it was
there that researchers bred the first tropical soybean variety suitable
for Brazil's vast interior. It was remarkable. The Embrapa seeds were
essentially super germplasm in the sense that they not only proved
viable in the harsh conditions of the Cerrado but also produced
higher yields than those of varieties grown in the temperate south.[15]

Over the next two decades, Embrapa helped to develop roughly
forty unique soybean varieties suited for the Brazilian environment.
It was an open-source model led by a public institution almost
entirely funded by the government. Embrapa shared its germplasm
with local breeders who in turn created seeds that year-after-year
outperformed older crosses, with yield per hectare productivity for
soybeans roughly doubling in the twenty years between the time
Embrapa created its soybean research center and 1995. That year,
roughly half of the country's soybeans came from the Cerrado.[16]

This was when Monsanto made its major move into the Brazilian
seed business. In 1997, the company purchased Agroceres and FT
Sementes, two of the largest seed companies in Brazil, both of which
contained seedstocks that in part benefited from years of Embrapa
research. The timing proved fortunate for Monsanto because it could
buy at bargain prices from struggling Brazilian seed firms that had to
deal with elevated interest rates implemented by the government in an
effort to curb years of destructively high inflation. Monsanto acquired
a goldmine: germplasm that had been perfected through decades of
public investments and open exchange between Brazilian breeders.[17]

Soybean seeds were just part of the bounty Monsanto got through
Brazilian acquisitions. By the start of the 2000s, the company also

controlled roughly 40 percent of the corn seed market in Brazil. This was a big deal because in the Cerrado, warm weather and year-round sunshine meant that farmers enjoyed two harvests each year, often rotating soybean and corn crops. There was no winter downtime like in the United States. The pause between harvesting a soybean crop and planting corn crops could be as short as fifteen minutes, with massive corn planters following right behind soybean combines in the field. Here was a seamless seeding system that stood to make GE companies billions.[18]

Monsanto put its Roundup Ready technology in germplasm that had been hand-selected by Brazilian breeders for decades, though it took credit for the yield potential offered by these varieties. Hugh Grant made expansive claims about this GE technology in Brazil, which by the end of the 2000s included Roundup Ready gene cassettes "stacked" atop "Bt" traits that allowed plants to produce their own insecticides. The firm claimed that these stacked varieties drove significant yield increases. In 2008, for instance, Grant even promised that Monsanto's biotechnology would effectively double corn, cotton, and soybean yields by the year 2030.[19]

Yet, despite Hugh Grant's claims, data collected by a leading national statistics agency within Brazil revealed that conventionally bred soybeans were actually outyielding GE varieties in certain areas of the Cerrado. The Mato Grosso Institute of Agricultural Economics (IMEA), the central source for agricultural yield and cost data in Brazil and an agency that supported both GE and non-GE agriculture, reported in 2016 that conventional soybean growers in the western farm country of Mato Grosso, deep in the Cerrado, harvested 60 sacks per hectare from their fields compared to GE growers' 55 sacks per hectare—five sacks per hectare less. Looking across the state, IMEA stated that three of the six Mato Grosso regions showed conventional growers with average productivity exceeding that of GE farmers. In the other three regions, IMEA reported identical yield per hectare results for conventional and transgenic crops.[20]

Nevertheless, as in Ohio, many farmers in Brazil purchased

Roundup Ready seeds because the technology made life easier and helped them save on costs—at least in the short run. To understand the scale of initial savings, one has to understand just how big some of these farms were in the Cerrado. As early as 1995, more than 60 percent of the farms in the Cerrado were bigger than 2,400 acres, and by the 2010s, some growers cultivated farms that were more than thirty times that size. It was an otherworldly landscape—commodity crops as far as the eye could see. This was the kind of big business that attracted investment types, including George Soros, who sunk money into what were called "pools de siembra," grower collectives that managed huge tracts of land in the Cerrado. Even TIAA-CREF channeled money from teachers' and professors' retirement funds into major investments in Cerrado soybean production.[21]

Roundup Ready technology helped these big farms turn profits for their big investors in the early years. Growers could simply set their GPS technology and sit back in their sprayers as satellites in outer space directed their path across a vast green sea of soybean pods, spraying cheap glyphosate on Roundup Ready crops. In 2015, herbicide costs for managing a GE soybean field in Mato Grosso ran about $76 per hectare, which was about $11 less than what farmers using conventional seeds paid. While this might not seem like a lot of money, savings added up fast for big growers. A massive 30,000-hectare farm could expect to save nearly $330,000 on herbicide costs every year using this system. Joining the Roundup Ready revolution seemed to make clear economic sense for many big Brazilian growers.[22]

But these savings only materialized if glyphosate worked effectively to kill weeds, and given what happened in the United States with the emergence of glyphosate-resistant species, there was reason to fear that this herbicide might not retain its potency indefinitely. As early as 2007, what had already played out in the United States began unfolding in Brazil. That year, scientists reported four different types of glyphosate-resistant weeds in the country and predicted then that these weeds "had great potential to become problems."[23]

Monsanto's Brazilian business, worth billions of dollars, was in jeopardy. By the end of the 2010s, company officials went to the Ministry of Agriculture to see if they could get approval for Monsanto's dicamba-tolerant seeds—the best tool they felt they had to beat back Roundup-resistant weeds they should have known would sprout up in a Cerrado now covered in glyphosate.

* * * *

"WE ARE AFRAID," said one senior official at Embrapa when asked about the introduction of dicamba-resistant crops in Brazil. In the United States, farmers had learned that hot temperatures, a commonality in the tropical Cerrado, turned dicamba into a volatile compound. Bill Bader, the Missouri peach farmer, was two years into a suit against Monsanto for damage caused by dicamba's volatility, and yet it seemed Brazil was headed toward the same trouble despite the warning signs from abroad. Some agricultural researchers feared the scale of destruction would be massive. Brazil was not the United States, João Paulo Capobianco had warned, and in this case that mattered. If drift was a problem in temperate Missouri, it promised to be worse in tropical Brazil.[24]

"I'm pretty scared about dicamba," confessed Dr. Rafael Pedroso, a UC Davis–trained weed scientist working at ESALQ, Brazil's leading agricultural university. Pedroso was certainly not opposed to GE technology, nor was he against using chemicals to combat weeds in Brazil. But he had watched what had happened in the United States with drift and he just could not see how Brazilians were going to avoid the same problems when Roundup Ready Xtend technology was introduced in his home country. "I think many growers are probably going to adopt it, not because they like it, but just to protect themselves against neighbors."[25]

And for Pedroso what was equally troubling was the fact that dicamba probably would not solve farmers' weed problems anyway. He noted that many of the weeds that farmers had the most trouble with in Brazil were actually grasses—which dicamba fails to control—as opposed to broadleaf weeds. In other words, here was another example of how technologies suited for one place were not the best technologies for another.[26]

Some people saw through Monsanto's marketing of miracles. Brazilian farmer José Soares said he was not going to buy into the Xtend system because it did not offer him any real benefit. "They are selling us a problem we don't have [in order] to be able to sell us the solution," he said. Monsanto's Xtend system simply was not going to work in Brazil, he reasoned, so why should he bother with it?[27]

Other growers joined Soja Livre, the "Free Soy" movement. Started in 2009 by soybean growers in Mato Grosso, Soja Livre's mission was to provide farmers an alternative to the GE seed system by developing and distributing high-yielding conventional varieties. As Embrapa explained, it was all about "keeping alive the freedom of choice of the rural producer." By the 2010s, nearly 40 percent of farmers in Mato Grosso were turning to non-GE soybean varieties.[28]

But dicamba remained a threat to these farmers, even if they chose not to buy into Monsanto's system. "Dicamba is dangerous," José Soares told a reporter. After all, dicamba drift could hurt his crops. For people like Soares and members of the Soja Livre movement who wanted nothing to do with dicamba-tolerant seeds, the only hope was for the government to ban the use of Monsanto's technology.[29]

And this was not going to happen. President Jair Bolsonaro, an admirer of Donald Trump and a conservative ideologue, put in place cabinet officials that were all about eliminating government regulations. The Ministry of Agriculture proposed one short year of Xtend trials in the 2019–2020 growing season—nothing like the five-year environmental review Capobianco had proposed for GE crops more than a decade before. And as this book went to press, Bayer announced that its INTACTA 2 Xtend seed—a "stacked" GE germplasm with two Bt genes, the Roundup Ready cassette, and a dicamba-tolerance trait—would "be commercialized in the 2021/2022 harvest." Whether farmers liked it or not, large volumes of dicamba were coming to Brazil.[30]

By the end of 2020, Brazilian farmers were being pulled into a seed system that was becoming increasingly more difficult to escape. But Monsanto had other frontiers to conquer. Halfway around the world, in another tropical ecosystem, Monsanto had been trying to pull off what some saw as an impossible feat: convincing people once ravaged by Agent Orange to buy Monsanto's seeds.

* * * *

IT WAS NOW FORTY YEARS after the Vietnam War, and former North Vietnamese colonel Dr. Trần Ngọc Tâm was beaming car-to-ear as he welcomed his guests to the headquarters of the Vietnam Association of Victims of Agent Orange (VAVA) on a hot June day in 2017. Founded in 2004 by Vietnamese war veterans and concerned citizens, VAVA had become over the course of more than a decade the main domestic charitable organization offering direct assistance to Agent Orange victims in Vietnam. Just how many Vietnamese citizens were harmed by Agent Orange is unclear, but estimates range from 1 million to more than 4 million people. VAVA officials, however, do not seek definitive proof that dioxin caused a person harm before administering aid. The organization pays for relief programs for a broad spectrum of people, soliciting donations from private citizens and international organizations, including UNICEF. Using these resources, VAVA set up schools for children suffering with disabilities allegedly linked to dioxin exposure.[31]

Next to Dr. Tâm was Dr. Trần Thị Tuyết Hạnh of Hanoi's University of Public Health. Dr. Hạnh had been studying dioxin contamination at two former US air bases: one in Bien Hoa and the other in the coastal town of Da Nang. Researchers in the 1990s had discovered that while dioxin had already undergone denaturing in many areas where Agent Orange had been sprayed, there were several hot spots around the country that needed immediate attention. They zeroed in on US air bases at Phu Cat, Bien Hoa, and Da Nang, the three primary locations where the US Air Force stored Agent Orange for use in Operation Ranch Hand. There, dioxin levels in soils and nearby waterbodies were hundreds of times higher than recommended safe standards set by the Vietnamese government. Blood and breast-milk samples from community members surrounding these bases showed that this contamination had spread into the bodies of Vietnamese citizens. At Da Nang and Bien Hoa, contamination was particularly bad, in part because this was where the US military stored thousands of barrels of Agent Orange as it carried out Operation Pacer IVY, the official campaign to remove all unused herbicides from Vietnam in the 1970s. As rusted and leaky barrels sat on-site for months prior to being shipped to a US military base at Johnston Atoll in the Pacific Ocean, the ground surrounding storage sites became contaminated with dioxin.[32]

In her studies on Da Nang and Bien Hoa, Hạnh showed how dioxin traveled from contaminated lakes to humans via fish, ducks, and other livestock. In a 2015 article published in an international environmental health journal, she found that in these hot-spot communities, consumption of "high-risk" foods, which included free-range chicken, fish, and fowl—popular items in Vietnamese cuisine—increased daily dioxin "intakes far above WHO recommended" levels.[33]

Dr. Hạnh's study corroborated findings made by American scientist Dr. Arnold Schecter, one of the world's leading experts on dioxin's effects on human health. Beginning in 1984, Schecter—then a professor of preventative medicine at SUNY Binghamton—traveled to Southeast Asia to collaborate with Vietnamese researchers. He worked on studies that examined avenues of dioxin transmission from contaminated sites to human bodies. After years of study, in 2005, he made the following conclusion: "Clearly, food . . . appears responsible for elevated TCDD [dioxin] of residents of Bien Hoa City, even though the original Agent Orange contamination occurred 30–40 years before sampling." Schecter's and Hạnh's research made clear that the dangers of dioxin seeped far beyond the borders of former US military bases. Dealing with this problem was going to cost a lot of money.[34]

VAVA officials in Hanoi had tried to force Monsanto to pay for relief programs. On January 30, 2004, the organization brought suit against Dow, Monsanto, and the other manufacturers of 2,4,5-T, seeking compensation for damages caused by Agent Orange during the Vietnam War. They invoked the Alien Tort Act, a law passed by the first Congress of the new American republic in 1789. Under this historic law, foreign nationals could file suit against US firms that violated international treaties. In this case, VAVA charged that Monsanto and Dow had sold the federal government a known poison that was later used against the Vietnamese people in violation of international agreements, such as the Geneva Protocol and the Biological Weapons Convention, that banned use of poisons during wartime.[35]

VAVA did not find a friendly ally in the Hon. Jack B. Weinstein, the federal judge assigned to hear the case in the US District Court for the Eastern District of New York. Now two decades removed from his settlement decision in the 1984 US veterans case that had Monsanto attorneys toasting with champagne in his chambers, Weinstein considered himself a legal expert on matters related to Agent Orange litigation, and he spoke with authority when issuing his opinion in this case. On March 30, 2005, he rejected the plaintiffs' complaint, first dismissing any domestic law claims against Monsanto on the grounds that the firm was protected by what is known as the "government contractor defense," a legal precedent that allows companies to avoid liability for harm caused by a product if that product was produced at the behest of the federal government. Then, Weinstein turned to the question of whether Monsanto and the other companies violated international law. On this question, Weinstein found that VAVA had not effectively demonstrated that Monsanto or Dow had committed war crimes. After all, he explained, the US government had not ratified the provisions of the Geneva Protocol and other international agreements barring the use of poisons in war at the time herbicides were deployed in Vietnam. Weinstein argued that the court could not be expected to retroactively enforce treaties that

were not ratified during the Vietnam War. More important, Weinstein argued that even if these treaties were binding on Monsanto
and Dow, Agent Orange was not a poison under the terms of international law: "Agent Orange and other agents . . . should be characterized as herbicides and not poisons. While their undesired effects
may have caused some results analogous to those of poisons in their
impact on people and land, such collateral consequences do not
change the character of the substance for present purposes." VAVA
appealed the decision, finally petitioning the US Supreme Court in
2009 to grant a writ of certiorari, but, in the end, they found no
relief.[36]

Monsanto also avoided paying for mounting costs that the US
Department of Veterans Affairs took on in the 1990s and 2000s to
compensate American veterans and their families for health problems caused by Agent Orange. In 1993, secretary of veterans affairs
Jesse Brown informed the US Senate that his agency would henceforward offer compensation to veterans suffering from a wide variety of ailments, including soft-tissue sarcoma, chloracne, Hodgkin's
disease, and other illnesses. Brown's announcement came two years
after Congress passed the Agent Orange Act of 1991, which called
on the VA to work with the National Academy of Sciences to review
the health hazards associated with dioxin exposure. In his 1993
remarks, Brown cited the 813-page National Academy of Sciences
report that resulted from the 1991 legislation, which showed that
dioxin exposure could increase risks for many of the ailments Brown
said the VA would now cover. According to the VA, servicemen did
not have to prove that dioxin caused their health problems because
the agency planned to apply the "presumptive policy," which meant
that there was a presumption that dioxin could have caused the diseases on the VA's Agent Orange list. The number of diseases eligible
for benefits under the new federal policy grew after the NAS revised
its reports in the years ahead. Tens of thousands of veterans ultimately received compensation under this program that was funded
by taxpayer dollars.[37]

With the government footing the bill, Monsanto, Dow, and other chemical manufacturers avoided big costs they might otherwise have been forced to take on. Their liability for US veterans' health issues associated with dioxin ended in 1997, when remaining funds from the $180 million settlement in 1984 were disbursed. Divided among seven firms, the settlement only made Monsanto liable for a portion of these costs. In the end, this lawsuit resulted in payouts to veterans and their relations that averaged around $3,900. Taking into account accrued interest, Monsanto was only responsible for roughly $98 million. Spread out over the course of more than a decade, this sum was a far cry from the $2.2 billion the Department of Veterans Affairs paid in just one year to more than 89,000 veterans that filed claims between August 2010 and August 2011 for three diseases on the VA's Agent Orange list.[38]

But if this multibillion-dollar firm sought to avoid the multibillion-dollar Agent Orange problem, a powerful congressional staffer in Washington, DC, made it his mission to see that justice was served to the people of Vietnam.

✳   ✳   ✳   ✳

TIM RIESER TOUCHED DOWN at the Hanoi airport in December 2006. He had come to find out more about Agent Orange issues in the country. A longtime aide to Senator Patrick Leahy of Vermont and an influential staffer on the Senate Appropriations Committee, Rieser was a brilliant dealmaker in Washington, DC. Bloggers for *The Hill* dubbed him one "of the most powerful staffers in Congress presiding over US foreign policy and US foreign assistance," while *Politico* listed him among the top fifty "thinkers, doers, and visionaries transforming American politics." Senators are often in the media spotlight, but aides like Rieser work out of the public eye to actually get things done on Capitol Hill.[39]

Now more than thirty years into his career, Tim Rieser had long worked with Senator Leahy to improve US–Vietnamese relations. Rieser had helped develop Senator Leahy's War Victims

Fund, which Congress authorized in 1989. This fund provides pros-
thetic limbs, wheelchairs, and other disability assistance for people
injured by landmines and other unexploded ordinance. Overseeing
these efforts for Senator Leahy, Rieser traveled multiple times to
Vietnam to meet with government officials on a range of issues,
and in virtually "every conversation, every meeting," he said in
2017, "the Vietnamese would raise this issue of Agent Orange and
the fact that it was continuing to be a serious health problem and
environmental problem." "It was really something that they clearly
resented enormously," Reiser recalled. Frustration grew worse as
billions of dollars flowed from the Department of Veterans Affairs
to American servicemen affected by Agent Orange. "The inconsis-
tency of our policy was so flagrant and indefensible," Rieser concluded. Something had to be done.[40]

Of course, Rieser was not alone. Others had been fighting for
change for some time, including Charles Bailey, the Ford Foundation's representative to Vietnam, who launched a special initiative
to address Agent Orange remediation. For Bailey, finding absolute proof linking dioxin exposure to specific claims of disability
was beside the point. Inaction had gone on for far too long, and it
remained a key talking point any time a Vietnamese official met with
a senior US diplomat. The Ford Foundation had money that could
seed transformative change in Vietnam, and Bailey was determined
to use it. Over the next ten years, in large part because of Bailey's
dogged determination, this foundation channeled more than $17
million toward Agent Orange relief.[41]

But despite the goodwill of men such as Bailey and organizations
such as the Ford Foundation, the problem demanded more substantive financial support. But neither the State Department nor the Pentagon wanted to touch this issue. Their major concern was liability.
What if the United States agreed to relief programs and millions of
Vietnamese citizens came forward seeking compensation? It could
cost a fortune.[42]

Reiser was not discouraged. He knew from studies funded by the Ford Foundation and the Vietnamese government where the worst dioxin hot spots were. With targeted funding, these sites could be cleaned up, which would go a long way to halting the spread of dioxin-related health problems in the country.[43]

In 2007, Senator Leahy worked with Rieser to secure $3 million in federal funds to initiate cleanup efforts and support medical programs at key dioxin hot-spot areas in Vietnam. As they considered ways to address the dioxin problem, Congress heard from prominent Vietnamese scientist Dr. Lê Kế Sơn, who traveled to Washington, DC, to explain the results of studies he worked on that showed persistence of dioxin in the Vietnamese environment. Five years later, USAID partnered with the Vietnamese Ministry of Defense to break ground on the first large-scale dioxin cleanup project in Vietnam at Da Nang airport.[44]

The scale of the cleanup was stunning. At the end of the main runway at Da Nang's international airport was a concrete structure a football field in length and approximately two stories high. Over the course of five years at Da Nang airport, USAID and the Vietnamese Ministry of Defense excavated roughly 90,000 cubic meters of dioxin-contaminated soil and cooked two separate 45,000 cubic-meter batches in this concrete container in a process known as "in-pile thermal desorption." Phase I, completed in 2015, involved heating 45,000 cubic meters of soil to approximately 335 degrees Celsius with the help of 1,250 heater wells drilled through a concrete cap. During cooking, the concrete structure got so hot that workers monitoring operating components on the structure were required to take frequent breaks every hour to have their blood pressure checked to ensure they did not collapse from heat exhaustion. Cooking proceeded for several months in order to ensure that the molecular bonds holding dioxin together were broken. Then, a prolonged cooling process began, water being added in small quantities over the course of several months to cool the soil. After completion of phase I, the concrete cap was cracked, soil removed, another batch of 45,000 cubic meters of soil added, and heating wells reinstalled all over again. By November 2018, USAID and the Vietnam Ministry of Defense declared the cleanup complete.[45]

This was a costly affair. As the cleanup approached conclusion, total expenses totaled about $116 million for the planning and execution of the thermal desorption. Additional money, roughly 20 percent of total appropriations approved for Agent Orange remediation in Vietnam, went toward direct assistance to families and communities dealing with dioxin-related health problems. These targeted health programs, however, represented only a fraction of the human costs associated with Agent Orange exposure in Vietnam. As dioxin expert Arnold Schecter put it, the federal government was spending a lot of money "to clean up dirt," while human health issues got inadequate attention. Real solutions would require more money.[46]

While US taxpayers paid the bill for the Da Nang airport cleanup, Monsanto, Dow, and the other companies that had sold the US government a contaminated product remained completely out of sight.

The Da Nang initiative, much lauded by US and Vietnamese politicians alike, was only a start. Bien Hoa air base lay ahead, and it was going to be an even bigger problem. People overseeing the cleanup estimated that more than five times as much soil was contaminated at Bien Hoa as at Da Nang, with that one property containing 85 percent of dioxin-laden soil found at all three of the major air-base hot spots. Estimates varied, but USAID's Chris Abrams guessed that this project could cost somewhere in the ballpark of $400 million.[47]

Tim Rieser considered the prospects for the future in 2017. "We need to find the money," he said, "and it's not easy." Leahy was going to need partners with bigger pockets, and he hoped that the Pentagon would help out with these efforts.[48]

Rieser noted that Leahy had never sought funds from Monsanto and other private contractors that had sold the US government a dirty product many years ago, but he admitted this might have been an oversight. "It seems to me that there are ways that they could" contribute "without opening themselves up to legal liability." But, he said that he wanted to learn more about the history of Monsanto. What did they know about dioxin contamination and Agent Orange before they sold it to the US government? Answering this question was important, but in the meantime, Rieser felt Monsanto should make a "gesture of goodwill" by helping to "mitigate the harm it helped to cause."[49]

One thing was clear: Monsanto had money that could seed change in Vietnam. In 2017, the firm's net income totaled more than $2.2 billion. Apportioning just a fraction of this income toward remediation at Bien Hoa, Monsanto could help to cover costs that Rieser and others were trying to squeeze out of Congress. But by the end of the 2010s, Monsanto gave no sign that it was interested in participating in such humanitarian relief. Instead, Monsanto offered the people of Vietnam something else: genetically engineered corn seeds.[50]

✳ ✳ ✳ ✳

IT WAS "AS MUCH fun as wedding parties," 64-year-old Vietnamese farmer Nguyen Hong Lam said when asked about a Monsanto corn seed "launch event" that took place sixty some miles outside Ho Chi Minh City in the mid-2010s. Lam detailed an extravagant affair, with tents capable of accommodating hundreds of farmers. Festivities went on for days as men like Lam learned about the magical qualities of Monsanto's patented genetically engineered seeds. As he spoke with a Vietnamese reporter, Lam wore a bright-colored polo shirt emblazoned with a "Dekalb Vietnam" logo and a catchy slogan, "Cùng Nhau Phát Triển," which loosely translates to "develop together." Dekalb was the Monsanto seed subsidiary now operating in Vietnam. The firm's promotional message was clear: Lam had become a partner in Monsanto's global empire, empowered to shape the firm's future. The clothes, the parties, the promise of partnership: all this must have seemed attractive to this aging farmer seeking a brighter future in the Vietnamese countryside.[51]

This was standard fare for Monsanto in developing countries around the world. CEO Narasimham Upadyayula, head of Monsanto's Vietnam subsidiary, said that the firm organized these promotional events all the time. The goal was to give farmers "a vision" of the future. "Seeing is believing," Upadyayula explained.[52]

Lam believed. In December 2014, he converted roughly 75,000 square feet of rice paddies into farmland devoted to cultivating Monsanto's genetically engineered corn. These were some of the first GE seeds to be planted in this Southeast Asian country. Monsanto was once again transforming the Vietnamese landscape in profound ways. After Lam, others followed. A local Vietnamese paper reported in October 2015 that "Dekalb Vietnam has trained more than 16,000 farmers in more than 200 locations" how to use GE corn seeds. Farmer Luu Van Tran echoed Lam's vote of approval for Monsanto's new product: "Now I have more peace of mind as I worry a lot less about my major weeds and pests in corn."[53]

This moment had been long in the making, and it involved more than just field parties. For years, Monsanto had worked with the USDA and State Department to promote its GE seed business in Vietnam. As early as 2007, the USDA's Foreign Agricultural Service helped bring "eight senior Vietnam officials" to Monsanto's headquarters as part of a "weeklong biotech study tour." A few months later, the USDA paid for "a delegation of high-ranking [Vietnamese] officials" to attend the 2007 BIO International Convention, the leading global conference focused on biotechnology and genetic engineering. Free international flights and other perks kept coming in the years ahead, with both the State Department and USDA funding trips around the world for powerful Vietnamese officials. Missouri, Monsanto's home state, remained a frequent stop on these government-financed trips.[54]

While government partners helped Monsanto make key political connections, the firm simultaneously sought influence in Vietnam's academic circles. In 2014, for example, the company offered the

Vietnam University of Agriculture a grant totaling more than $70,000 to fund graduate work in genetic engineering. Monsanto gave similar donations to other Vietnamese research institutions. Just as Monsanto had done back in the United States, the company was becoming an important financial partner for colleges and universities in Vietnam.[55]

The USDA was also helping Monsanto make important connections to media outlets in Southeast Asia. The agency, for example, financed training workshops for Vietnamese journalists on "How to Educate Farmers about Biotech Crops." One of the training sessions occurred in Monsanto's home state. The USDA also offered funding to several Vietnamese journalists to take a course in Missouri titled "Training Journalists on Benefits [of] Biotechnology."[56]

All these efforts laid the groundwork for Monsanto's GE launch in 2014. That year, the Vietnamese government approved the use of two genetically engineered corn seeds distributed by Dekalb Vietnam. One seed included a gene capable of producing the Bt pesticide that would repel insects, while another conferred Roundup tolerance to corn. The Roundup Ready revolution had come to Vietnam.[57]

✳   ✳   ✳   ✳

MONSANTO HAD PULLED OFF a remarkable feat. The firm had avoided paying any costs for Agent Orange remediation in Vietnam, despite having been the largest producer by volume of the chemical in the world. Back home, the firm had also escaped major financial liability, securing a generous settlement in the Vietnam veterans' case that allowed it to avoid billions of dollars in health expenses ultimately covered by the VA. In short, the federal government played a big role in protecting the cash reserves that eventually enabled Monsanto to build a seed empire overseas. When it came to Agent Orange, Monsanto never lost big.

And in Vietnam, state aid also proved essential in securing Monsanto's access to commercial markets. There the company benefited from a government bailout, as American taxpayers covered the costs

of multimillion-dollar cleanup campaigns that helped cool hostili-
ties over the dioxin contamination Monsanto had helped create in
Vietnam. In addition, the USDA and State Department doled out
cash to woo scientists, journalists, and Vietnamese politicians to the
benefits of a biotechnology.

When corn grower Nguyen Hong Lam was asked how he felt
about buying seeds from a company that had such a sordid history,
Lam, who neither spoke nor read English, confessed that he did not
know about Dekalb Vietnam's connection to Agent Orange. All he
knew was that these seeds promised to bring money that would help
him build a better life. He pointed to a 20 percent bump in profits as proof that his GE gamble was paying off. Accepting Monsanto's offer did not induce moral anxiety for folks like Lam; it simply
seemed smart business.[58]

But recent history hinted at trouble ahead. Just four years after
the Vietnamese government approved Monsanto's Roundup Ready
seeds for use in the country, Californian Lee Johnson won a case in
San Francisco, when a jury found that Roundup contributed to the
cancer that threatened to take Johnson's life. Meanwhile, Missourian
Bill Bader and his lawyers had worked through the pre-trial discovery process in their dicamba case, securing corporate documents that
showed there were real problems with Monsanto's GE system. And
trouble was afoot in Brazil, where weed scientists waited anxiously
to see what would happen when a volatile chemical called dicamba
was dropped in the tropical Cerrado to beat back weeds Monsanto
had once claimed its Roundup Ready system would annihilate.

No doubt, Lam believed he was taking part in the future of agriculture by buying into the Roundup Ready seed system. He was
now in his mid-sixties, and he had surely seen nothing like what
Monsanto had to offer. But the truth was he was spraying a chemical on his farm, Roundup, that had been created half a century ago
to replace the 2,4,5-T in Agent Orange, a chemical whose legacy
still lingered in Vietnamese soil. And if things played out as they
had halfway around the world, Lam would soon have to purchase

larger volumes of other chemicals first commercialized in the mid-twentieth century—perhaps dicamba or even 2,4-D, once used in Agent Orange—to deal with herbicide-tolerant weeds created by Roundup Ready crops. Ultimately, Lam's GE seeds connected him to a chemical past that, in so many ways, was not even past in Vietnam.

At Bayer's April 2019 annual shareholders' meeting, one can see the company's concerted efforts to focus on new "digital farming" technologies—products such as drones and tablets. But despite this new emphasis on data mining and information selling, the firm was still deeply wedded to chemicals such as glyphosate and dicamba, packaged with the Roundup Ready Xtend system.

PART V

HARVEST

"Malicious Code"

WHEN BAYER EXECUTIVES MET FOR THE 2020 SHAREHOLDERS'
meeting in Leverkusen, Germany, the COVID pandemic gripped the
globe, forcing the board to stream the meeting over the Internet from
a room without any shareholders. Six suited executives sat at bleach-
white desks, positioned more than six feet apart from one another, in
a sterile, glass walled room at company headquarters. Before Werner
Baumann, Bayer's CEO, got up to speak, a woman wearing a mask
and rubber gloves approached the podium, wiping the microphone
down with alcohol. There was an eerie silence in the room.[1]

Bayer had been here before. At the end of World War I, the German firm sold its popular aspirin, invented by the company in the
1890s, to reduce fevers caused by the influenza commonly referred to
as the Spanish flu. People feared for their lives as the pandemic killed
millions. The world was anxious.[2]

Back then, across the Atlantic, a still-infant Monsanto was just
trying to keep up. Company founder John Queeny worked feverishly to develop his own line of aspirin and a drug called phenacetin
that could help quell the fevers of the infected. In the heat of development, Monsanto workers on the phenacetin production line col-
lapsed on the factory floor as their blood cells burst—a consequence
of exposure to toxic chemical compounds. Queeny lamented the
loss but framed his work as a matter of life and death given the fact
that Americans had few sources of supply for key flu-fighting drugs
beyond German and European firms. When World War I broke out,

he said the Germans told him they were going to cut off their supplies to St. Louis unless Monsanto ceased chemical sales to allies in England. He would not let that happen. "I told them to go to blazes," Queeny quipped. He would liberate the United States from the death grip of German companies so he could heal sick soldiers fighting to make the world safe for democracy.[3]

Now, more than a century later, as Werner Baumann took to the microphone, Germans once again controlled Queeny's company. In June 2018, Baumann successfully orchestrated a Monsanto buyout for $63 billion—the largest acquisition by a German corporation in history. Bayer owned all of Monsanto's assets, including its Roundup Ready Xtend seed technology. Queeny's quest for independence from German firms had ended in failure.[4]

Big deals were happening throughout the genetically engineered (GE) seed business. Dow and DuPont combined their assets in a 2017 mega-merger, eventually spinning off their agricultural divisions into a new company called Corteva. A year later, ChemChina, headquartered in Beijing, acquired the Swiss biotechnology giant Syngenta. By 2019, Bayer, Corteva, and ChemChina, along with BASF, controlled roughly 60 percent of the entire global seed market.[5]

In Leverkusen, not everyone was celebrating; some Bayer employees believed that the aspirin maker had just swallowed a poison pill. Among them was Werner Baumann's predecessor, Marijn Dekkers, who feared that Monsanto legacies would sully the Bayer brand. Dekkers, a Dutchman, ran the company from 2010 to 2016 and tried, like Monsanto, to eliminate many old chemical product lines from the company portfolio, focusing on core areas in the life sciences, including extensive crop-technology investments. The German DAX exchange had rewarded his efforts, which is why he was reluctant to take on Monsanto's toxic assets.[6]

Baumann was less cautious. He had started working on the Monsanto merger in 2015, and when he became CEO a year later, he moved aggressively to get the deal done, hoping to take advantage of the fact that low commodity crop prices and a failed Monsanto bid for Syngenta dropped the St. Louis firm's selling price. Now was the right time to buy.[7]

Only it wasn't. Just a month after the buyout, Lee Johnson got his $289 million Roundup verdict in court, and Bayer's stock price plummeted, the firm losing nearly $30 billion in market capitalization by November 2018. The following March, another plaintiff against Bayer, Californian Edwin Hardeman, won an $80 million verdict, the jury in his case finding that Roundup exposure had been a "substantial factor" contributing to his non-Hodgkin's lymphoma. By the next shareholders' meeting, a third of the firm's value was gone, making it only worth the entire amount it had paid to purchase Monsanto in the first place. Investors were livid. In a historic moment, a majority of shareholders issued a vote of no confidence in Werner Baumann and the board of management. This had never happened before in the history of the DAX exchange.[8]

Baumann worked for the next twelve months to gain back investors' trust, even taking off on a "roadshow" to calm shareholders' fears, but he kept getting hit with bad news. Right after the 2019 shareholders' meeting, yet another Roundup case went against Bayer, when Albert and Alva Pilliod, a husband and wife claiming glyphosate caused their cancer, won an astounding $2 billion verdict in the San Francisco Bay Area. By this point, several countries, including Belgium, Denmark, and France, among many others, had banned or were moving to phase out the use of glyphosate. (Vietnam prohibited the use of glyphosate in 2019, but then, facing pressure from the US, temporarily lifted that ban in 2020. As this book went to press, glyphosate was still approved for use in Vietnam, though an import and manufacturing ban remained in place.)[9]

In February 2020, Judge Limbaugh read the $265 million verdict against Bayer in the *Bader Farms* case. Limbaugh had allowed attorneys Bev and Billy Randles to seek punitive damages in the case because he believed the evidence showed that Monsanto acted with "reckless indifference" when it rolled out its dicamba-tolerant Xtend seeds. Dozens of other farmers awaiting trial rejoiced at the prospects of redemption. A month later, Europe went into coronavirus lock-down, and the global economy seized. Stock prices continued to drop, from a high of $135 per share in June 2017 to $68 per share when investors met in Leverkusen in April 2020.[10]

Shareholders did not issue a no-confidence vote this time, many perhaps heartened by news that Bayer was working on a big settlement to end pending Roundup lawsuits—now well over 120,000, including filed and unfiled claims. Still, it was clear that Monsanto's legacy issues were very much on the minds of those gathered. For almost four hours, Baumann and the other board members read and responded to a flood of questions about the dicamba and Roundup cases, answering more than 240 inquiries without stopping for a single break. Baumann seemed unfazed, a signature demeanor the *Financial Times* dubbed "slightly robotic." At the end of the day, he still had his job, which was more than many Bayer employees could say, as the firm laid off thousands of workers to keep its costs down after the Monsanto merger.[11]

Despite this maneuver, the problems of Monsanto's past continued to haunt Bayer's future. In June 2020, the US Court of Appeals for the Ninth Circuit issued an injunction against the further use of Bayer's dicamba herbicide, holding that "dicamba had caused substantial and undisputed damage" across the United States. The Center for Food Safety, the National Family Farm Coalition, Pesticide Action Network, and the Center for Biological Diversity partnered with several nonprofit organizations to bring this case to trial, after the EPA failed to address dicamba-drift problems. Bayer had asked the court not to revoke registrations for use of dicamba in the upcoming growing season, even if the court found in favor of the plaintiffs, but the judge refused to honor this request. In a forceful summary statement, the three-judge panel said that doing so would further "tear the social fabric of farming communities" across the country.[12]

President Trump's EPA issued an order immediately after the case telling farmers that they could use remaining stocks of dicamba in the upcoming growing season, but plaintiffs in the Ninth Circuit filed a motion to block this exemption. Growers across the country rushed to get dicamba on their fields before there was resolution. Ohio State weed scientist Mark Loux said he talked to a friend in Iowa that said dicamba was everywhere: "You could about taste it in the air." In October 2020, the EPA issued new registrations for dicamba formulations, but the Center for Food Safety immediately declared its intentions to challenge the agency's actions. And in March 2021, a new acting assistant administrator within President Biden's EPA issued a jarring statement acknowledging that "political interference" had "compromised the integrity" of the 2018 dicamba registration process. Given the legal uncertainties, many farmers did not know what to do.[13]

This was not just an American issue. As this book went to press, farmers deep in the Brazilian Cerrado were preparing for the first large-scale release of Bayer's Xtend seed system. Though dicamba-drift issues had not been resolved in the United States, and despite weed scientists' warnings that the Cerrado was just the kind of ecosystem that made Bayer's herbicide particularly volatile, Bayer was now exporting its technology to a hot, tropical environment.[14]

Dicamba and Roundup were just part of the trouble. Three weeks before the 2020 Bayer shareholders' meeting, a US district court judge allowed the city of Baltimore to seek damages against Bayer for PCB contamination that Monsanto never cleaned up. The Maryland metropolis joined Seattle, Cincinnati, and San Diego, which all filed similar cases a few years earlier. Soon other cities and states followed. Werner Baumann never mentioned these PCB liabilities in his comments to shareholders in April 2020, but they nevertheless represented grave financial risks. After all, PCBs, like Roundup and dicamba, could still be found around the world. In Seattle, Attorney General Bob Ferguson said his office had identified roughly 600 PCB hot spots stretching from Puget Sound to Spokane and noted that there were numerous fishing advisories on the Columbia River and other tributaries because of PCB pollution. Mike Dewine, then serving as Ohio's attorney general, echoed these concerns in 2016. He cited an EPA study, which concluded that "PCB contamination in fish is the cause of most of the human health impairments" of Ohio waters. If Bayer really wanted to feed the world, these cities were saying, the company could play a role in cleaning up streams that could provide sustenance to millions of people.[15]

In June 2020, Baumann tried to suppress all these issues—the dicamba trouble, the Roundup litigation, and the lingering PCB legacies—once and for all. He announced that Bayer was putting aside $400 million to settle all remaining dicamba claims and $820 million to deal with any outstanding PCB litigation. He also said that he had met with several of the law firms in the Roundup case and agreed to a $10 billion settlement that would cover some 95,000 cases of the 125,000 or so still pending against the firm. Bayer lawyers had tried to include a provision in the agreement that called for the creation of an independent scientific advisory panel tasked with issuing summary judgment on whether Roundup causes cancer. Bayer demanded that if this panel found that Roundup was safe, future class action claims related to this product should not be allowed to go forward. But this $10 billion deal ultimately fell through, and settlement talks associated with the dicamba, PCB, and Roundup litigations had yet to be finalized as of the time of this writing.[16]

And then there was Agent Orange. At Bien Hoa air base in southern Vietnam, USAID began work on a decade-long dioxin cleanup campaign in December 2019, estimating that total costs would amount to more than $300 million—all of it coming from public funds and none of it from the corporations that had sold these contaminated products to the US government. There was a big celebration, Vietnamese and US officials shaking hands in an uplifting moment of reconciliation. But some activists, including Nguyen Thi Lan Anh of the Action to the Community Development Center, were not satisfied. "We get funds for working with disabled people," she said, "but we know that around us, there are people who have been affected by dioxin and need special care." She needed more funding: "We cannot wait for another 40 years, it will be too late."[17]

Back in West Virginia, Charleston attorney Stuart Calwell kept fighting, more than forty years after he had taken the first case of the 2,4,5-T workers in Nitro. He organized a class action lawsuit, *Bibb v. Monsanto*, that involved hundreds of litigants, charging that Monsanto had polluted the entire town of Nitro. His clients demanded that the company pay to monitor the health of more than 5,000 Nitro residents and pay to clean up the disaster it created. The case never went to trial because Monsanto agreed to a $93 million settlement in 2012 that would help pay for medical monitoring over the course of three decades as well as residential cleanup programs. In addition, Monsanto agreed to allow plaintiffs to retain the right to file suit against the chemical company if blood samples did indeed reveal elevated dioxin levels. Here was another liability, inherited from Monsanto, hanging over Bayer.[18]

* * * *

IN THIS MOMENT of controversy, Bayer worked hard to rebrand itself as a high-tech, life-sciences company no longer tethered to its polluted past as it focused on its role as a pioneer in "digital farming." This "Fourth Agricultural Revolution," as some boosters dubbed it, first emerged in the late 1990s, when agricultural scientists began promoting the idea of using advanced computer technology to enhance "precision agriculture" techniques. The idea was to apply new and emerging tools—soil sensors, satellite imagery, drones, and global positioning systems—to help farmers make precise herbicide applications, monitor field nutrients, and track weather patterns, thereby boosting yields via data mining.[19]

Some of the new technologies coming out of the digital-farming pipeline were eye-catching, including the "WEED-IT," an herbicide applicator featuring LED optical sensors capable of detecting weeds under a sprayer boom traveling up to twenty-five miles an hour. Bayer did not own the WEED-IT (a Dutch start-up launched the product in the late 1990s), but Bayer invested in similar computerized gadgets, including unmanned aerial vehicles (UAVs) that could deliver targeted doses of herbicide. Bayer also invested in artificial intelligence (AI) that would allow the company to aggregate data extracted from digital sensors, monitors, and drones in fields across the world and provide analysis for farmers so that they could make "critical, timely, in-field decisions."[20]

The part about farmers making the decision was important. Bayer held that its big-data tools would "empower farmers through actionable insights," allowing them to get the "most out of their fields while using less." It was all about freedom. Data mining would help farmers break their dependence on costly petrochemical inputs mined from the ground.[21]

This was a strange thing for a company like Bayer to say, especially as it simultaneously assured investors that its herbicide sales would remain sound. A board member made this clear in the 2020 shareholders' meeting when asked a question about Bayer's development of "alternatives for glyphosate." "In coming years," he said, "glyphosate will continue to be a good tool for controlling weeds in agriculture worldwide," adding, "the product features and traits of glyphosate have not been met or exceeded by any other herbicide." Roundup Ready technology was a killer app that Bayer could not abandon in this new age of digital farming. Bayer, in other words, was still very much a chemical company born of a chemical economy, no matter how many drones it owned.[22]

To seed its future, the German firm recycled feed-the-world messaging from the past. In the 1990s, Monsanto's Bob Shapiro had promised that "DNA-encoded information," a kind of genetic software built into GE seeds, would allow the company to replace "stuff with information," thereby increasing food production without placing heavy demands on the environment. Now, Bayer made the same appeal. "The world's population is growing," Bayer said on a company website about digital farming. "Agricultural productivity will have to increase if we want to safeguard our food supply in the long term." Bayer claimed that farmers would need everything from "drones, sensors and other digital technologies to trusted herbicides like glyphosate" in order "to shape a healthy and sustainable future for agriculture."[23]

GE firms had always been good problem sellers—defining the dilemma and positioning themselves as the solution. But history showed they were often bad problem solvers. When Monsanto launched the GE revolution in 1996, it said it was doing so in part to address famine around the world. Who could argue with that? Yet, twenty-five years later, historical data analyzed by the USDA and the National Academy of Sciences, among other renowned scientific bodies, showed that yield gains from Monsanto's GE technology were nonexistent in some cases or marginal at best. Yes, productivity had increased over the years, and Bt technology had helped many growers increase yields in particularly pest-ridden parts of the world, but productivity gains for non-GE crops tracked the yield increases for GE varieties.[24]

And genetically engineered seeds did not eliminate the weeds they were sold to annihilate. The claim, for example, that these technologies would reduce chemical dependence proved false. Glyphosate use on American farms increased exponentially between 1995 and 2015, and herbicides that Monsanto had promised farmers they would never again need in large volumes—such as 2,4-D and dicamba—came flooding back into fields as weed resistance ravaged farms. Monsanto profited from these new problems because they allowed the company to sell new "solutions"—GE seeds 2.0 and 3.0 packaged with "new" chemicals, the next round in the commercial cycle.[25]

It is true that Monsanto's Bt traits dramatically decreased the use of many toxic insecticides on corn and cotton farms around the world in the early years of GE seed adoption. But by the 2010s, several insect species had developed resistance to Bt, causing scientists to call for better management practices. And between 2015 and 2018, the total volume of synthetic insecticides used on corn and cotton farms in the United States began to tick back up. Still, many Bt farmers were using fewer insecticides than they had before GE seed adoption, but those chemical savings were dwarfed by the astronomical rise in herbicide usage. By the end of the 2010s, the total chemical pesticide inputs in the United States—meaning herbicides and insecticides combined—had grown considerably for commodity crops such as corn, soybeans, and cotton since the introduction of GE seeds in 1996.[26]

This is perhaps the most important thing we have learned about the GE food revolution on the occasion of its silver anniversary: that the recombinant DNA Monsanto sold did not produce the dramatic yield increases promised, nor did it free farmers from their deep dependency on chemicals developed decades ago that came from the remains of the dead buried deep in the ground.

Monsanto had tried to escape this fossil-fueled economy in the 1970s and 1980s. Then, executives looked at the firm's chemical portfolio and realized that roughly 80 percent of Monsanto's products ultimately came from petroleum and natural gas feedstocks. They watched with grave concern as powerful oil and gas companies—Exxon, Chevron, among others—invested in chemical production facilities, using up petrochemical resources that they once offered Monsanto on the cheap. The company felt trapped, which is one of the reasons why it made the bold move to start investing in seeds and biotechnology and begin divesting from most commodity chemical production altogether.[27]

But the firm hung onto blockbuster brands, such as Roundup, bringing the chemical past into the biotechnology beyond. Though executives liked to talk about how the "new" Monsanto was very different than the old, the truth was that the firm's future was still tethered to its chemical origins—to scavenger capitalism. Roundup derived from phosphate ore mined from ancient seabeds that drained millions of years ago produced half of the firm's revenue by 2001. Monsanto, in other words, could never fully break free of the chemical economy it helped create.

* * * *

THIS BOOK IS NOT an indictment of genetic engineering *in toto*. After all, emerging gene-editing tools, such as CRISPR, might well—in the right hands—unleash new yield potentials in the decades ahead. And genetic engineering has also led to the development of important drugs and vaccines, including ones that proved critical in the fight against COVID-19. But the first twenty-five years of GE seed development shows that consolidated corporate management of GE technologies by companies that sold chemicals yielded a farming system plagued with problems. The possibilities for GE innovation were circumscribed by the product pipelines that narrowed Monsanto's vision of where to channel capital investments. Drought-resistant seeds, so often touted by Monsanto, never gained the same attention as Roundup Ready technology, in large part because it was difficult to create the same kind of returns on investment for research and development in this area. As it turns out, there are a lot of different factors that make a crop drought-tolerant. It is hardly something that can be easily distilled into a short gene sequence. As a result, the company focused more on selling seeds that could sell its chemicals because that was how it could earn money to cover costs associated with all the production and marketing infrastructure it had built around its herbicide business. With Bayer announcing in 2020 that it had a "new candidate as an active substance for an innovative herbicide" in its "product pipeline," the firm indicated that it still saw the world's food future through a tunnel vision bounded by the chemical channels that still pulsed profits through its corporate body.

There are many people featured in this book who are proponents of genetic engineering and others who are less enthusiastic about the seeds Monsanto created. But their collective stories reveal that GE technology was erroneously deployed over the past two decades and was more about selling chemicals than investing in real solutions to our food problems, which has resulted in wasted opportunities and wasted resources.

By 2020, Bob Shapiro's sincere conviction in the 1990s that GE technology would replace "stuff with information" was essentially being turned on its head. In the era of digital farming, GE firms such as Bayer were using information mined from farmers to sell more stuff—sensors and drones, satellites and chemicals. And all these big machines and branded herbicides cost money, money that many cash-strapped farmers did not have.[28]

The Department of Homeland Security was worried. In 2018, it issued a report titled "Threats to Precision Agriculture," in which it outlined the national security concerns associated with the digital-farming revolution. The agency pointed out that most unmanned aerial systems were "foreign built," creating vulnerabilities that could allow enemy states to gain access to critical data related to American food production. The report also explained that foreign countries could remotely disable this computerized equipment with "malicious code . . . during times of crisis or during key planting or harvesting windows."[29]

This did not sound like the kind of liberation and freedom Bayer hawked in its big-data branding. Here was a new, digital pest problem capable of disrupting US food production that farmers just a half-century earlier could have never envisioned.

*　*　*　*

BAYER'S BOARD MEMBERS UNDERSTOOD the dangers of a virus. As they sat in the 2020 shareholders' meeting, the world economy was spiraling into a recession caused by a single strand of rapidly replicating RNA that was sweeping swiftly across the globe. Everything they had planned for was put on hold.

And yet, the entire message of the meeting was that Bayer had things under control. For more than six hours, Baumann and the board managers assured investors that Bayer had a handle on all the problems of the past. It was moving confidently into the future.[30]

But should farmers, and the people they feed, believe Bayer? The history offered here shows that Monsanto's technology did not work the way the company said it would. For years, Monsanto and Bayer both challenged this assertion, saying that the proof was in farmers' purchases. But by the end of the 2010s, this logic was no longer so sound. In 2016, the *Wall Street Journal* ran a major story, "Behind the Monsanto Deal, Doubts About the GMO Revolution," detailing how many farmers were abandoning biotech products because high prices for GE seeds were "harder to justify." Since 1997, the price for GE soybean and corn seeds has roughly quadrupled, even though they did not always produce dramatically more grain than that yielded by comparable non-GE varietals. The explosion in resistant weeds also made the herbicide savings Monsanto customers once enjoyed less attractive.[31]

Bayer is willing to spend billions of dollars to settle cases so that it can keep selling Roundup, an indication of just how dependent the firm still is on chemicals developed half a century ago. Meanwhile, the evidence from the past twenty-five years shows that farmers do not need Roundup in the quantity that the industry has wanted to sell. Non-GE growers that do not spray Roundup year-round have been able to produce the same yields (or more) as those of other farmers using Monsanto GE seed varietals.

This should be good news for farmers looking to escape rising costs associated with GE technology. The door for an alternative food future has not closed. Fortunately, there are still seed merchants selling non-GE seeds for commodity crops, and there are many foods—especially fruits and vegetables—that are still grown outside the GE system Monsanto created.

But gaining independence from this system may become harder in the future. The dicamba Xtend story shows that genetic software can function like a kind of malware, spreading like a virus across farms, forcing growers into a GE system, whether they like it or not. And Bayer has little incentive to stop this from happening. After all, the company is anxious to assure jittery investors shell-shocked by recent litigation that the firm can still make big profits from herbicide sales.

So it is that Bayer seeks to make farmers see old chemicals as the future of agriculture. The only question is whether farmers will continue to buy what Bayer is selling.

Acknowledgments

I WRITE THIS IN THE MIDDLE OF A GLOBAL PANDEMIC. IN SUCH times, one values, more than ever, the support of institutions and the love and kindness of colleagues, friends, and family who have volunteered their time and energy to assist in a project that took the better part of a decade to finish.

In 2012, the History Department at the University of Alabama in Tuscaloosa, led by talented chair Kari Frederickson, took a chance on an all-but-dissertation graduate student, offering me a tenure-track job in environmental history. I am eternally grateful to the people that served on that search committee—Teresa Cribelli, Andrew Huebner, George McClure, and Josh Rothman. They brought me into a community that supported my early research on this book. I am also indebted to my Tuscaloosa friends that kept me sane by venturing into the woods, paddling down streams, and traveling to the majestic Gulf Coast. Friday trips to the Alcove also stand out prominently in my mind.

Four years later, I headed north, to The Ohio State University, where a new academic family welcomed me in. In that department, deftly directed by compassionate and thoughtful chairs Nathan Rosenstein and Scott Levi, I joined a talented team of terrific scholars focusing on the history of the environment, health, technology, and society (EHTS), including Nick Breyfogle, John Brooke, Phil Brown, Kip Curtis, Jennifer Eaglin, Hieu Phung, Chris Otter, Tina Sessa, Dave Staley, and Sam White. I also had the pleasure of working closely with a great group of modern US historians, including Paula Baker, Clay Howard, Hasan Jeffries, Joseph Parrott, Randy Roth, Daniel Rivers, David Stebenne, and David Steigerwald. We were fortunate to have several EHTS graduate students—Dylan Cahn, Mike Corsi, James Esposito, Jim Harris, Ives Lux Hartman, Neil Humphrey, Katie Lang, Dustin Meier, and Cody Patton—who added so much to our environmental history "lab" meetings. These scholars were core collaborators as I worked on this project, and I am especially thankful to the graduate teaching assistants at Alabama and Ohio State, who put in so much hard work to make the courses I taught over the past four years a success. The Sustainability Institute also offered me opportunities for interdisciplinary exchange, and I'm especially grateful to Elena Irwin for helping me connect with scholars outside my department.

Several Ohio State undergraduate and graduate students volunteered their time to work on this project as research assistants. Sasha Zborovosky poured over documents related to Monsanto's PCB business in the Poison Papers archives, while Matthew Bonner flew to St. Louis where he found key documents related to Monsanto's public affairs strategies. Callia Tellez traveled with me to Cape Girardeau, Missouri, to observe the *Bader Farms* case, and Johnathon Hendy, Meagan Kellis, Megan LaFrance, Elliot Ping, Markus Schoof, Maddie Sisk, and Arrie Zimmerman all researched and assisted with specific aspects of this book.

The Ohio State University Brazil Gateway program helped make arrangements for my field research in São Paulo, Londrina, and beyond. Gateway director Jane K. Aparecido was incredible, connecting me with Brazilian agricultural specialists. One of those experts was Felipe Sartori, then a PhD student at Luiz de Queiroz College of Agriculture (ESALQ), University of São Paulo, who served as an indispensable guide, driving all over the south of Brazil and connecting me to farmers and people that could speak to Monsanto's activity in the country. I am also thankful for the time I had with many ESALQ affiliates, especially Professor Rafael Munhoz Pedroso and

Professor André Froes da Borja Reis, who both sat down with me for an extended conversation about weed-control programs in Brazil. Officials at Embrapa's headquarters in Londrina and Brasilia were extremely helpful, as was Professor Cláudio Vaz Di Mambro Ribeiro of the Universidade Federal da Bahia, who helped me organize a trip to Monsanto's glyphosate plant in Camaçari. I owe thanks as well to Karine Peschard, a postdoctoral fellow who put me in touch with Embrapa officials in Brasilia.

In Vietnam, I was grateful to US army veteran Chuck Searcy, who connected me with officials at VAVA and helped me plan my trip to Hanoi, Da Nang, and Ho Chi Minh City. Chris Abrams at the US Agency for International Development office in Da Nang was also generous with his time, meeting with me to discuss the ongoing Agent Orange cleanup operations. While in Hanoi, I met with Dr. Lê Kế Sơn, who worked within the Vietnamese government on dioxin-remediation programs. Charles Bailey, formerly of the Ford Foundation, and Tim Rieser, senior policy aide to Senator Patrick Leahy of Vermont, discussed the history of Agent Orange cleanup campaigns in Vietnam. At VAVA headquarters in Hanoi, I met with Dr. Trần Ngọc Tâm of VAVA and Dr. Trần Thị Tuyết Hạnh of Hanoi's University of Public Health and others to talk about the long history of dioxin pollution in Vietnam. I also met with VAVA affiliates in Da Nang. In southern Vietnam, I worked closely with reporter Dien Luong, whose work on Agent Orange and Monsanto's present-day seed business proved instrumental to Chapter 13. In addition, I am indebted to Madame Tôn Nữ Thị Ninh, former Vietnamese ambassador to the United Nations, for her insights into the concerns about dioxin contamination in Vietnam.

In Soda Springs, I was fortunate to be able to meet with several people, including former Monsanto engineer and city council president Mitch Hart, who discussed topics covered in Chapter 7 and Chapter 9. Robert Gunnell also sat down for an interview in Provo, Utah, to talk about his negotiations with Monsanto in the wake of EPA Superfund investigations in Soda Springs. Matthew Cheramie, a former Soda Springs resident, shared important resources about ongoing pollution problems in this Idaho town, while Professor Emeritus Thomas Gesell at Idaho State University offered critical insights into the radiological studies conducted around Monsanto's plant in Idaho. EPA officials also met with me in Pocatello and spoke over the phone to discuss ongoing remediation efforts at phosphate mines and facilities in Idaho. Lastly, chairman Blaine Edmo and other members of the Shoshone-Bannock tribal business council provided critical perspective on the shortcomings of EPA remediation efforts in Soda Springs.

Through all these travels, I had my trusted friend, Jon Zadra, by my side. He not only served as a field photographer and valuable campground mate but also used his skills as a data scientist to help with statistical analysis. It would have been a far-lonelier research journey but for his companionship.

This book would not have been possible without the financial support of New America, a nonpartisan think tank based in Washington, DC, that funds research on a variety of public policy issues. In 2016, New America offered me a year-long fellowship that was funded by the Carnegie Corporation of New York. I stayed on as an Eric and Wendy Schmidt Fellow during the 2017–2018 academic year.

Columbia's Journalism School and Harvard University's Nieman Foundation for Journalism also offered generous funding for the book. In 2020, they gave this project the J. Anthony Lukas Work-in-Progress Award. Considering the financial strictures many historians faced as the pandemic set in, this funding proved critical as I moved toward completion of the book.

The Ohio State University's Office of the Vice President for Research in the College of Arts and Sciences offered me a Large Grant Award, which helped cover much of the cost associated with Brazilian field research. The Rachel Carson Center in Munich offered funding for a two-week trip to Germany, and the Julian Edison Department of Special Collections at Washington University in

St. Louis also provided funding, which allowed me to work through Monsanto corporate records housed at that facility.

Archivist Miranda Rectenwald at Washington University was an incredible ally, helping me get permission from Monsanto to access the company's records housed in special collections. I would also like to thank Kathy Shoemaker at Emory University, who assisted me when I traveled to the Stuart A. Rose Manuscript, Archives, and Rare Book Library. The law office of Stuart Calwell also handed over key court documents related to the Nitro trial, and archivists at the National Archives helped me find a variety of primary sources featured in this book.

The Business History Conference, the journals *Agricultural History* and *Enterprise and Society*, and the American Society for Environmental History accepted papers based on my Monsanto research. At conference meetings, I received invaluable feedback on key chapters. I am also grateful to Chris Jones and the Julie Ann Wrigley Global Institute of Sustainability at Arizona State University and Brenden Resnick at Brigham Young University's Charles Redd Center for Western Studies for hosting talks based on my Roundup research. The Redd Center event was particularly fruitful because I met historian Brian Cannon, who told me he was neighbors with Robert Gunnell and soon put me in touch. The STEAM factory at The Ohio State University also proved a valuable place for cross-disciplinary collaboration. There I met weed scientist Mark Loux, who later introduced me to farmers and other Ohio State scientists who helped with this project.

I conducted dozens of interviews with scientists, lawyers, Monsanto employees, farmers, and journalists over the course of seven years for this book. Some of these interviews were with people the reader now knows—Linda Birnbaum, Stuart Calwell, Fred Gould, Stephen Duke, Fred Pond, Bev and Billy Randles, Marc Vanacht, and others. But some individuals, including people who worked for Monsanto, spoke to me off the record, providing invaluable information for the historical account offered in these pages. Others offered valuable background information but are not featured in the text. Though I do not include all of their names here in the acknowledgments, I am deeply grateful for their willingness to speak with me.

I filed several Freedom of Information Act requests with various public agencies and am indebted to the men and women who worked tirelessly to produce documents that had not yet been released to the public. I'd especially like to thank Scott Hainer with The Ohio State University's Office of Public Records, who helped with an FOIA request that shed light on Monsanto's financial contributions to the university's College of Food, Agricultural, and Environmental Sciences.

This global environmental history of Monsanto built on the work of several talented historians and journalists. PCB chapters leaned heavily on the scholarship of my good friend Ellen Griffith Spears, while sections detailing recent Roundup litigation drew on Carey Gillam's journalism. For a bird's-eye view of the chemical industry, Fred Aftalion's *A History of the International Chemical Industry* and Alfred Chandler's *Shaping the Industrial Century* were touchstone texts. Marie-Monique Robin's book on Monsanto and Dan Charles's *Lords of the Harvest* were also never far away from my desk as I pounded away on the computer.

So many wonderful people read sections as well as full drafts of the manuscript. Readers included Ed Ayers, Brian Balogh, Kate Brown, Brent Cebul, Jennifer Eaglin, Jerome Elmore, Joya Elmore, Shane Hamilton, Mark Hersey, William Thomas Okie, Tore Olsson, Furl Mercer, Stephen Macekura, Catherine McNeur, Andy Robichaud, Chris Rosen, Ed Russell, Ellen Griffith Spears, Mark Stoll, Steven Stoll, Bert Way, Joanna Zadra, and Jon Zadra. I'd also like to thank my agent, Geri Thoma, who read early iterations of chapters and helped me prepare for publication.

Of course, there was one reader that put in more effort than anyone, and that was my editor at Norton, Justin Cahill. He read draft after draft of this book. He is a dream to work with. Thank you for sticking with me for another wild global ride and for reigning me in when the prose ventured off into wandering backroads.

So many friends helped get this book to the finish line. Some did so by offering direct aid, including Jesse Pappas, who came up with the name for this book. Thomas Altman, Tony Buhr, Lauren Cardon, Eric Courchesne, Kathryn Drago, Josh Eyer, Patrick Frantom, Elliot Panek, Nigel Seaman, Sarah Steinbock-Pratt, and Rachel Stephens tried hard to distract me in Tuscaloosa. Joe Bayer, Chen Chen, Erin Crotty, Duncan Forbes, James Forbes, Pete and Jen Kaser, Craig Kent, and Sarah Lenkay did the same in Columbus. Erin and Tom Cochran put me up at their house in DC during my trips to New America and the National Archives, and Erin worked tirelessly as the communications director for the project. Justin Storbeck and Christine LaDuca were my kind hosts when I had work in New York City. Zubin Desai designed and managed my website and helped strategize how to make my scholarship available to a broad audience. Matt Rahn stayed up "late," coaching me through the trials of writing while being a parent. My college buddies—Kyle Hatridge, Hemant Joshi, Kristian Lau, and Joe Thistle—provided key psychological support. I'm also eternally grateful to Bhavani Lev, who offered up the "Blue House" as a place for our family to escape the rush and whir of the city, and for Joy Leblanc, who kept us happy and well fed out on the farm. Other friends offered support that went beyond editorial advice—Frisbee golf sessions, whitewater escapades, mountain bike rides, delicious dinners, North Carolina hikes, music by the campfire, and banter on the phone and Zoom. I cannot name everyone here but know that I love you.

I lost my mother in the course of writing this book, but I welcomed two wonderful people into this world as well: River Lamar and Blue Boyd Wistik Elmore. Without them, I know some of the hard times we faced as a family would have been even harder. Thank you River and Blue for bringing so much joy to your dad's life. I hope I can be as good of a father to you as my dad was to me.

To those I hold so dear, Collins, Scott, and Lisa, I could not take on these big projects without the confidence you have shown in me. We are on the cusp of this pandemic being over, and I know we are all ready for that big group hug soon. To my in-laws, John Puts, Raquel Jacobs, Gio McMurray, Maya Van Dyck, Amanda McMurray, and Noah McMurray, and my nephews Owen and Griffin, thank you for treating me as if we were born under the same roof. You have offered timely distractions, moments of jest and laughter that eased burdens in those final stages of writing.

And, lastly, to my wife, Joya. This book is dedicated to you because I could not have finished it without you. I have never known someone so selfless, so empathic, so committed to what is right and just. I hope this book is worthy of all the sacrifices you have made to get us to this point.

Notes

INTRODUCTION: **"Don't Do It. Expect Lawsuits."**

1. Bev and Billy Randles (plaintiff's attorneys in *Bader Farms, Inc. v. Monsanto Co., and BASF Corporation*, MDL No. 1:18-md-2820-SNLJ, Case No. 1:16-CV-299-SNLJ, US District Court, E. D. Missouri, Southeastern Division [2019]) [hereinafter *Bader Farms v. Monsanto*], interview by the author, May 29, 2020. The author also attended the trial on January 27, 2020.
2. Randles, interview.
3. Randles, interview.
4. Randles, interview; "Randles, Pierson Enter Race for Missouri Lieutenant Governor," *Columbia Daily Tribune*, July 2, 2012, https://www.columbia tribune.com/article/20150702/news/307029914. Bev Randles, email correspondence with author, April 5, 2021.
5. On the University of Nebraska and dicamba resistance, see Transcript of Testimony at 1702–1703, 2428, *Bader Farms v. Monsanto*.
6. On fruit orchards and sycamore trees, see Transcript of Testimony at 448, 550, 1325–37, *Bader Farms v. Monsanto*.
7. Randles, interview.
8. Transcript of Testimony at 53–55, *Bader Farms v. Monsanto*.
9. Transcript of Testimony at 144, 260–61, *Bader Farms v. Monsanto*; Johnathan Hettinger, "'Buy It Or Else': Inside Monsanto and BASF's Move to Force Dicamba on Farmers," *St. Louis Post-Dispatch*, December 6, 2020, https://www.stltoday.com/news/local/state-and-regional/buy-it-or-else-inside-monsanto-and-basf-s-moves-to-force-dicamba-on-farmers/article_002f5e83-004d-52de-a686-eef5cb108192.html; The author would like to thank Hettinger and St. Louis Public Radio's Corinne Ruff for discussing the case with me when we all attended the *Bader* trial in 2020. Plaintiff's Exhibit 13, Letter from Kimberly Magin, Monsanto, to Steve Smith, July 24, 2012. These exhibits are available at the US Right to Know (USRTK) website. USRTK research director and journalist Carey Gillam organized documents and wrote articles about the dicamba and Roundup trials since they started and helped to curate

an invaluable collection of documents related to these cases. Her two books, *Whitewash: The Story of a Weed Killer, Cancer, and the Corruption of Science* (Washington, DC: Island Press, 2017) and *Monsanto Papers: Deadly Secrets, Corporate Corruption, and One Man's Search for Justice* (Washington, DC: Island Press, 2021), offer detailed analysis of the Dewayne "Lee" Johnson Roundup case and Monsanto's efforts to influence the debate about Roundup's carcinogenicity. I would like to thank Carey Gillam for corresponding with me about these cases. Dicamba documents from the trial can be found here: https://usrtk.org/pesticides/dicamba-papers/.

10. Transcript of Testimony at 157, 2479, *Bader Farms v. Monsanto*; Plaintiff's Exhibit 22, Roundup Ready Xtend Crop System, Reflections on Building FTO [Freedom to Operate], marked "Monsanto Company Confidential," 31, USRTK.

11. Transcript of Testimony at 147, 444, 2056, *Bader Farms v. Monsanto*.

12. Transcript of Testimony at 2454, 2455, *Bader Farms v. Monsanto*; Randles, interview.

13. Transcript of Testimony at 150, 430–32, 742, *Bader Farms v. Monsanto*; Plaintiff's Exhibit 292, Email from Joseph Sandbrink to Jeff Travers, Re: Norsworthy Visit, April 11, 2015, USRTK.

14. Author observations during *Bader Farms* trial, January 27, 2020; Transcript of Testimony at 983, 994–98, *Bader Farms v. Monsanto*.

15. Transcript of Testimony at 155, 542–44, *Bader Farms v. Monsanto*; Plaintiff's Exhibit 178, Email from Boyd Carey to Melanie Knaak-Guyer et al., Re: Xtend Inquiry Support—Grower List, July 28, 2017, USRTK. Plaintiff's Exhibit 179 also indicated that Monsanto would only inspect the farms of "good MON customers." "I will need a purchase history," Monsanto's Sara Allen wrote her superior, "to determine if [complaining farmers] are 'good MON customers' and thus need a dispatch." Email from Sara Allen to Ty Witten, July 3, 2017, USRTK.

16. Transcript of Testimony at 156–57, 160–61, *Bader Farms v. Monsanto*.

17. Transcript of Testimony at 2080, *Bader Farms v. Monsanto*.

18. Transcript of Testimony at 158, *Bader Farms v. Monsanto*; Plaintiff's Exhibit 177, Email from John Cantwell to Tony White, August 7, 2017, USRTK.

19. Transcript of Testimony at 156, 160–61, *Bader Farms v. Monsanto*.

20. Dewayne "Lee" Johnson, interview by journalist Carey Gillam, 2018. Gillam made this interview available to the author; "Jurors Give $289 Million to Man They Say Got Cancer from Monsanto's Roundup Weedkiller," *CNN*, August 11, 2018, https://www.cnn.com/2018/08/10/health/monsanto-johnson-trial-verdict/index.html; "Bayer loses California appeal of Roundup verdict, but damages are reduced," *Reuters*, July 21, 2020, https://www.reuters.com/article/us-bayer-glyphosate-lawsuit/bayer-loses-california-appeal-of-roundup-verdict-but-damages-are-reduced-idUSKCN24M2BT; "Roundup's Maker Agrees to Pay More Than $10 Billion to Settle Thousands of Claims that the Weedkiller Causes Cancer," *New York Times*, June 24, 2020, B1. For a summary of the Roundup cases and key facts on the Dewayne Johnson case, see the US Right to Know webpage, "Glyphosate Fact Sheet: Cancer and Other

Health Concerns," https://usrtk.org/pesticides/glyphosate-health-concerns/. See also Carey Gillam, *Whitewash*, and *Monsanto Papers* for detailed histories of the Lee Johnson case and other Roundup litigation.

21. "Jurors Give $289 Million to Man They Say Got Cancer from Monsanto's Roundup Weedkiller," *CNN*, August 11, 2018, https://www.cnn .com/2018/08/10/health/monsanto-johnson-trial-verdict/index.html; "Chemical Giant Monsanto Ordered to Pay $289 Million in Roundup Cancer Trial," *HuffPost*, August 10, 2018, https://www.huffpost.com/entry/ monsanto-to-pay-roundup-cancer-trial_n_5b6e14f1e4b0bdd062095477.

22. Videotaped Deposition of Monsanto toxicologist Donna Farmer, January 11, 2017, referencing Donna Farmer email from September 21, 2009; Email from Stephen Adams to Gary Klopf, December 14, 2010, Monsanto Papers, made available by the Baum Hedlund Aristei & Goldman law firm. Files can be accessed here: https://www.baumhedlundlaw.com/toxic-tort -law/monsanto-roundup-lawsuit/monsanto-secret-documents/ [hereinafter Monsanto Papers]. See also Carey Gillam, "Formulations of Glyphosate-Based Weedkillers are Toxic, Tests Show," *Guardian*, January 23, 2020, https://www.theguardian.com/business/2020/jan/23/formulations-gly phosate-based-weedkillers-toxic-tests.

23. For example, see Robin Mesnage, Charles Benbrook, and Michael N. Antoniou, "Insight into the Confusion over Surfactant Co-formulants in Glyphosate-Based Herbicides," *Food and Chemical Toxicology* 128 (2019): 138; Robin Mesnage et al., "Potential Toxic Effects of Glyphosate and Its Commercial Formulations," *Food and Chemical Toxicology* 84 (2015): 135; Christophe Gustin, Mark Martens, and C. Bates [scientists at Monsanto], "Clustering Glyphosate Formulations with Regard to the Testing for Dermal Uptake," confidential draft dated July 2001, released in trial; Email from William Heydens to Charles Healey, April 2, 2002, Monsanto Papers.

24. Email from Donna Farmer to John Acquavella, September 18, 2014; Email from William Heydens to Richard Garnett, October 15, 2014, Monsanto Papers.

25. Email from Mark A. Martens to Larry D. Kier, William F. Heydens, et al., April 19, 1999; Email from William F. Heydens to Mark A. Martens, Larry D. Kier, et al., September 16, 1999, Monsanto Papers.

26. Proposal for Post-IARC Meeting Scientific Projects, Draft, 5, May 11, 2015; Email from William Heydens to Michael S. Koch, Donna Farmer, et al., May 11, 2015; Email from John Acquavella to William F. Heydens, November 3, 2015; Dr. David Saltmiras "custodial file," "Glyphosate activities," August 4, 2015, Monsanto Papers.

27. EPA Office of Pesticide Programs, *Glyphosate Issue Paper: Evaluation of Carcinogenic Potential*, 140, September 12, 2016, https://www.epa.gov/sites/ production/files/2016–09/documents/glyphosate_issue_paper_evaluation_ of_carcincogenic_potential.pdf; International Agency for Research on Cancer, *IARC Monographs Volume 112: Evaluation of Five Organophosphate Insecticides and Herbicides*, published by the World Health Organization (2015), https://monographs.iarc.fr/wp-content/uploads/2018/06/mono112– 10.pdf; Agency for Toxic Substances and Disease Registry, "Toxicological

Profile for Glyphosate," Draft for Public Comment (April 2019), 5, https://www.atsdr.cdc.gov/toxprofiles/tp214.pdf.

28. "'We Did The Right Thing': Jurors Urge Judge to Uphold Monsanto Cancer Ruling," *Guardian*, October 18, 2018, https://www.theguardian.com/business/2018/oct/18/monsanto-verdict-jurors-judge-dewayne-johnson.

29. "EPA Takes Next Step in Review Process for Glyphosate, Reaffirms No Risk to Public Health," EPA Press Release, April 30, 2019, https://www.epa.gov/newsreleases/epa-takes-next-step-review-process-herbicide-glyphosate-reaffirms-no-risk-public-health.

30. The author visited Monsanto's headquarters at the Centec Tower, Ho Chi Minh City, Vietnam, June 14, 2017. See the International Service for the Acquisition of Agri-Biotech Applications (ISAA) website for dates of GE corn seed adoption in Vietnam and beyond, https://www.isaaa.org/gmapprovaldatabase/event/default.asp?EventID=86.

31. On the size of Monsanto's herbicide contracts with the US military during the Vietnam War, see Peter H. Schuck, *Agent Orange on Trial: Mass Toxic Disasters in the Courts* (Cambridge, MA: Belknap Press of Harvard University Press, 1987), 156.

32. This book builds on environmental historian Edmund P. Russell's pioneering work in evolutionary history, a field of research which considers the ways in which plants, animals, and humans have coevolved throughout history. For more on this area of historical inquiry, see Edmund P. Russell's *Evolutionary History: Uniting History and Biology to Understand Life on Earth* (Cambridge: Cambridge University Press, 2011); "Man Awarded $80M in Lawsuit Claiming Roundup Causes Cancer," *USA Today*, March 27, 2019, https://www.usatoday.com/story/money/2019/03/27/monsanto-roundup-cancer-lawsuit-california-man-awarded-80–million/3293824002/; "Findings Released as Major Scientific Study Shows Eating Organic Lowers Cancer Risk," Environmental Working Group Press Release, October 24, 2018, https://www.ewg.org/release/roundup-breakfast-part-2–new-tests-weed-killer-found-all-kids-cereals-sampled.

33. Roundup Ready Canola, FarmCentral.com [a Monsanto-sponsored website], https://web.archive.org/web/19981202105106/http://www.farmcentral.com/s/rr/s3rrzzzzz.htm; "Another Great Option—XtendiMax® Herbicide with VaporGrip® Technology," Bayer product website, https://traits.bayer.ca/en/soybeans/roundup-ready-xtend-crop-system/chemistry-options/; Roundup Ready Xtend Crop System Farmer Testimonial Video, featuring Kentucky farmer Brian Shouse, https://www.youtube.com/watch?v=aSDKO50KriQ; "Innovation for Generations," *Today's Acre* (Winter 2019) [Monsanto's online magazine], https://www.roundupreadyxtend.com/todays-acre/Pages/winter2019–innovation-for-the-generations.aspx.

34. Philip H. Howard, "Visualizing Consolidation in the Global Seed Industry: 1995–2008," *Sustainability* 1, no. 4 (2009), 1274.

35. This book seeks to bring together environmental histories of the factory with environmental histories of the farm. In 2003, agricultural historian Deborah Fitzgerald penned the award-winning book, *Every Farm a Factory: The Industrial Ideal in American Agriculture* (New Haven: Yale University Press, 2003),

examining how farmers adopted business practices from the industrial sector, thereby modernizing American agriculture in the 1920s and beyond. Fitzgerald's factory–farm theme is one that needs more attention. By putting the histories of industrial plants alongside the histories of agricultural plants, this book on Monsanto shows more clearly how chemical company executives, industrial plant workers, farmers, and flora and fauna interacted to shape the food system we depend on today. This represents an important departure from the way most historians have written about the ecological history of industrial farming and agricultural chemicals. Many studies on this topic have focused on the end-use environmental effects of synthetic herbicide and insecticide use. See, for example, Angus Wright, *The Death of Ramón González: The Modern Agricultural Dilemma* (Austin: University of Texas Press, 1990); Carol Van Strum, *A Bitter Fog: Herbicides and Human Rights* (San Francisco: Sierra Club Books, 1983); Clinton L. Evans, *The War on Weeds in the Prairie: An Environmental History* (Calgary: University of Calgary Press, 2002); Pete Daniel, *Toxic Drift: Pesticides and Health in the Post–World War II South* (Baton Rouge: Louisiana State University Press, 2007); J. L. Anderson, *Industrializing the Corn Belt: Agriculture, Technology, and Environment, 1945– 1972* (Dekalb: Northern Illinois University Press, 2009); Frederick Rowe Davis, *Banned: A History of Pesticides and the Science of Toxicology* (New Haven: Yale University Press, 2014); David D. Vail, *Chemical Lands: Pesticides, Aerial Spraying, and Health in North America* (Tuscaloosa: University of Alabama Press, 2018). For a cultural history of pesticides in America, see Michelle Mart, *Pesticides, a Love Story* (Lawrence: University of Kansas Press, 2018). Building on these works but venturing into new territory, this book weaves together more intimately the lives of industrial workers that manufactured herbicides and the lives of farmers that used these chemicals on rural land, ultimately explaining how decisions made by managers and workers in the factory connected to decisions being made by growers and businessmen working in the field. In addition, this book offers a deeper historical account of the Roundup Ready Revolution, a topic that needs much more attention from historians. Most sweeping histories of the emergence of genetically engineered (GE) crops have been written by journalists, not agricultural or environmental historians. Here it is worth mentioning journalist Daniel Charles's *The Lords of the Harvest: Biotech, Big Money, and the Future of Food* (Cambridge, MA: Perseus Publishing, 2001), which provides a deeply researched account of the birth of GE crop cultivation. Sociologist Jack Ralph Kloppenburg Jr. also discusses the rise of GE seed companies in his excellent *First the Seed: The Political Economy of Plant Biotechnology, 1492–2000* (1988; repr., Madison: University of Wisconsin Press, 2004). But both of these books take the story up to the early 2000s, before the Roundup-resistant weed problem really became a major concern for farmers. A 2017 survey of *Environmental History* and *Agricultural History*, two leading academic journals, revealed that there were virtually no articles published on Roundup or the emergence of GE crops, even though agricultural and environmental historians have unique skills that can be deployed to digest historical data related to GE crop cultivation. Only four articles in *Agricultural History* briefly mentioned GE crops introduced since the 1990s. For those studies, see "Agricultural History Talks to Karen-Beth G. Scholthof," *Agricultural History* 87, no. 2 (Spring 2013): 194–200; Melissa Walker, "Contemporary Agrarianism: A Reality Check," *Agricultural History* 86, no. 1 (Winter 2012): 1–25; Louis Ferleger, "Arming American Agriculture for the Twentieth Century: How the USDA's Top Managers Promoted Agricultural Development," *Agricultural History* 74, no. 2 (Spring 2000): 211–26; Peter A. Coclanis, "Food Chains: The Burdens of the (Re)Past," *Agricultural History* 72, no. 4 (Autumn 1998): 661–74. This book seeks to address this lacuna, bringing together environmental, agricultural, and business histories to show how the GE crops that feed us today are in fact tethered to an aging industrial chemical infrastructure built long ago.

36. For Monsanto's listing on a ranking of "most-hated" companies, see "Bad Reputation: America's Top 20 Most-Hated Companies," *USA Today*, February 1, 2018, https://www.usatoday.com/story/money/business/2018/02/01/bad-reputation-americas-top-20-most-hated-companies/1058718001/. For an analysis of how Monsanto executives wrestled with this "Monsatan" label, see Shane Hamilton and Beatrice D'Ippolito, "From Monsanto to 'Monsatan': Ownership and Control of History as a Strategic Resource," *Business History* (2020), https://www.tandfonline.com/doi/full/10.1080/00076791.2020.1838487 (published online).

37. Daniel Charles, an NPR reporter who spent some time among Monsanto scientists and researchers when working on his excellent book *The Lords of the Harvest* was also convinced that many Monsanto employees "dream, as all of us do, of doing something significant, of making a difference in the world." See "Prologue," xiv. In *Seeds of Science: Why We Got It So Wrong on GMOs* (London: Bloomsbury Sigma, 2018; 2020), Mark Lynas cites a 2010 sociological study involving "Monsanto insiders" who said they earnestly believed that they "were doing something good for the world" (89). Lynas' book details his conversion from being an anti-GE activist to being a proponent of the benefits of GE technology. References to Lynas' work in this book are to the 2020 edition.

CHAPTER 1: "You Are Getting into Chemistry Now, Senator, on Which Subject I Am Rather Weak"

1. Document titled "John F. Queeny," based on interview with Samuel Allender authored by Francis J. Curtis, December 7, 1950, series 14, box 22, folder: Queeny, John F. (Misc.) (Folder 1), Monsanto Company Records, Washington University in St. Louis, Julian Edison Department of Special Collections, St. Louis, Missouri [hereinafter MCR]. Louis Veillon's diary entries indicate that on February 6, 1902, "the inauguration of the works with the engine under steam took place in the presence of Mr. & Mrs. Q." Veillon also states that "actual manufacture started on Febr. 14." See document titled "Reminiscences about Monsanto's Beginnings," featuring excerpts from Dr. Louis Veillon's diary, series 10, box 5, folder: Monsanto Company History (Historical Accounts [Veillon, Louis]), MCR; On Queeny's mustache and height,

see Monsanto History, by Hubert Kay, 1st Draft, Edgar M. Queeny's Copy, Part II, "John F. Queeny and the Founding of American Industry," F-9, series 6, box 2, folder: Monsanto and the American Idea (Hubert Kay) (1st Draft) (Queeny's Copy); "Early Days," document authored by Francis J. Curtis, series 10, box 5, folder: Monsanto Company History (Historical Accounts [Kernon, Jules])—Oversaw Personnel, MCR. For weather in St. Louis in February, see the National Oceanic & Atmospheric Administration's National Center for Environmental Information, which contains records on climatological observations for St. Louis dating back to 1902. The database can be accessed here: https://www.ncdc.noaa.gov/cdo-web/search.

2. Deer Creek Tapes, November 17, 1951, 10–11, series 10, box 5, folder: Monsanto Company History (Historical Accounts [Bebie, DuBois, Veillon Reunion, 1951]), MCR.

3. Deer Creek Tapes, November 17, 1951, 10–11; Document titled "John F. Queeny and the Early Days," based on interview with Bert Langreck, authored by Francis J. Curtis, February 12, 1951, series 10, box 4, folder: Monsanto Company History (Historical Accounts [Misc.] [Folder 2]), MCR.

4. Document titled "John F. Queeny," based on interview with Samuel Allender authored by Francis J. Curtis, December 7, 1950, series 14, box 22, folder: Queeny, John F. (Misc.) (Folder 1), MCR; On Olga Monsanto's concern for Queeny's drinking, see Deer Creek Tapes, November 17, 1951, 10; Document titled "John F. Queeny," based on interview with Gaston DuBois/ Louis Veillon authored by Francis J. Curtis, April 3, 1951, series 14, box 22, folder: Queeny, John F. (Misc.) (Folder 1), MCR; On Queeny's "fondness for drink," see Monsanto History, by Hubert Kay, 1st Draft, Edgar M. Queeny's Copy, Part II, "John F. Queeny and the Founding of American Industry," F-32, series 6, box 2, folder: Monsanto and the American Idea (Hubert Kay) (1st Draft) (Queeny's Copy); "Early Days," document authored by Francis J. Curtis, series 10, box 5, folder: Monsanto Company History (Historical Accounts [Kernon, Jules])—Oversaw Personnel, MCR.

5. Document titled "John F. Queeny," based on interview with Samuel Allender authored by Francis J. Curtis, December 7, 1950, series 14, box 22, folder: Queeny, John F. (Misc.) (Folder 1), MCR.

6. "John F. Queeny," based on interview with Samuel Allender authored by Francis J. Curtis, December 7, 1950; Dan Forrestal, *Faith, Hope & $5,000: The Story of Monsanto* (New York: Simon & Schuster, 1977), 18.

7. Document titled "John F. Queeny," based on interview with Charles Belknap authored by Francis J. Curtis, December 29, 1950; Document titled "John F. Queeny," based on interview with Charles Huisking authored by Francis J. Curtis, January 15, 1951; Document titled "Early Days: John F. Queeny," based on interview with Miss Fitzpatrick authored by Francis J. Curtis, December 28, 1950, series 14, box 22, folder: Queeny, John F. (Misc.) (Folder 1); Forrestal, *Faith, Hope & $5,000*, 13.

8. Leonard A. Paris, "Monsanto: The First 75 Years," Corporate Public Relations article (1976), series 10, box 5, folder: Monsanto Company History (Historical Accounts [Misc.]); "President's Message," *Monsanto Current Events* 15, no. 1 (February 1936): 3–6, series 8, box 10, folder: Monsanto

Current Events (1936–37); John Queeny timeline (June 1966), series 14, box 22, folder: Queeny, John F. (Obituaries); "It's Dangerous to be *Too* Good a Loser," *Monsanto Current Events* 4, no. 1 (May 1925): 1–2, 14, series 14, box 22, folder: Queeny, John F. (Misc.) (Folder 1); "John Queeny," in *The Book of St. Louisans*, 2nd ed. (St. Louis: St. Louis Republic, 1912), 486–87, series 14, box 22, folder: Queeny, John F. (Misc.) (Folder 1), MCR; Forrestal, *Faith, Hope & $5,000*, 11.

9. Queeny offered a succinct summary of his early life during testimony in the 1911 *United States v. Forty Barrels and Twenty Kegs of Coca-Cola* case. See Transcript of Testimony at 1062–63, *United States v. Forty Barrels and Twenty Kegs of Coca-Cola*, 191 F. 431 (E. D. Tenn. 1911); Forrestal, *Faith, Hope & $5,000*, 12; "Little Stories of Big Successes," reprint from Greater St. Louis, *Monsanto Current Events* 4, no. 4 (November 1925): 1–2; "Mr. John Queeny Celebrates His Fiftieth Anniversary in the Drug and Chemical Business," *Monsanto Current Events* 3, no. 3 (May 1922): 5, series 8, box 9, folder: Monsanto Current Events, 1920–26, MCR.

10. Document titled "John F. Queeny," from interview with Charles Belknap authored by Francis J. Curtis, December 29, 1950, series 14, box 22, folder: Queeny, John F. (Misc.) (Folder 1); William Haynes, *Chemical Pioneers: The Founders of the American Chemical Industry* 1 (New York: D. Van Nostrand, 1939), 230–31, series 14, box 22, folder: John F. (Misc.) (Folder 2), MCR; "Drugs: Is There Any Adulteration in Those Unpleasant Articles?," *Chicago Daily Tribune*, July 12, 1874, 10; "The Manitoba or 'Tolu Wave' Is Now Upon Us," *Chicago Daily Tribune*, October 17, 1880, 1; "Brunker's Carminative Balsam," *Herald and Review* (Decatur, IL), August 4, 1881, 2.

11. "Drugs," *Chicago Daily Tribune*, 10.

12. William Haynes, *Chemical Pioneers*, 231; Boss quoted in Forrestal, *Faith, Hope & $5,000*, 12; "John Queeny," in *The Book of St. Louisans*, 2nd ed., 486–87.

13. For a concise global history of the chemical and pharmaceutical industries, see Alfred D. Chandler Jr., *Shaping the Industrial Century: The Remarkable Story of the Evolution of the Modern Chemical and Pharmaceutical Industries* (Cambridge, MA: Harvard University Press, 2005), 114–15, 117, 178, 291. Also see Fred Aftalion, *A History of the International Chemical Industry: From the 'Early Days' to 2000* (Philadelphia: Chemical Heritage Foundation, 2002), 39–47.

14. Chandler Jr., *Shaping the Industrial Century*, 125, 127.

15. Chandler Jr., *Shaping the Industrial Century*, 32, 42–43, 183, 189, 193–94, 197–98. For more on the early history of DuPont, see Alfred Chandler, "Du Pont—Creating the Autonomous Division," in *Strategy and Structure: Chapters in the History of The Industrial Enterprise* (Cambridge, MA: MIT Press, 2013, 1962), 52–113. For more on the history of Dow Chemical Company, see the work of Dow public relations director E. N. Brandt, *Growth Company: Dow Chemical's First Century* (East Lansing: Michigan State University Press, 1997).

16. "Drunkenness," *Times-Democrat* (New Orleans, LA), May 24, 1892, 10; "Bladon Springs," *Times-Picayune* (New Orleans, LA), July 12, 1885, 9.

17. "Boneset Bourbon Tonic," *Times-Picayune* (New Orleans, LA), June 25,

1881, 8. Evidence suggests that John Pemberton, the Atlanta pharmacist who created Coca-Cola, looked to cocaine as one source of the cure for "morphinism," an addiction that he apparently suffered from, according to close associates. See Mark Pendergrast, *For God, Country and Coca-Cola: The Definitive History of the Great American Soft Drink and the Company That Makes It* (New York: Basic Books, 2000, 1993), 25.

18. "John Queeny," in *The Book of St. Louisans*, 2nd ed., 486–87; "It's Dangerous to be *Too* Good a Loser," *Monsanto Current Events*; Document titled "Monsanto Chemical Company," dated 1941, series 10, box 5, folder: Monsanto Company History (Historical Accounts [Misc.]), MCR. Records related to Queeny's day-to-day work at Meyer is sparse, in part because the firm did not maintain extensive archival materials from the period. A letter to historian Hubert Kay contained in the Monsanto Records at Washington University explained: "Carl Meyer reports that Meyer Brothers Drug Company destroy their records every five years and that there is no way of finding out anything about what went on in 1901 with regard to saccharin or Queeny's salary." See letter to Hubert Kay, April 27, 1951, series 10, box 5, folder: Monsanto Company History (Historical Accounts [Misc.] [Folder 2]), MCR. Most accounts of Queeny's life suggest he went straight to St. Louis from New Orleans, but in his testimony for the 1911 *United States v. Forty Barrels and Twenty Kegs of Coca-Cola* case, Queeny noted that he went first to "Los Angeles, and remained there a year," before heading to St. Louis. See Transcript of Testimony at 1063, *United States v. Forty Barrels and Twenty Kegs of Coca-Cola*. On the early history of Merck's subsidiary in the United States, see Chandler Jr., *Shaping the Industrial Century*, 183.

19. Letter to Hubert Kay, Re: Monsanto Book, September 22, 1952, series 14, box 22, folder: Queeny, John F. (Misc.) (Folder 1), MCR. According to this document, Don Manuel was knighted by Queen Isabella of Spain for his services to the crown in New Grenada (present-day Colombia) and later moved to St. Thomas in the Virgin Islands. His son and Olga's father, Maurice, lived a privileged life, traveling to a polytechnic university to study civil engineering in Hanover, Germany, where he met his wife, Emma Cleeves, whose father was secretary to King George IV of Hanover. Maurice and Emma returned to St. Thomas in the mid-1800s and gave birth to their daughter Olga, but soon moved to New York after a hurricane devastated their island home in 1877. Deer Creek Tapes, November 17, 1951, 10–11; John Queeny timeline (June 1966), series 14, box 22, folder: Queeny, John F. (Obituaries), MCR; Forrestal, *Faith, Hope & $5,000*, 13; "The Wedding Bells Rang Merrily," *Merck's Market Report*, February 15, 1896, 98, series 14, box 22, folder: John F. Queeny (Misc.) (Folder 1), MCR; "Hoboken" [Queeny wedding announcement], *New-York Daily Tribune*, February 9, 1896, 18.

20. John Queeny timeline (June 1966), series 14, box 22, folder: Queeny, John F. (Obituaries); "Unique Double Birthday Party," *St. Louis Republic Sun*, August 19, 1900, 4. On Olga Monsanto as a wealthy woman, see for example the description of Olga Monsanto offered in document titled "To Hubert Kay, Reference: Monsanto Book," September 22, 1952, series 14, box 22, folder: Queeny, John F. (Misc.), Folder 1, MCR.

21. William Haynes, *Chemical Pioneers*, 232–33.

22. "It's Dangerous to be *Too* Good a Loser," *Monsanto Current Events*; Document titled "John F. Queeny," from interview with Charles Huisking authored by Francis J. Curtis, January 15, 1951, series 14, box 22, folder: Queeny, John F. (Misc.) (Folder 1), MCR; Forrestal, *Faith, Hope, and $5,000*, 14.

23. For press speaking to Mrs. Queeny's musical talents and professional responsibilities, see "Hoboken," [Queeny wedding announcement], *New-York Daily Tribune*, February 9, 1896, 18; "Music Teacher's Convention," *St. Louis Post-Dispatch*, May 30, 1897, 5; "Heathen Concert in a Christian Country," *Star Tribune* (Minneapolis), June 20, 1897, 4; "Music Teachers Convention," *The Inter Ocean* (Chicago), June 20, 1897, 24; "Heathen Concerts in a Christian Country," *Philadelphia Inquirer*, June 20, 1897, 26; "Where Music Will Charm," *Pittsburgh Post*, June 20, 1897, 21; On the "porch-climber" incident, see "Chicago Man is Held," *St. Louis Post-Dispatch*, March 20, 1899, 7; On John Queeny's father's death, see "Deaths," *Chicago Tribune*, September 15, 1898, 5.

24. Document titled "Early Days," based on interview with Carl Meyer, Meyer Brothers executive, and Mr. Vaughn, treasurer, authored by Francis J. Curtis, December 27, 1950, series 14, box 22, folder: Queeny, John F. (Misc.) (Folder 1), MCR.

25. Document titled "Early Days: John Queeny and Pure Food and Drug Laws," authored by Francis J. Curtis, June 19, 1951, series 14, box 22, folder: John F. (Civic Issues), MCR.

26. "Early Days: John Queeny and Pure Food and Drug Laws," authored by Francis J. Curtis, June 19, 1951.

27. "Hydroleine," *Meyer Brothers Druggist* 10, no. 1 (January 1888), xxxvi; John F. Queeny, *Minority Report of the Committee on Adulteration*, National Wholesale Druggists Association (October 4, 1899), series 14, box 22, folder: Queeny, John F. (on Civic Issues), MCR. Queeny's support for federal regulation fits with what scholars of the Progressive Era have said about big business and government agencies in this period. Beyond Monsanto, many firms sought the support of regulatory agencies in order to legitimize their businesses at a time when there was growing anti-corporate sentiment within the country and overseas. On this history, see for example Gabriel Kolko, *The Triumph of Conservatism: A Reinterpretation of American History, 1900–1916* (New York: Free Press of Glencoe, 1963); Thomas K. McCraw, *Prophets of Regulation: Charles Francis Adams, Louis D. Brandeis, James M. Landis, Alfred E. Kahn* (Cambridge, MA: Belknap Press of Harvard University, 1984); Martin J. Sklar, *The Corporate Reconstruction of American Capitalism, 1890–1916* (Cambridge, UK: Cambridge University Press, 1988); James Weinstein, *The Corporate Ideal in the Liberal State, 1900–1918* (Boston: Beacon Press, 1968).

28. Queeny, *Minority Report of the Committee on Adulteration*.

29. Letter from Minneapolis wholesaler to John F. Queeny, October 6, 1899; Letter from Clayton F. Shoemaker, Miers Busch, May 9, 1899, regarding Pure Food and Drug Act, series 14, box 22, folder: Queeny, John F. (on Civic Issues), MCR.

30. Letter from Harvey W. Wiley to John F. Queeny, October 29, 1900; John F. Queeny, "Hearing Before the Committee on Manufactures, United States Senate, On the Bill (S. 198) For Preventing the Adulteration, Misbranding, and Imitation of Foods, Beverages, Candies, Drugs, and Condiments, in the District of Columbia and the Territories, and for Other Purposes, and the Bill (H.R. 6295) For Preventing the Adulteration of Misbranding of Foods or Drugs And for Regulating Traffic Therein, and for Other Purposes," February 24, 1904, series 14, box 22, folder: Queeny, John F. (on Civic Issues), MCR.

31. Letter from Alvin G. Lazen to George Roush Jr., September 13, 1979; Letter from Helen M. Vaden to Mr. M. C. Throdahl, February 27, 1980, series 3, box 3, folder: Saccharin (History); "About Saccharin," *Monsanto Current Events* 17, no. 2 (June–July 1938): 20–21, series 8, box 10, folder: Monsanto Current Events (1938–39), MCR; Carolyn De La Peña, *Empty Pleasures: The Story of Artificial Sweeteners from Saccharin to Splenda* (Chapel Hill: University of North Carolina Press, 2010), 16, 19.

32. Interview with Leo Hertling (June 1956), series 10, box 5, folder: Monsanto Company History (Historical Accounts [Hertling, Leo]), MCR; Carolyn de la Peña, *Empty Pleasures*, 19–20; Bartow J. Elmore, *Citizen Coke: The Making of Coca-Cola Capitalism* (New York: Norton, 2014), 77; Frederick Allen, *Secret Formula: How Brilliant Marketing and Relentless Salesmanship Made Coca-Cola the Best-Known Product in the World* (New York: HarperBusiness, 1994), 104; Deborah Jean Warner, *Sweet Stuff: An American History of Sweeteners from Sugar to Sucralose* (Washington, DC: Rowman & Littlefield, 2011), 182.

33. Swiss chemist Jules Bebie, one of the founding members of Monsanto, once quipped that Queeny "was no technical man" in the early years. See Deer Creek Tapes, November 17, 1951, 10–11; US Senate Committee on Finance, *Tariff Act of 1921: Volume II, Schedule 1—Chemicals, Oils, and Paints*, 67th Cong., 2nd Sess., 1922, 887.

34. Monsanto History, by Hubert Kay, 1st Draft, Edgar M. Queeny's Copy, Part II, "John Queeny and the Founding of American Industry," F-17; Proceedings of a Special Meeting of the Board of Directors of the Monsanto Chemical Works, November 30, 1902, series 14, box 22, folder: Queeny, John F. (Misc.) (Folder 1), MCR.

35. Document titled "Section 1, Monsanto Chemical to 1947, series 10, box 5, folder: Monsanto Company History (Historical Accounts [Misc.]), MCR; Ellen Griffith Spears, *Baptized in PCBs: Race, Pollution, and Justice in an All-American Town* (Chapel Hill: University of North Carolina Press, 2014), 2; Alfred Chandler, *Scale and Scope: The Dynamics of Industrial Capitalism* (Cambridge, MA: Belknap Press of Harvard University Press, 1990), 170.

36. Document titled "Early Days," based on interview with Gaston DuBois, authored by Francis J. Curtis, series 10, box 5, folder: Monsanto Company History (Historical Accounts [DuBois, Gaston]); Leonard A. Paris, "Monsanto: The First 75 Years," 3, Corporate Public Relations article (1976), series 10, box 5, folder: Monsanto Company History (Historical Accounts [Misc.]); Document titled "Early Days, Some of Edgar Queeny's Recollections," authored by Francis J. Curtis, December 25, 1950, series 10, box 5,

folder: Monsanto Company History (Historical Accounts [Misc.] [Folder 2]); Document titled "Reminiscences about Monsanto's Beginnings," featuring excerpts from Dr. Louis Veillon's diary, series 10, box 5, folder: Monsanto Company History (Historical Accounts [Veillon, Louis]), MCR.

37. Monsanto History, by Hubert Kay, 1st Draft, Edgar M. Queeny's Copy, Part II, "John F. Queeny and the Founding of American Industry," F-20; "Reminiscences about Monsanto's Beginnings."

38. Gaston DuBois would later recall that "everything was built, it seemed, in a very great hurry." Gaston DuBois, "How Monsanto Grew Chemically," speech before Monsanto Study Group, February 6, 1940, series 10, box 5, folder: Monsanto Company History (Historical Accounts [Misc.] [Folder 2]), MCR; Monsanto History, by Hubert Kay, 1st Draft, Edgar M. Queeny's Copy, Part II, "John F. Queeny and the Founding of American Industry," F-21.

39. Proceedings of Meeting of the Stockholders of the Monsanto Chemical Works, August 22nd, 1902, at the office of the company, no. 1812 South Second Street, in the City of St. Louis, Missouri, series 14, box 22; Document titled "John F. Queeny," based on interview with Samuel Allender authored by Francis J. Curtis, December 7, 1950, series 14, box 22, folder: Queeny, John F. (Misc.) (Folder 1), MCR.

40. Document titled "Monsanto Growth," based on interview with J. W. Livingston, January 8, 1951, series 14, box 15, folder: Queeny E. M. (Character + Outlook [Business Concerns] [2]), MCR.

41. Document titled "Early Days, Veillon Comes to Monsanto," based on interview with Louis Veillon, Gaston DuBois, and Jules Bebie authored by Francis J. Curtis, April 1, 1951, series 10, box 05, folder: Monsanto Company History (Historical Accounts [Veillon, Louis]), MCR; US Government Industrial Inventory (1916), marked confidential, series 10, box 6, folder: Monsanto Company (WWI), Monsanto Company Archives; On Queeny only speaking English, see document titled "Early Days," authored by Francis J. Curtis, February 19, 1951, series 10, box 5, folder: Monsanto Company History (Historical Accounts [Kernon, Jules]), MCR; "Magdalene Emma Kithon Cleeves," Ancestory.com, https://www.ancestry.com/genealogy/records/magdalene-emma-kithon-cleeves-24-zvln1v.

42. "Reminiscences about Monsanto's Beginnings."

43. "Early Days, Veillon Comes to Monsanto"; Monsanto History, by Hubert Kay, 1st Draft, Edgar M. Queeny's Copy, Part II, "John F. Queeny and the Founding of American Industry," F-25.

44. "Reminiscences about Monsanto's Beginnings."

45. "The Accomplishments of Gaston DuBois," speech by Dr. C. A. Thomas, Dayton, Ohio, in presentation of the Perkins Prize, January 7, 1944, series 14, box 3, folder: Gaston F. DuBois (Correspondence and Speeches), MCR; Document titled "John F. Queeny," interview with Gaston DuBois/Louis Veillon authored by Francis J. Curtis, April 3, 1951, series 14, box 22, folder: Queeny, John F. (Misc.) (Folder 1), MCR; "Reminiscences about Monsanto's Beginnings"; On revenues from caffeine sales, see document titled "Monsanto Chemical Company," dated 1941, series 10, box 5, folder: Monsanto Company History (Historical Accounts [Misc.]); "How Monsanto Grew

Chemically," speech before Monsanto Study Group by Gaston DuBois, February 6, 1940, series 10, box 5, folder: Monsanto Company History (Historical Accounts [Misc.] [Folder 2]), MCR.

46. Document titled "Saccharin (very early history), Very old notes on Saccharin found in Monsanto library. Some written by Howard McDonough and others by [Gaston F.] DuBois," series 3, box 3, folder: Saccharin (History), MCR.

47. "Reminiscences about Monsanto's Beginnings"; Document titled "Early Days, Some of Edgar Queeny's Recollections," authored by Francis J. Curtis, December 25, 1950, series 10, box 5, folder: Monsanto Company History (Historical Accounts [Misc.] [Folder 2]), MCR.

48. Financial History of Monsanto Chemical Company, document dated 1957 series 10, sub-series 1, box 4, folder: Finance (History), MCR; Document titled "Monsanto Chemical Company," dated 1941, series 10, box 5, folder: Monsanto Company History (Historical Accounts [Misc.]), MCR; "Caffeine," *Meyer Brothers Druggist* 29, no. 3 (March 1908), 16, box 216, Harvey Wiley Papers, Library of Congress, Washington, DC [hereinafter Wiley Papers]; "Reminiscences about Monsanto's Beginnings"; Notes on Monsanto from a diary of Jules Bebie (1910–1911), series 10, box 5, folder: Monsanto Company History (Historical Accounts [Bebie, Jules]), MCR.

49. "Prelude to a Start-Up: A Century of Changes," *Monsanto Magazine* 2 (1997), 14, series 10, box 6, folder: Monsanto Company History (Timeline/Milestones); Document titled "Miscellaneous Stuff on Early Days," authored by Francis J. Curtis, n.d., series 10, box 5, folder: Monsanto Company History (Historical Accounts [Misc.] [Folder 2]); "Reminiscences about Monsanto's Beginnings"; Haynes, *Chemical Pioneers*, 237–38; "Early Days, Some of Edgar Queeny's Recollections"; Document titled "Bebie's Recollections of Early Problems," from interview with Dr. Bebie, authored by Francis J. Curtis, December 27, 1950, series 10, box 5, folder: Monsanto Company History (Historical Accounts [Bebie, Jules]), MCR.

50. Haynes, *Chemical Pioneers*, 238; "Bebie's Recollections of Early Problems"; Deer Creek Tapes, November 17, 1951, 10–11, "Reminiscences about Monsanto's Beginnings."

51. Monsanto History, by Hubert Kay, 1st Draft, Edgar M. Queeny's Copy, Part II, "John F. Queeny and the Founding of American Industry," F-30; "Bebie's Recollections of Early Problems"; "Reminiscences about Monsanto's Beginnings"; Document titled "Saccharin (very early history), Very old notes on Saccharin found in Monsanto library. Some written by Howard McDonough and others by [Gaston F.] DuBois," series 3, box 3, folder: Saccharin (History), MCR.

52. "Monsanto—A Missouri Story," *Missouri News Magazine* (December 1956), series 10, box 6, folder: Monsanto Company History (Publications About Monsanto 1933–1969); "John Queeny," in *The Book of St. Louisans*, 2nd ed., 486–87; Letter from John Queeny to L. F. DuBois [Gaston DuBois's father], LeLocle, Switzerland, 1908, series 14, box 3, folder: Gaston F. DuBois (Correspondence and Speeches), MCR; "Monsanto Chemical Works," *St. Louis Daily Globe-Democrat*, January 18, 1910, 13; Financial

History of Monsanto Chemical Company, document dated 1957; Monsanto History, Significant Events (October 1955), series 10, sub-series 1, box 4, folder: Finance (History); Interview with Leo Hertling, June 1956, series 10, box 5, folder: Monsanto Company History (Historical Accounts [Hertling, Leo]); Document titled "Early Days, Veillon Comes to Monsanto," based on interview with Louis Veillon, Gaston DuBois, and Jules Bebie authored by Francis J. Curtis, April 1, 1951, series 10, box 05, folder: Monsanto Company History (Historical Accounts [Veillon, Louis]); History of Monsanto, filed by company librarian Mr. Marple, July 20, 1944, series 10, box 5, folder: Monsanto Company History (Historical Accounts [Misc.]), MCR; On further expansion in 1912, see "Chemical Work Buys Land," *St. Louis Daily Globe-Democrat*, November 28, 1912, 13; "Building Records of 1911 Will Be Broken This Year," *St. Louis Star* , December 1, 1912, 4; "Bebie's Recollections of Early Problems"; "Reminiscences about Monsanto's Beginnings"; Deer Creek Tapes, November 17, 1951, 10–11.

CHAPTER 2: **"A Coal-Tar War"**

1. US Department of Agriculture, Office of the Secretary, Food Inspection Decision 135, issued April 29, 1911, in US Bureau of Chemistry, *Food Inspection Decisions*, 1–212 (Washington, DC: GPO, 1905–34); "The New 'Poison Squad,'" *New York Times*, November 1, 1907, 8; On caffeine and Coca-Cola, see Bartow J. Elmore, *Citizen Coke: The Making of Coca-Cola Capitalism* (New York: Norton, 2014), 58–75; As Wiley put it, "The extraction of caffein [sic] from any of its natural sources and the use of it in beverages which by their manner of use give no suggestion of containing this product, appear to me to be an objectionable practice, irrespective of any opinion regarding the injurious qualities of this alkaloid." See "Is the Drinking of Tea or Coffee Harmful to Health," *New York Times*, September, 15, 1912, SM11; "Fight on Coca-Cola is Waxing Warm," *Atlanta Georgian*, March 16, 1911, in Scrapbook marked "Coca-Cola Trial, March 13–March 24, 1911, box 200, folder: Caffein [sic], Harvey Wiley Papers, Library of Congress, Washington, DC [hereinafter Wiley Papers]; On the tariff, see "The Chemist Steps Out of the Laboratory," speech by Gaston F. DuBois, Vice President, Monsanto Chemical Company, January 7, 1944, on the occasion of Perkin Medal award, series 14, box 3, folder: Gaston F. DuBois (Correspondence and Speeches), Monsanto Company Records, Washington University in St. Louis, Julian Edison Department of Special Collections, St. Louis, Missouri [hereinafter MCR].
2. Notes on Monsanto from Jules Bebie diary, 1910–11, series 10, box 5, folder: Monsanto Company History (Historical Accounts [Bebie, Jules]), MCR.
3. Note dated March 13, 1911, in scrapbook titled "Coca-Cola Trial, Book 1 March 13–March 24, 1911"; "Candler Cursed Me Says the Inspector," *The Atlanta Georgian*, March 14, 1911, in scrapbook titled "Coca-Cola Trial, Book 1, March 13–March 24, 1911," box 200, folder: Caffein [sic]; "Government Spies Busy For Over Two Years," *Chattanooga Times*, March 16,

1911, in scrapbook titled "Coca-Cola Trial, Book 1, March 13–March 24, 1911," box 200, folder: Caffein [sic], Wiley Papers; "The Caffeine Content in Eight Coca-Colas Would Kill If Concentrated in One Dose, Says Expert on the Witness Stand," *Columbus Daily Enquirer* (Columbus, GA), March 19, 1911, 1; "Dangerous 'Soft Drinks,'" *Our Messenger* (Wichita, KS) 25, no. 8 (July 1, 1910), 1; "Coca-Cola Contains Caffeine, Deadly to Interior Organisms," *The Weekly Tribune* (Tampa, FL), March 23, 1911, 10. See also Elmore, *Citizen Coke*, 58–75, 60–62.

4. "Rebuttal Evidence in the Coca-Cola Case," *Chattanooga News*, April 1, 1911, in scrapbook titled "Coca-Cola Trial, Book 2, March 24–, 1911," box 200, folder: Caffein [sic], Wiley Papers.

5. Transcript of Testimony at 499 and 562, *United States v. Forty Barrels and Twenty Kegs of Coca-Cola*, 191 F. 431 (E. D. Tenn. 1911); See also Elmore, *Citizen Coke*, 65–66.

6. Opinion in *United States v. Forty Barrels and Twenty Kegs of Coca-Cola*, 191 F. 431 (E. D. Tenn. 1911), in US Department of Agriculture, *Decisions of Courts in Cases Under the Federal Food and Drugs Act* (Washington, DC: GPO, 1934), 242, 245, 256.

7. Regulators scrutinized margarine, milk, and other processed foods in addition to caffeine and saccharin. On the history of debates about the naturalness of processed foods in the Gilded Age and Progressive Era, see Benjamin R. Cohen, *Pure Adulteration: Cheating on Nature in the Age of Manufactured Food* (Chicago: University of Chicago Press, 2019); Kendra Smith-Howard, *Pure and Modern Milk: An Environmental History since 1900* (New York: Oxford University Press, 2013); Ai Hisano, *Visualizing Taste: How Business Changed the Look of What We Eat* (Cambridge, MA: Harvard University Press, 2019); Anna Zeide, *Canned: The Rise and Fall of Consumer Confidence in the American Food Industry* (Berkeley: University of California Press, 2018).

8. *U.S. v. Forty Barrels and Twenty Kegs of Coca-Cola*, 241 U.S. 265, 284–85, 276 (1916).

9. Frederick Allen, *Secret Formula: How Brilliant Marketing and Relentless Salesmanship Made Coca-Cola the Best-Known Product in the World* (New York: HarperCollins, 1994), 90.

10. Letter from John F. Queeny to Theodore Roosevelt, July 3, 1911, series 3, box 3, folder: Saccharin (Correspondence) (1906–19), MCR; Letter from President Theodore Roosevelt to John F. Queeny, July 7, 1911, series 3A, vols. 20–23, May 29–July 19, 1911, reel 367, Theodore Roosevelt Papers, Library of Congress.

11. Warwick Hough pointed out that he had been sharing the Roosevelt letter widely: "I have been using it for some time." See USDA, Office of the Secretary, *Saccharin: Under the Food and Drugs Act of June 30, 1906*, Issued December 16, 1911; Letter from Warwick M. Hough to Ira Remsen, November 18, 1911, Record Group 16: Records of the Office of the Secretary of Agriculture, General Correspondence of the Office of the Secretary, 1906–70, Reports-Retirement, 1911–12, box 37, folder: Saccharin [2 of 2], National Archives, College Park, Maryland [hereafter NARA].

12. Letter from Warwick M. Hough to editor, January 2, 1912, Record Group

16: Records of the Office of the Secretary of Agriculture, General Correspondence of the Office of the Secretary, 1906–70, Reports-Retirement,
1911–12, box 37, folder: Saccharin [1 of 2], NARA.

13. Warwick M. Hough, "A Regulation on Saccharin: Brief on the Interpre-
tation of the Conclusions of the Referee Board of Consulting Scientific
Experts on Saccharin with Suggestions as to a Regulation to be Adopted
by the Three Secretaries to give Legal Effect Thereto," published by Nixon-
Jones Press (1911), Record Group 16: Records of the Office of the Secretary of Agriculture, General Correspondence of the Office of the Secretary,
1906–70, Reports-Retirement, 1911–12, box 37, folder: Saccharin [2 of 2],
NARA; William Haynes, *Chemical Pioneers: The Founders of the American Chemical Industry* 1 (New York: D. Van Nostrand, 1939), 239, series
14, box 22, folder: John F. (Misc.) (Folder 2), MCR.

14. Letter from John Queeny to Frank L. McCartney, War Department, Purchase, Storage & Traffic Division, Office of the Director of Purchase &
Storage, April 8, 1919, series 3, box 3, folder: Saccharin (Correspondence)
1906–19, MCR; Letter from John F. Queeny, October 12, 1912, series 3,
box 3, folder: Saccharin (History), MCR.

15. Letter from Warwick Hough, April 2, 1912, series 3, box 3, folder: Saccharin (Correspondence) 1906–19, MCR.

16. History of Monsanto, filed by company librarian Mr. Marple, July 20, 1944,
series 10, box 5, folder: Monsanto Company History (Historical Accounts
[Misc.]); "Government Drops Saccharin Suit," *Monsanto Current Events*
4, no. 1 (May 1925): 5, series 14, box 22, folder: Queeny, John F. (Misc.),
Folder 1, MCR.

17. Document titled "Accounting notes," series 10, box 5, folder: Monsanto
Company History (Historical Accounts [DuBois, Gaston]); Document
titled "Monsanto Chemical Company," dated 1941, series 10, box 5, folder:
Monsanto Company History (Historical Accounts [Misc.], MCR.

18. Letter from John F. Queeny to Claude Kitchin, Tariff Conference Committee, House of Representatives, September 12, 1913, series 14, box 22, folder:
Queeny, John F. (Facility Planning—Chlorine), MCR; US Senate Committee on Finance, *Tariff: American Valuations*, 67th Cong., 1st Sess., July 25,
1921, 310.

19. Fred Aftalion, *A History of the International Chemical Industry: From the
'Early Days' to 2000* (Philadelphia: Chemical Heritage Foundation, 2002),
78, 125.

20. On Monsanto's reliance on Europe at the start of World War I, see document titled "Monsanto in World War I," based on interview with Louis Veillon, Gaston DuBois, and Jules Bebie, authored by Francis J. Curtis, 1951,
series 10, sox 5, folder: Monsanto Company History (Historical Accounts
[Veillon, Louis]), MCR; "It's Dangerous to be Too Good a Loser," *American Magazine* (May 1925), 196, series 14, box 22, folder: Queeny, John F.
(Misc.) (Folder 1), MCR; Document titled "Saccharin (very early history),
Very old notes on Saccharin found in Monsanto library. Some written by
Howard McDonough and others by [Gaston F.] DuBois," series 3, box 3,
folder: Saccharin (History), MCR.

21. "Monsanto in World War I"; Gaston DuBois, "How Monsanto Grew Chemically," talk before Monsanto Study Group, February 6, 1940, series 10, box 5, folder: Monsanto Company History (Historical Accounts [Misc.] [Folder 2]), MCR.

22. On Queeny's frugality, see document titled "Early Days: John F. Queeny," based on interview with Miss Fitzpatrick, authored by Francis J. Curtis, December 28, 1950, series 14, box 22, folder: Queeny, John F. (Facility Planning-Chlorine); Monsanto History, by Hubert Kay, 1st Draft, Edgar M. Queeny's Copy, Part II, "John F. Queeny and the Founding of American Industry," F-69, series 6, box 2, folder: Monsanto and the American Idea (Hubert Kay) (1st Draft) (Queeny's Copy), MCR; On broom incident, see document titled "Early Days," based on interview with Bill Wendt and Leo Hertling, authored by Francis J. Curtis, February 26, 1951, series 10, box 5, folder: Monsanto Company History (Historical Accounts [Wendt, Leo]), MCR.

23. Harry A. Stewart, "It's Dangerous to be Too Good a Loser!," *American Magazine*, 196–97, series 14, box ??, folder: Queeny, John F. (Misc.) (Folder 1), MCR.

24. DuBois, "How Monsanto Grew Chemically"; Letter from John Queeny to W. F. Gephart, Chairman of the United States Food Administration, October 25, 1918, series 3, box 3, folder: Saccharin (Correspondence) 1906–19; "Monsanto Division on Parade," 1965, series 1, box 4, folder: International Division (History) (Folder 1); Monsanto Press Release from 1962, series 3, box 3, folder: Saccharin (History); Document titled "Early Days," based on interview with J. F. Stickley, Plant B, authored by Francis J. Curtis, December 5, 1950, series 3, box 3, folder: Saccharin (History); Excerpt on saccharin taken from Gaston DuBois "Statement Regarding the Value of Processes Used by Monsanto Chemical Works for the Manufacture of Glycerophosphates, Saccharin, Vanillin, Phenacetin, Phenolphthalein," dated April 26, 1920, series 3, box 3, folder: Saccharin (History), MCR.

25. Stewart, "It's Dangerous to be Too Good a Loser!"; Ellen Griffith Spears, *Baptized in PCBs: Race, Pollution, and Justice in an All-American Town* (Chapel Hill: University of North Carolina Press, 2014), 62–63.

26. Stewart, "It's Dangerous to be Too Good a Loser!"; Spears, *Baptized in PCBs*, 62–63.

27. On bank debts, see Financial History of Monsanto Company, 1957, series 10, sub-series 1, box 4, folder: Finance (History) (Folder 6); Document titled "Monsanto Chemical Company," dated 1941, series 10, box 5, folder: Monsanto Company History (Historical Accounts [Misc.]); Stewart, "It's Dangerous to be Too Good a Loser!"; Deer Creek Tapes, November 17, 1951, 10–11, series 10, box 5, folder: Monsanto Company History (Historical Accounts [Bebie, DuBois, Veillon Reunion, 1951]), MCR.

28. Ferdinand Zienty, "Roots, Origins of the Monsanto Agricultural Chemical Enterprise," presented to the Technical Community of Monsanto, April 26, 1989, St. Louis, Missouri, series 10, box 5, folder: Monsanto Company History (Historical Accounts [Misc.] [Folder 2]); "Speech: Industrial Nurses Association," no author, November 16, 1967, series 1, box 4, folder: Medical Department, MCR.

29. Document titled "John F. Queeny and the Early Days," based on interview with Bert Langreck, authored by Francis J. Curtis, February 12, 1951, series 14, box 22, folder: Queeny, John F. (Misc.) (Folder 1), MCR.

30. Document titled "Early Days," based on interview with Jules Kernon, authored by Francis C. Curtis, February 19, 1951, series 10, box 5, folder: Monsanto Company History (Historical Accounts [Kernon, Jules]—Oversaw Personnel); US Government Industrial Inventory (1916), marked confidential, series 10, box 6, folder: Monsanto Company (WWI); Document titled "Early Days," authored by Francis J. Curtis, February 19, 1951, series 10, box 5, folder: Monsanto Company History (Historical Accounts [Kernon, Jules]—Oversaw Personnel).

31. Deer Creek Tapes, November 17, 1951, 10–11; In his history of St. Louis, Walter Johnson writes about the long history of racial violence in St. Louis in *The Broken Heart of America: St. Louis and the Violent History of the United States* (New York: Basic Books, 2020), 217–19, 225, 253.

32. Monsanto History, by Hubert Kay, 1st Draft, Edgar M. Queeny's copy, Part II, "John F. Queeny and the Founding of American Industry," F-48; Document titled "Monsanto Chemical Company," dated 1941, series 10, box 5, folder: Monsanto Company History (Historical Accounts [Misc.]); John F. Queeny, "The West at Work: Filling the Gaps in the St. Louis Factory Line," n.d., series 14, box 22, folder: Queeny, John F. (Misc.) (Folder 2), MCR.

33. "A Romance," published by Monsanto, 1931, series 10, box 5, folder: Monsanto Company History (Historical Accounts [Misc.] [Folder 2]); See also "Heavy Chemical Making in the Mississippi Valley Metropolis," April 8, 1920, series 10, sub-series 1, box 2, folder: Chemical Industry (1900–50); John F. Queeny, "The West at Work: Filling the Gaps in the St. Louis Factory Line," n.d., series 14, box 22, folder: Queeny, John F. (Misc.) (Folder 2), MCR.

34. "A whole arm of this industry starts in coal," Gaston DuBois exclaimed in 1922, "which is the greatest organic deposit known to man." Gaston DuBois, "Politics, Statesmanship and Organic Chemicals," *Monsanto Current Events* 4, no. 3 (September 1925), 2, series 8, box 9, folder: Monsanto Current Events, 1920–26, MCR.

35. Untitled document dated December 1916, series 3, box 1, folder: Coal Tar, MCR; See also Bartow J. Elmore, "The Commercial Ecology of Scavenger Capitalism: Monsanto, Fossil Fuels, and the Remaking of a Chemical Giant," *Enterprise and Society* 19, no. 1 (March 2018), 160.

36. US Tariff Commission, *Report on Dyes and Related Coal-Tar Chemicals, 1918* (Washington, DC: Government Printing Office, 1919), 15; This is confirmed in a study reported by Francis J. Curtis. See Curtis, "Twenty Years of the American Chemical Industry," *Monsanto Current Events* (June 1937), series 8, box 9, folder: Monsanto Current Events, 1920–26, MCR.

37. Untitled document titled "Coal tar!" dated December 1916, potentially authored by Dr. Nickel, series 3, box 1, folder: Coal Tar; For more on scavenger capitalism, see Elmore, "The Commercial Ecology of Scavenger Capitalism: Monsanto, Fossil Fuels, and the Remaking of a Chemical Giant," 153–78. The author began thinking about this concept of scavenger

capitalism after reading the work of Van Jones, who talked about an economy reliant on "dead things" in his book *The Green Collar Economy*. See
Van Jones, "The Green Economy," Center for American Progress website, October 7, 2008, https://www.americanprogress.org/issues/green/
news/2008/10/07/5063/the-green-collar-economy/. Van Jones, *The Green
Collar Economy: How One Solution Can Fix Our Two Biggest Problems*
(New York: HarperOne, 2008).

38. Dan J. Forrestal, *Faith, Hope & 5,000: The Story of Monsanto* (New York:
Simon & Schuster, 1977), 29; Letter from Juanita McCarthy to H. A. Marple, May 2, 1961, series 10, box 6, folder: Monsanto Company History
(WWI); Document titled "Monsanto in World War I," based on interview
with Louis Veillon, Gaston DuBois, and Jules Bebie, authored by Francis
J. Curtis, 1951, series 10, box 5, folder: Monsanto Company History (Historical Accounts [Veillon, Louis]); "The Coal-Tar Industry After the War,"
interview with John F. Queeny and Gaston DuBois, October 14, 1920,
series 10, sub-series 1, box 2, folder: Chemical Industry (1900–50); Document titled "Monsanto Chemical Company," dated 1941, series 10, box 5,
folder: Monsanto Company History (Historical Accounts [Misc.]), MCR;
John Barry, *The Great Influenza: The Epic Story of the Deadliest Plague in
History* (New York: Penguin Books, 2005), 354.

39. "Ruabon's Century of Progress," *Autoclave*, a Monsanto Chemicals Limited publication, 1–11, series 1, box 4, folder: Monsanto Chemical Limited (History); Document titled "A Romance," produced by the Monsanto
Chemical Company in 1931, series 10, box 5, folder: Monsanto Company
History (Historical Accounts [Misc.] [Folder 2]); "Monsanto in England,"
Chemical and Metallurgical Engineering 44, no. 4 (April 1937), 191, series
1, box 4, folder: Monsanto Chemical Limited (History), MCR.

40. Forrestal, *Faith, Hope & $5,000*, 38.

41. Monsanto History, by Hubert Kay, 1st Draft, Edgar M. Queeny's copy, Part
II, "John F. Queeny and the Founding of American Industry," F-51; Letter
from Edgar M. Queeny to Joseph Pulitzer, February 6, 1941, series 14, box 16,
folder: Queeny, E. M. (Correspondence, 1938–43); "New Year's Greeting,"
Monsanto Current Events (January 1923), series 14, box 22, folder: Queeny,
John F. (Misc.), Folder 1; Jules Kernan, "Evolution personnel Function, Medical Services," n.d., series 1, box 4, folder: Medical Department; "Prelude to
a Start-Up: A Century of Changes," *Monsanto Magazine* 2 (1997), 15, series
10, box 6, folder: Monsanto Company History (Timeline/Milestones), MCR.

42. "Financial History of Monsanto Chemical Company," 1957, series 10, sub-
series 1, box 4, folder 6, Finance (History); Monsanto History, by Hubert
Kay, 1st Draft, Edgar M. Queeny's copy, Part II, "John F. Queeny and the
Founding of American Industry," F-54; On the acquisitions strategy, see
also Alfred D. Chandler Jr., *Shaping the Industrial Century: The Remarkable Story of the Evolution of the Modern Chemical and Pharmaceutical
Industries* (Cambridge, MA: Harvard University Press, 2005), 63; Aftalion,
A History of the International Chemical Industry, 175; On the strength of
financial markets, and the rise in the number of households with equity in
the stock market at this time, see Julia Ott, *When Wall Street Met Main*

Street: *The Quest for an Investor's Democracy* (Cambridge, MA: Harvard University Press, 2011).

43. Monsanto History, by Hubert Kay, 1st Draft, Edgar M. Queeny's copy, Part II, "John F. Queeny and the Founding of American Industry," F-69; On Queeny's stroke, see document titled "John F. Queeny," based on interview with Charles Huisking, authored by Francis J. Curtis, January 15, 1951; On tongue cancer, see document titled, "John F. Queeny," based on interview with Samuel Allender, authored by Francis J. Curtis, December 7, 1950, series 14, box 22, folder: Queeny, John F. (Misc.), Folder 1, MCR; Forrestal, *Faith, Hope & $5,000*, 40.

CHAPTER 3: "A Die-Hard Admirer of the Tooth-and-Claw"

1. Interview with Ralph C. Piper conducted by James E. McKee Jr., September 11, 1981, 23, series 10, box 5, folder: Monsanto Company History (Oral History Project, 1981, Folder 2), Monsanto Company Records, Washington University in St. Louis, Julian Edison Department of Special Collections, St. Louis, Missouri [hereinafter MCR].

2. Interview with Ralph C. Piper, 24.

3. Interview with Ralph C. Piper, 24. In his memoir, Ralph Piper discusses his time as a Monsanto pilot. See Ralph E. Piper, *Point of No Return: An Aviator's Story* (Ames: Iowa State University Press, 1990).

4. Interview with Ralph C. Piper, 32–33.

5. Interview with Ralph C. Piper, 33–34; Letter from Edgar M. Queeny to Pietro Crespi, Crespi Cotton Company, July 31, 1958, series 14, box 16, folder: Queeny, E. M. (1958), MCR.

6. Interview with Ralph C. Piper, 24.

7. Interview with Ralph C. Piper.

8. Interview with Ralph C. Piper.

9. Interview with Ralph C. Piper, 24–25.

10. Letter from Edgar Queeny to Richard E. Bishop, July 27, 1942, series 14, box 16, folder: Queeny, E. M. (Correspondence, 1938–43), MCR; Leonard A. Paris, "Monsanto: The First 75 Years," 4, Corporate Public Relations article (1976), series 10, box 5, folder: Monsanto Company History (Historical Accounts [Misc.]), MCR; On piano and Edgar's mother, see document titled "Bebie's Recollections of Early Problems," from interview with Dr. Bebie, authored by Francis J. Curtis, December 27, 1950, series 10, box 5, folder: Monsanto Company History (Historical Accounts [Bebie, Jules]); Document titled "John F. Queeny," authored by Clayton Wolfe, n.d.; Document titled "Early Days," interview with Carl Meyer, Meyer Brothers executive, and Mr. Vaughn, treasurer, authored by Francis J. Curtis, December 27, 1950; Document titled "John F. Queeny and the Early Days, Transition and Growth," authored by William M. Rand, February 8, 1951, series 14, box 22, folder: Queeny, John F. (Misc.), Folder 1, MCR.

11. "His interest in the chemical business was slight," explained one company chronicler. See Hubert Kay, "Monsanto and the American Dream," draft

copy, "Part II, Chapter 3, The Boss's Son," E-14, series 6, box 1, folder: Monsanto and the American Idea, by Hubert Kay, 1st Draft (Forrestal's Copy), Folder 1, MCR.

12. Monsanto Chemical Works 1928 Annual Report, 6; Hubert Kay, "Monsanto and the American Dream," draft copy, "Part II, Chapter I: The Two Selves of Edgar Queeny," E-14, series 6, box 1, folder: Monsanto and the American Idea, by Hubert Kay, 1st Draft (Forrestal's Copy), Folder 1, MCR.

13. "Tailors Say Businessmen Best Dressed Astaire Is Lone Actor on List of Ten," *Bergen Evening Record* (Hackensack, NJ), November 29, 1935, 2; Letter from Edgar Queeny to Robert Woodruff, December 10, 1935; Letter to Robert W. Woodruff, December 3, 1935; Letter from Robert W. Woodruff to Edgar Queeny, December 4, 1935; Letter from Edgar Queeny to Robert W. Woodruff, March 21, 1945, series 1, box 257, folder: Queeny, Edgar M., 1935–46, Robert Winship Woodruff Papers, Stuart A. Rose Manuscript, Archives, and Rare Book Library, Emory University [hereinafter RWW Papers]; Letter from Edgar Queeny to Garet Garett, October 23, 1945, series 14, box 16, folder: Queeny, E. M. (Correspondence, 1944–45); Letter from Edgar Queeny to J. P. Sprang Jr., Gillette Razor Company, March 28, 1960, series 14, box 16, folder: Queeny, E. M. (Correspondence, 1959–60), MCR.

14. Edgar Queeny, *The Spirit of Enterprise* (New York: Charles Scribner's Sons, 1943), ix, 17, 19–21, 34, 147, 150; Letter from Edgar Queeny to Rose Wilder Lane, September 9, 1943, series 14, box 16, folder: Queeny, E. M. (Correspondence, 1938–43); Letter from Edgar Queeny to Charles J. Graham, September 1, 1943; Letter from Edgar Queeny to Rose Wilder Lane, July 29, 1943, series 14, box 16, folder: Queeny, E. M. (Correspondence, 1938–43); Letter from Edgar Queeny to the editor of the *St. Louis Post-Dispatch*, April 22, 1943, series 14, box 15, folder: Queeny, E. M. (Character + Outlook [Business Concerns] [2]), MCR.

15. Queeny, *Spirit of Enterprise*, 32–33; Letter from Edgar Queeny to Herbert Hoover, April 8, 1943; In one letter, Queeny did say that he realized "the contributions that have been made by immigrants from Eastern Europe, but," he added, "I think the facts substantiate the assertion that the Communist movement in this country is composed very largely of Eastern Europeans." See letter from Edgar Queeny to Elisha E. Friedman, August 11, 1943, series 14, box 16, folder: Queeny, E. M. (Correspondence, 1938–43), MCR.

16. Letter from Edgar M. Queeny to Ben Reese, *St. Louis Post-Dispatch*, April 1, 1943, series 14, box 16, folder: Queeny, E. M. (Correspondence, 1938–43), MCR; Queeny, *Spirit of Enterprise*, ix; Leonard A. Paris, "Monsanto: The First 75 Years."

17. "Prelude to a Start-Up: A Century of Changes," *Monsanto Magazine* 2 (1997), 15, series 10, box 6, folder: Monsanto Company History (Timeline/ Milestones), MCR.

18. "Merrimac is Oldest and Largest New England Chemical Company," *Monsanto Current Events* 8, no. 7, December 4, 1929, series 8, box 10, folder: Monsanto Current Events (1927–29); Document titled "Monsanto Chemical Company," dated 1941, series 10, box 5, folder: Monsanto Company History (Historical Accounts [Misc.]), MCR.

19. Letter from Gaston DuBois to Edgar M. Queeny, March 22, 1932, series 14, box 9, folder: Livingston, William G., MCR.

20. Letter from Charles Belknap to Edgar M. Queeny, March 14, 1932; Letter from G. Lee Camp to Edgar M. Queeny, March 10, 1932, series 14, box 9, folder: Livingston, William G., MCR; On bad blood between Queeny and DuBois, see "Allender's Comments on First Outline," series 14, box 15, folder: Queeny E. M. (Character + Outlook [Business Concerns] [2]), MCR.

21. Memorandum from Edgar M. Queeny to the Board of Directors, March 8, 1932, series 14, box 9, folder: Livingston, William G., MCR.

22. Hubert Kay, "Monsanto and the American Dream," draft copy, "Part II, Chapter 5, How Monsanto Grew," E-39, series 6, box 1, folder: Monsanto and the America Idea, by Hubert Kay, 1st Draft (Forrestal's Copy); "Swann Corporation Has Shown Diversity in Its Development," *Monsanto Current Events* 8, no. 1 (February 1934), 7, 15, series 8, box 10, folder: Monsanto Current Events (1934); "Monsanto at a Glance," *Monsanto Magazine* 3 (1995), 32, series 10, box 6, folder: Monsanto Company History (Timeline/Milestones); "President's Message, *Monsanto Current Events* 14, no. 2 (April 1935), 3, series 8, box 10, folder: Monsanto Current Events (1935), MCR; For a list of some of the compounds derived from phosphates, see Monsanto Chemical Company 1939 Annual Report, 13; Ellen Griffith Spears, *Baptized in PCBs: Race, Pollution, and Justice in an All-American Town* (Chapel Hill: University of North Carolina Press, 2014), 57; Gerald Markowitz and David Rosner, "Monsanto, PCBs, and the Creation of a 'World-Wide Ecological Problem," *Journal of Public Health Policy* 39 (2018), 467.

23. "Electro Chemical Operations Feature Monsanto's Newest Subsidiary," *Monsanto Current Events* 14, no. 3 (June–July 1935): 4–6; On the $3 million price tag for this investment, see document titled "Monsanto Chemical Company, 1941," series 10, box 5, folder: Monsanto Company History (Historical Accounts [Misc.]); Letter from J. Harris, Dayton Laboratories, Monsanto, to Curtis, November 6, 1951, series 3, box 2, folder: Detergents, MCR; Dan Forrestal, *Faith, Hope, & $5,000: The Story of Monsanto* (New York: Simon & Schuster, 1977), 80–81.

24. Hubert Kay, "Monsanto and the American Idea," draft copy, "Part II, Chapter 5, How Monsanto Grew," E-39, June 1958, series 6, box 1, folder: Monsanto and the America Idea, by Hubert Kay, 1st Draft (Forrestal's Copy), MCR; Paul H. Sisco, "Population Changes in Memphis, 1950–1958," *Southeastern Geographer* 1 (1961), 24.

25. Edgar M. Queeny, "The Chemical Industry Turns to the South," *Manufacturer's Record*, August 1933, series 10, sub-series 1, box 2, folder: Chemical Industry (1900–1950); "President's Message," *Monsanto Current Events* 14, no. 4 (September 1935), 3, series 8, box 10, folder: Monsanto Current Events (1935), MCR; Queeny, *Spirit of Enterprise*, 139. For other Queeny critiques of the TVA, see letter to the editor authored by Edgar Queeny, *St. Louis Post-Dispatch*, April 22, 1943, series 14, box 15, folder: Queeny, E. M., MCR.

26. Report on Idaho–Utah Visits, July 11 and 12, 1951; "Idaho Phosphate Furnace Promises Record Flow," *Lake Telegram* (Salt Lake City, UT), July 13, 1951; Pamphlet on Monsanto, Soda Springs, Idaho, undated; "Power Rates

OKed by Idaho PUC," n.d., 1951, series 2, box 6, folder: USA (Soda Springs, Idaho), MCR; The firm expressed concern about phosphate supplies as early as 1936 in its annual report: "In order to fulfill our part, we have projected a comprehensive program for 1936. Involved are the acquisition of phosphate mines and reserves to secure and lower the cost of this important raw material." See Monsanto Chemical Company 1935 Annual Report, 13.

27. Monsanto Chemical Company 1934 Annual Report, 13.

28. Alfred Chandler, *Shaping the Industrial Century: The Remarkable Story of the Evolution of the Modern Chemical and Pharmaceutical Industries* (Cambridge, MA: Harvard University Press, 2005), 27, 55–56, 69, 72–73, 113, 146; Mark Fiege, *Republic of Nature: An Environmental History of the United States* (Seattle: University of Washington Press, 2014), 372.

29. Memorandum from W. G. Livingston to H. C. Greer regarding Chemstrand Corporation, March 18, 1958, series 14, box 17, folder: Queeny, E. M. (Correspondence by Subject: Board of Directors, 1935–64), MCR; Alfred D. Chandler Jr., *Shaping the Industrial Century*, 42–46; Fred Aftalion, *A History of the International Chemical Industry: From the 'Early Days' to 2000* (Philadelphia: Chemical Heritage Foundation, 2002), 155. For more on DuPont's history, see Pap A. Ndiaye, *Nylon and Bombs: DuPont and the March of the Modern World* (Baltimore: Johns Hopkins University Press, 2007); Alfred Chandler, "Du Pont—Creating the Autonomous Division," in *Strategy and Structure: Chapters in the History of the Industrial Enterprise* (Cambridge, MA: M.I.T. Press, 1962), 52–113; On the history of how DuPont's synthetic fibers changed American fashion, see, for example, Regina Lee Blaszczyk, "Styling Synthetics: DuPont's Marketing of Fabrics and Fashions in Postwar America," *Business History Review* 80, no. 3 (2006): 485–528. Blaszczyk was working on a new project related to DuPont's fibers as this book went to press.

30. Chandler Jr., *Shaping the Industrial Century*, 56–57. For more on Dow's history, see E. N. Brandt, *Growth Company: Dow Chemical's First Century* (East Lansing: Michigan State University Press, 1997).

31. Chandler Jr., *Shaping the Industrial Century*, 71–75.

32. Monsanto Chemical Company 1939 Annual Report, 3–4; "Prelude to a Start-Up: A Century of Changes," *Monsanto Magazine* 2 (1997): 15–16, series 10, box 6, folder: Monsanto Company History (Timeline/Milestones), MCR.

33. Monsanto Chemical Company 1939 Annual Report, 15; "President's Message," *Monsanto Current Events* 17, no. 2 (June–July 1938), 3, series 8, box 10, folder: Monsanto Current Events (1938–39), MCR; US Bureau of Labor Statistics, "Labor Force Statistics from the Current Population Survey, 1929–1939," https://data.bls.gov/timeseries/ LFU21000100&series_id=LFU22000100&from_year=1929&to_ year=1939&periods_option=specific_periods&periods=Annual+Data.

34. "Charles Allen Thomas, Monsanto's New President," *Monsanto Magazine* 30, no. 3 (April 1951), 4–7, series 8, box 12, folder: Monsanto Magazine (1950–51), MCR; For a discussion of Thomas Midgley's infamy, see J. R. McNeill, *Something New Under the Sun: An Environmental*

History of the Twentieth-Century World (New York: W. W. Norton, 2001), 111–13.

35. "Prelude to a Start-Up: A Century of Changes," 15–16; Document titled "Historical Background of Monsanto," n.d., series 1, box 4, folder: Monsanto Chemical Limited (History); "From the Hub of the 'Middlewest' to the World's Chemical Marketplace," *Chemical Markets* (August 1930), series 10, box 5, folder: Monsanto Company History (Incorporation Statements, Missouri); Document titled "Monsanto Canada Limited, History," n.d., series 10, box 5, folder: Monsanto Company History (Canada) (Folder 1); Letter from Edgar M. Queeny to Monsanto's Board of Directors, May 11, 1938, series 14, box 17, folder: Queeny, E. M. (Correspondence by Subject: Board of Directors, 1935–64), MCR.

36. Letter from Edgar M. Queeny to Lammot du Pont, October 3, 1938, series 14, box 17, folder: Queeny, E. M. (Correspondence by Subject: Board of Directors, 1935–64); Letter from Lammot du Pont, President of Du Pont, to Edgar M. Queeny, October 6, 1938, series 14, box 17, folder: Queeny, E. M. (Correspondence by Subject: Board of Directors, 1935–64), MCR.

37. Letter from Edgar M. Queeny to Lammot du Pont, October 18, 1938, series 14, box 17, folder: Queeny, E. M. (Correspondence by Subject: Board of Directors, 1935–64), MCR.

38. E. I. du Pont de Nemours & Company 1939 Annual Report, 12; "How Recent and Prospective Chemical Developments Will Affect Business," speech by Francis J. Curtis at the National Association of Purchasing Agents conference, San Francisco, California, May 22, 1939, series 14, box 3, folder: Francis J. Curtis, MCR.

39. Curtis, "How Recent and Prospective Chemical Developments Will Affect Business." Curtis's words would have found favor with a growing number of agricultural scientists and business leaders in the "chemurgy" movement that emerged in the 1920s and 1930s. American chemurgists promoted the idea of turning domestically produced agricultural crops toward industrial use, eschewing technologies that tended to create dependencies on foreign-derived raw materials and goods. For a succinct summary of the chemurgy movement, see, for example, Mark R. Finlay, "Old Efforts at New Uses: A Brief History of Chemurgy and the American Search for Biobased Materials," *Journal of Industrial Ecology* 7, no. 3–4 (2004), 33–46.

40. Curtis, "How Recent and Prospective Chemical Developments Will Affect Business"; "Fill 'Er Up," *Monsanto Magazine* 26, no. 4 (September 1946), 20, series 8, box 12, folder: Monsanto Magazine (1946–47), MCR.

41. Rubber Situation, 1942, series 10, box 7, folder: Monsanto Company History (WWII [Misc.]), Folder 2, MCR.

42. "A Synthetic Rubber Plant in Less Than a Year," *Monsanto Magazine* 22, no. 3 (June–July 1943), 4–6, series 8, box 11, folder: Monsanto Magazine (1942–1943), MCR; Hubert Kay, "Monsanto and the American Dream," draft copy, "Part V—Challenge and Response in Texas City," TC-4, series 6, box 2, folder: Monsanto and the American Idea (Hubert Kay) (1st Draft) (Forrestal's Copy) (Folder 3), MCR.

43. Monsanto in World War II, Summary of Divisions and Plant Reports Written in 1945, authored by Francis J. Curtis, April 12, 1951, series 10, box 7, folder: Monsanto Company History (WWII [Misc.] [Folder 2]); Spears, *Baptized in PCBs*, 72, 161; Recommended Procedures for the Disposal of PCB-Containing Wastes (Industrial Facilities), February 27, 1976; 1972(?) US Government Printing Office (GPO) Report on PCBs, Poison Papers archive, managed by the Bioscience Research Project and the Center for Media and Democracy and largely based on papers collected by environmentalist and journalist Carol Van Strum, https://www.poisonpapers.org/the-poison-papers/ [hereinafter PPA]; Markowitz and Rosner, "Monsanto, PCBs, and the Creation of a 'World-Wide Ecological Problem,'" 468–69.

44. Keith V. Gilbert, "History of the Dayton Project" (June 1969), series 2, box 2, folder: USA (Dayton, Ohio) Central Research Department, MCR.

45. Letter from J. Robert Oppenheimer to Dr. Charles A. Thomas, September 8, 1945, series 2, box 2, folder: USA (Dayton, Ohio) Central Research Department, MCR; For more on the history of the "Dayton Project," see declassified secret document titled "Manhattan District History, Book VIII, Los Alamos Project (Y) – Volume 3, Auxiliary Activities," 4.1-11.1, posted by the US Department of Energy Office of Scientific and Technical Information on the agency's website, https://www.osti.gov/includes/opennet/includes/MED_scans/Book%20VIII%20-%20%20Volume%203%20-%20Auxiliary%20Activities%20-%20Chapter%204,%20Da.pdf.

46. Document titled "Development Projects Which were Either Completed By or Received The Attention of Monsanto During the War," series 10, box 7, folder: Monsanto Company History (WWII [Misc.] [Folder 2]); Document titled "Summary of Operations of Government-Owned Plants by Monsanto," series 10, box 7, folder: Monsanto Company History (WWII [Misc.] [Folder 2]), MCR.

47. "Monsanto in World War II, Summary of Divisions and Plant reports Written in 1945," authored by Francis J. Curtis, April 12, 1951, series 10, box 7, folder: Monsanto Company History (WWII [Misc.] [Folder 2]), MCR.

48. On Pepsi's displeasure with Coca-Cola's wartime contracts, see Letter from Walter Mack to Chester Bowles, Director of the Office of Price Administration, October 9, 1944, Record Group 188: Records of Office of Price Administration, box 927, folder: Sugar Problems, National Archives, College Park, Maryland; Bartow J. Elmore, *Citizen Coke: The Making of Coca-Cola Capitalism* (New York: Norton, 2014), 160; Constance Hays, *The Real Thing: Truth and Power at the Coca-Cola Company* (New York: Random House, 2004), 81–82; For monthly GI demand in 1943, see also Classified Message from Eisenhower's Headquarters in North Africa, June 29, 1943, box 85, folder 2, RWW Papers; Memorandum from Ralph Hayes to Mr. W. J. Hobbs, September 12, 1947, box 49, folder: Caffeine, 1927–51, RWW Papers.

49. Document titled "Section 1, Monsanto Chemical to 1947," series 10, box 5, folder: Monsanto Company History (Historical Accounts [Misc.]), MCR; Monsanto Chemical Company 1943 Annual Report, 2.

50. Chandler Jr., *Shaping the Industrial Century*, 26–27, 83, 113, 120–21,

129; Michael J. Kelly, *Prosecuting Corporations for Genocide* (New York: Oxford University Press, 2016), 32; Fred Aftalion, *A History of the International Chemical Industry*, 168.

51. Letter from Edgar Queeny to J. M. O. Monasterio, August 3, 1943, series 14, box 16, folder: Queeny, E. M. (Correspondence, 1938–43); "Meet These Officers and Executives of Merrimac Chemical Company," *Monsanto Current Events* 8, no. 7 (December 4, 1929), 14, series 8, box 10, folder: Monsanto Current Events (1927–29); "Prelude to a Start-Up: A Century of Changes," *Monsanto Magazine* 2 (1997), 16, series 10, box 5, folder: Monsanto Company History (Timeline/Milestones), MCR; Queeny, *Spirit of Enterprise*; Letter from Edgar Queeny to L. G. Ryan, August 4, 1943, series 145, box 16, folder: Queeny, E. M. (Correspondence, 1938–43), MCR; Letter from Herbert Hoover to Edgar Queeny, March 16, 1961; Letter from Herbert Hoover to Edgar Queeny, December 29, 1942, box 182, folder: Queeny, Edgar Monsanto, Correspondence—1946–63, Herbert Hoover Presidential Library, West Branch, Iowa.

52. Leonard A. Paris (Corporate Public Relations), "Monsanto: The First 75 Years," 5, 1976, series 10, box 5, folder: Monsanto Company History (Historical Accounts [Misc.]), MCR.

53. Letter from Edgar Queeny to Walt Disney, July 25, 1958, series 14, box 16, folder: Queeny, E. M. (1958), MCR; Letter from Edgar Queeny to Robert W. Woodruff, December 21, 1942; Letter from Robert W. Woodruff to Edgar Queeny, November 8, 1946, box 257, folder: Queeny, Edgar M., 1935–46, RWW Papers.

54. Letter from Edgar Queeny to Honorable John W. Bricker, March 30, 1956, series 14, box 15, folder: Queeny E. M. (1953–56); Edgar Queeny, "Tariffs or Socialism," speech delivered before the Chamber of Commerce Annual Dinner, Little Rock, Arkansas, January 13, 1956, series 14, box 15, folder: Queeny, E. M. (Character + Outlook [Business Concerns]), MCR.

55. Letter from Edgar M. Queeny to Edward Mallinckrodt, May 24, 1954, series 14, box 16, folder: Queeny, E. M. (1953–56), MCR.

56. Document titled "Monsanto Products Used in World War II," series 10, box 7, folder: Monsanto Company History (WWII [Products]), MCR; When Monsanto stopped selling DDT, Edgar Queeny wrote company president Charles H. Sommer: "I am sorry to read in the reports that we have given up the production of DDT. You undoubtedly had good reasons, but I would like to see the reasons for it." Letter from Edgar M. Queeny to Charles H. Sommer, February 21, 1956, series 14, box 16, folder: Queeny, E. M. (1953–56), MCR. For an excellent history of DDT and other pesticides developed in World War II, see Edmund P. Russell, *War and Nature: Fighting Humans and Insects with Chemicals from World War I to* Silent Spring (Cambridge, UK: Cambridge University Press, 2001). On DDT, see also David Kinkela, *DDT and the American Century: Global Health, Environmental Politics, and the Pesticide That Changed the World* (Chapel Hill: University of North Carolina Press, 2011); Thomas Dunlap, *DDT: Scientists, Citizens, and Public Policy* (Princeton: Princeton University Press, 2016).

57. Letter from Edgar M. Queeny to Mr. Louis Reinhart, August 6, 1945, series 14, box 16, folder: Queeny, E. M. (Correspondence, 1944–45), MCR.

58. Paul K. Conklin, *Revolution Down on the Farm: The Transformation of American Agriculture Since 1929* (Lexington: University of Kentucky Press, 2009), 11, 60–76; Deborah Fitzgerald, *Every Farm a Factory: The Industrial Ideal in American Agriculture* (New Haven: Yale University Press, 2003), 93; For more on New Deal agricultural reforms, see Sarah T. Philips, *This Land, This Nation: Conservation, Rural America, and the New Deal* (New York: Cambridge University Press, 2007).

59. Paul K. Conklin, *Revolution Down on the Farm*, 23, 60–69, 72–73, 75, 79–80, 87, 99; Fitzgerald, *Every Farm a Factory*, 6–8, 53, 73; On the development of high-yielding hybrid corn seeds in the early twentieth century, see Deborah Fitzgerald, *The Business of Breeding: Hybrid Corn in Illinois, 1890–1940* (Ithaca: Cornell University Press, 1990).

60. Tore Olsson, *Agrarian Crossings: Reformers and the Remaking of the US and Mexican Countryside* (Princeton: Princeton University Press, 2017), 10, 41, 99, 132–33, 153, 155; Nick Cullather, *The Hungry World: America's Cold War Battle Against Poverty in Asia* (Cambridge, MA: Harvard University Press, 2013), 45, 61, 68; John Perkins, *Geopolitics and the Green Revolution: Wheat, Genes, and the Cold War* (New York: Oxford University Press, 1997), vi, 103, 115, 187, 259.

61. For an excellent corporate summary of how Monsanto got into agricultural chemicals, see Ferdinand Zienty, "Roots: Origins of the Monsanto Agricultural Chemical Enterprise," presentation to the Technical Community of Monsanto, April 26, 1989, St. Louis, Missouri, series 10, box 5, folder: Monsanto Company History (Historical Accounts [Misc.] [Folder 2]); "We Cannot Waste Our Land," *Monsanto Magazine* 28, no. 5 (October 1948): 18–23, series 8, box 12, folder: Monsanto Magazine (1948–49); Frederick W. Hatch, Consolidating to Compete in the Agricultural Products Markets, document dated June 1960, series 1, box 5, folder: Organic Chemical Divisions (Product Development), MCR. Parathion was extremely dangerous to humans, but in the mid-twentieth century regulators incorrectly assumed that parathion and other organophosphates were less dangerous than some chlorinated hydrocarbons like DDT because they seemed to degrade rapidly in the environment. Organophosphates remained in use well after the EPA banned DDT in 1972, though these chemicals "caused thousands of poisonings among farm workers and huge mortality in wildlife inadvertently exposed." See Frederick Rowe Davis, *Banned: A History of Pesticides and the Science of Toxicology* (New Haven; London, Yale University Press, 2014), 104–106, 159–161, 185, 217, 223.

62. "Our Dwindling Resources," *Monsanto Magazine* 27, no. 3 (June 1948), 4–5, series 8, box 12, folder: Monsanto Magazine (1948–49), MCR.

63. Minutes of the Research Directors Meeting, November 14–15, 1940, series 10, sub-series 1, box 1, folder: Board Committee and Meetings, MCR.

64. On the development of industrial hygiene as a profession in the early twentieth century, see Christopher C. Sellers, *Hazards of the Job: From Industrial Disease to Environmental Health Science* (Chapel Hill: University of

North Carolina Press, 1997); Quoted in Spears, *Baptized in PCBs*, 70–71; Markowitz and Rosner, "Monsanto, PCBs, and the Creation of a 'World-Wide Ecological Problem,'" 467–68. Frederick Rowe Davis discusses other advancements in toxicology at this time, including at the University of Chicago where pharmacologist E. M. K. Geiling established the Toxicity Laboratory. Davis also notes that DuPont, Dow, and Union Carbide all established their own in-house toxicology labs in the mid-1930s. See Frederick Rowe Davis, *Banned*, 8, 214–215.

65. Quoted in Spears, *Baptized in PCBs*, 72; Speech to the Industrial Nurses Association, author unknown, November 15, 1967, series 1, box 4, folder: Medical Department, MCR; Markowitz and Rosner, "Monsanto, PCBs, and the Creation of a 'World-Wide Ecological Problem,'" 469.

66. Company-Wide Medical Problems, n.d., series 14, box 8A, folder: Kelly, R. E., MCR.

CHAPTER 4: "Wonderful Stuff, This 2,4,5-T!"

1. Transcript of Direct Examination of James Ray Boggess in *James R. Boggess, et. al., v. Monsanto Company*, submitted with errata sheet on July 20, 1984, 6137–38, 6162, The Calwell Practice law office files, Charleston, West Virginia [hereinafter TCP files]. West Virginia personal injury attorney Stuart Calwell represented the Nitro workers in their first case against Monsanto in the 1980s. He has been fighting for them and their families ever since. Calwell gave me access to the law firm's extensive case records, which include briefs, depositions, confidential corporate files released during discovery, and court transcripts.

2. "How Industry Handled the Dioxin Question," *St. Louis Post-Dispatch*, November 14, 1983, 80–81; Transcript of Direct Examination of James Ray Boggess in *James R. Boggess, et. al., v. Monsanto Company*, 6137–38, 6162.

3. Transcript of Direct Examination of James Ray Boggess in *James R. Boggess, et. al., v. Monsanto Company*, 6166, 6168; "How Industry Handled the Dioxin Question," *St. Louis Post-Dispatch*, November 14, 1983, 80–81.

4. Deposition of Chester A. Jeffers in *James M. Adkins v. Monsanto Company*, Civil Action Nos. 81–2098, 81–2239, 81–2504, 83–2199, United States District Court for the Southern District of West Virginia (1983), 3–5, 8, 13, 21–22, 25, 37, TCP files; For the 200-employee statistic, see Marion Moses et. al. "Health Status of Workers With Past Exposure to 2,3,7,8-Tetrachlorodibenzo-p-dioxin in the Manufacture of 2,4,5-Trichlorophenoxyacetic Acid: Comparison Findings With and Without Chloracne," *American Journal of Industrial Medicine* 5 (1984), 167.

5. Plaintiff's Response to Defendant's Motion for Summary Judgment—Statute of Limitations in *Jeffers v. Monsanto*, Civil Action No. 81–2239, Docket No. 225, 73–74, TCP files.

6. "Neighbors Defend Safety at Carbide," *New York Times*, April 6, 1986, 23.

7. Bill Wintz, *Nitro: The World War I Boom Town: An Illustrated History of Nitro West Virginia and the Land on Which It Stands* (Charleston, WV:

Jalamap Publications, 1985), 3–4; For a description of Nitro in the 1910s, see Letter from Ward C. Griffing, infantryman in Nitro, to Mrs. Hattie P. Griffing, his mother in Manhattan, Kansas, January 27, 1919, http://griff-wjg .blogspot.com/2009/03/letter-119–december-21–1918.html. See also Nathan Cantrell, "West Virginia's Chemical Industry," *West Virginia Historical Society Quarterly* 18, no. 2 (April 2004): 1–15, https://web.archive.org/web /20100707161835/http://www.wvculture.org/history/wvhs1821.pdf.

8. "India, Falls Are to Enter Merger," *Akron Beacon Journal* (Akrin, OH), May 29, 1929, 15; "J. Q. Dickinson 'Oldest,'" *Sunday Gazette-Mail* (Charleston, WV), October 9, 1977; Monsanto Chemical Works 1929 Annual Report, 4–5; Monsanto Chemical Company 1951 Annual Report, 28; Kelly Hill, *Cases in Corporate Acquisitions, Buyouts, Mergers, and Takeovers* (Detroit: Gale Group, 1999), 59.

9. Carol Van Strum, *A Bitter Fog: Herbicides and Human Rights* (San Francisco: Sierra Club Books, 1983), 11; John Lewallen, *Ecology of Devastation: Indochina* (Baltimore: Penguin Books, 1971), 62; Marie-Monique Robin, *The World According to Monsanto* (New York: New Press, 2010), 40; Russell, *War and Nature*, 225.

10. "2,4-D Action Remains a Mystery," *Washington Post*, October 21, 1961, C13; Lewallen, *Ecology of Devastation*, 62; Robin, *The World According to Monsanto*, 35–36.

11. Gale E. Peterson, "The Discovery and Development of 2,4-D," *Agricultural History* 41, no. 3 (July 1967): 247–48; Van Strum, *A Bitter Fog*, 11, 75; Lewallen, *Ecology of Devastation*, 63; Michelle Mart, *Pesticides, A Love Story: America's Enduring Embrace of Dangerous Chemicals* (Lawrence: University Press of Kansas, 2015), 94.

12. American Chemical Paint Company advertisement, *Fort Worth Star-Telegram* (Fort Worth, TX), June 7, 1953, 16; For more on chemicals and the American lawn, see Ted Steinberg, *American Green: The Obsessive Quest for the Perfect Lawn* (New York: Norton, 2006).

13. Peter H. Schuck, *Agent Orange on Trial: Mass Toxic Disasters in the Courts* (Cambridge, MA: Belknap Press of Harvard University Press, 1987), 85; Monsanto Chemical Company 1949 Annual Report, 7; "Company History," Monsanto corporate website preserved in Internet Archive, https://web.archive .org/web/20170602185911/http://www.monsanto.com/whoweare/pages/ monsanto-history.aspx; C. J. Burns et al., "Mortality in Chemical Workers Potentially Exposed to 2,4-Dichlorophenozyacetic Acid (2,4-D) 1945-94: An Update," *Occupational & Environmental Medicine* 58, no. 1 (2001), 24.

14. Van Strum, *A Bitter Fog*, 12; Thomas Whiteside, *Defoliation: What Are Our Herbicides Doing to Us?* (New York: Ballantine Books, 1970), 5; "Country Diary: Wonderful Stuff, This 2,4,5-T!" *Daily Boston Globe*, December 14, 1953, 15; "The Garden Doctor," *Los Angeles Times*, May 27, 1956, N59. For more on how Americans "fell in love" with synthetic herbicides in the postwar period, see Michelle Mart, *Pesticides: A Love Story* (Lexington: University Press of Kentucky, 2015), 11–30.

15. Transcript of Direct Examination of James Ray Boggess in *James R. Boggess, et. al., v. Monsanto Company*, 6137–6138.

16. On the "slow violence" often inflicted on disadvantaged people working in dirty environments, see Rob Nixon, *Slow Violence and the Environmentalism of the Poor* (Cambridge, MA: Harvard University Press, 2011); Ellen Griffith Spears, *Baptized in PCBs: Race, Pollution, and Justice in an All-American Town* (Chapel Hill: University of North Carolina Press, 2014), 15; Transcript of Direct Examination of James Ray Boggess in *James R. Boggess, et. al., v. Monsanto Company*, 6167–68, 6388; "How Industry Handled the Dioxin Question," *St. Louis Post-Dispatch*, November 14, 1983, 32.

17. Transcript of Direct Examination of James Ray Boggess in *James R. Boggess, et. al., v. Monsanto Company*, 6170–71.

18. *Jeffers v. Monsanto*, Civil Action No. 81–2239, Docket No. 225, Plaintiff's Response to Defendant's Motion for Summary Judgment—Statute of Limitations, 50, TCP files; Raymond Suskind, "Report on Clinical and Environmental Survey, Monsanto Chemical Company, Nitro, West Virginia," report commissioned by Monsanto for internal use, 1953, 2, TCP files.

19. Steven Higgs, *Eternal Vigilance: Nine Tales of Environmental Heroism in Indiana* (Bloomington: Indiana University Press, 1995), 139; *Jeffers v. Monsanto*, Civil Action No. 81–2239, Docket No. 225, Plaintiff's Response to Defendant's Motion for Summary Judgment—Statute of Limitations, 46, TCP files.

20. Raymond R. Suskind et al., "Progress Report. Patients from Monsanto Chemical Company, Nitro, West Virginia," report commissioned by Monsanto for internal use, July 20, 1950, TCP files.

21. Suskind et al., "Progress Report," 10–17.

22. Suskind et al., "Progress Report," 23; *Jeffers v. Monsanto*, Civil Action No. 81–2239, Docket No. 225, Plaintiff's Response to Defendant's Motion for Summary Judgment—Statute of Limitations, 55, TCP files.

23. In 2010, only about 12 percent of West Virginia workers belonged to a union, but over half a century earlier more than 36 percent of laborers in the state paid union dues; Barry T. Hirsch, David A. Macpherson, and Wayne G. Vroman, "Union Density Estimates by State, 1964–2015," http://unionstats.gsu.edu/MonthlyLaborReviewArticle.htm; Bureau of Labor Statistics, "Union Affiliation of Employed Wage and Salary Workers by State," https://www.bls.gov/webapps/legacy/cpslutab5.htm; John Skaggs (attorney, Stuart Calwell Practice), interview by the author, February 23, 2017; "Decline of the Unions," *Washington Post*, June 5, 1983, C6.

24. Note that in 1913, West Virginia politicians used "workmen" not "workers," as did many other state legislatures. I use the term "workmen" only in reference to the specific laws and commissions that emerged in the Progressive Era. *Jeffers v. Monsanto*, Civil Action No. 81–2239, Docket No. 225, Plaintiff's Response to Defendant's Motion for Summary Judgment—Statute of Limitations, 48, TCP files; Gregory P. Guyton, "A Brief History of Workers' Compensation," *Iowa Orthopedic Journal* 19 (1999): 106–8; "Personal Injury Within the Meaning of the West Virginia Workmen's Compensation Act," *West Virginia Law Quarterly* 43 (1936–37): 154–56; James Weinstein, "Big Business and the Origins of Workmen's Compensation," in

United States Constitutional and Legal History: A Twenty Volume Series Reproducing Over 450 of the Most Important Articles on the Topic, ed. Kermit L. Hall (New York: Garland Publishing, 1987), 158, 166–67, 174; John Fabian Witt, *The Accidental Republic: Crippled Workingmen, Destitute Widows, and the Remaking of American Law* (Cambridge, MA: Harvard University Press, 2004), 11, 127.

25. *Jeffers v. Monsanto*, Civil Action No. 81–2239, Docket No. 225, Plaintiff's Response to Defendant's Motion for Summary Judgment—Statute of Limitations, 48-51, TCP files.

26. William F. Ashe and Raymond R. Suskind, "Progress Report—Patients from Monsanto Chemical Company, Nitro, West Virginia," n.p., report commissioned by Monsanto, April 1950, TCP files. On the lack of communication with workers about health study findings, see also Plaintiff's Response to Monsanto's Motion for Directed Verdict and Brief in Support Thereof, filed by Stuart Calwell in the case of *James R. Boggess, et. al., v. Monsanto Company*, November 9, 1984, 62–73, TCP files.

27. *Jeffers v. Monsanto*, Civil Action No. 81–2239, Docket No. 225, Plaintiff's Response to Defendant's Motion for Summary Judgment—Statute of Limitations, 52, 54, 57, TCP files; Raymond Suskind, "Report on a Clinical and Environmental Survey, Monsanto Chemical, Co., Nitro, W. VA.," report commissioned by Monsanto, 1953, 2, TCP files.

28. Raymond Suskind, "Chloracne and Associated Health Problems in Manufacture of 2,4,5T," report commissioned by Monsanto, n.d., 2; *Jeffers v. Monsanto*, Civil Action No. 81–2239, Docket No. 225, Plaintiff's Response to Defendant's Motion for Summary Judgment—Statute of Limitations, 53–55, TCP files.

29. Suskind, "Report on a Clinical and Environmental Survey, Monsanto Chemical, Co., Nitro, W. VA," 2.

30. Transcript of Direct Examination of James Ray Boggess in *James R. Boggess, et. al., v. Monsanto Company*, 6185–87.

31. Transcript of Direct Examination of James Ray Boggess in *James R. Boggess, et. al., v. Monsanto Company*, 6185–87; John Skaggs, an attorney who represented the Nitro workers in their case against Monsanto in the 1980s, described his clients similarly: "Most of them were born and raised in that area" and "none of them had a lot of formal education." Skaggs, interview; A local Nitro historian who graduated from Nitro High School in 1936 recounted that 90 percent of his graduating class worked in the chemical industry and noted that Monsanto had an established practice of hiring young men coming right out of high school to operate the company's expanding facilities. See William B. Wintz, *Nitro*, 106–7.

32. "Workers Sue Monsanto, Claiming Chemical Company Ignored Health Risk," *Baltimore Sun*, July 8, 1984, 11A.

33. Skaggs, interview; "How Industry Handled the Dioxin Question," *St. Louis Post-Dispatch*, November 14, 1983, 34; Union contracts for the Nitro workers contain details about this shoveling and shuffling work. Local 12610, District 50, U.M.W.A., and Monsanto Chemical Company, Nitro, West Virginia, Amended Raw Material Handlers Agreement, September 29,

1963, in Confidential Contract and Grievance History of Nitro Plant, prepared by Dwight E. Harding, TCP files; In *Plutopia: Nuclear Families,
Atomic Cities, and the Great Soviet and American Plutonium Disasters*
(New York: Oxford University Press, 2013), historian Kate Brown describes
how workers in Richland, Washington, exchanged their "biological rights
for consumer rights," agreeing to put their bodies in highly irradiated environments in exchange for middle-class, suburban accommodations and
accoutrements (p. 5). Here we see a similar story of compromise.

34. Agreement dated March 14, 1956, outlining the wage rate for Mr. Paul Willard, Confidential Contract and Grievance History of Nitro Plant, prepared
by Dwight E. Harding, TCP files.

35. Document titled "Monsanto's Progress," offering financial statistics on
the firm and comparisons with other chemical companies, series 10, box 5,
folder: Monsanto Company Historical Accounts ([Misc.] [Folder 2]), MCR.

36. Hubert Kay, "Monsanto and the American Idea," draft copy, "Part II,
Chapter 5: How Monsanto Grew," E-38, series 6, box 1, folder: Monsanto
and the American Idea (Hubert Kay), 1st Draft (Forrestal's Copy), Folder
1; Organizational Chronology of Monsanto Domestic Operating Units
Including Major Acquisitions and Divestitures, prepared by H. C. Godt Jr.,
PhD, private consultant for Monsanto, August 1997, series 10, box 7, folder:
Organization, 1928–97; "Prelude to a Start-Up: A Century of Changes,"
Monsanto Magazine 2 (1997), 17, series 10, box 6, folder: Monsanto Company History (Timeline/Milestones), MCR; Alfred D. Chandler Jr., *Shap-
ing the Industrial Century: The Remarkable Story of the Evolution of the
Modern Chemical and Pharmaceutical Industries* (Cambridge, MA: Har-
vard University Press, 2005), 144–54.

37. Frederick W. Hatch, "Consolidating to Compete in the Agricultural Products
Market," confidential Monsanto document, series 1, box. 5, folder: Organic
Chemical Divisions (Product Development); Ferdinand Zienty, "Roots: Origins of the Monsanto Agricultural Chemical Enterprise," presentation given
to the Technical Community of Monsanto, April 26, 1989, series 10, box
5, folder: Monsanto Company History (Historical Accounts [Misc.] [Folder
2]), MCR; Speaking of the 1950s, Fred Aftalion notes, "Curiously enough,
however, Monsanto, whose beginnings were in fine chemicals with aspirin,
vanillin, coumarin, and saccharin, now mainly focused on large-scale inor-
ganic and organic production, leaving aside pharmaceuticals." Fred Aftalion,
*A History of the International Chemical Industry: From the 'Early Days' to
2000* (Philadelphia: Chemical Heritage Foundation, 2002), 252.

38. It is worth noting here that what made Monsanto's Roundup so attractive
to company scientists when they first commercialized it in the 1970s was
not only the fact that it was much more effective at killing weeds than earlier
Monsanto brands, but also the fact that initial studies suggested Roundup
was more safe for human health and ecosystems compared to these earlier
herbicides. "Charles Allen Thomas, Monsanto's New President," *Monsanto Magazine* 30, no. 2 (April 1951), 4–7, series 8, box 12, folder: Monsanto Magazine (1950–51); Interview with Bob Rumer, February 11, 2007,
series 10, box 5, folder: Monsanto Company History (Historical Accounts

[Phillion, Lee]); Zienty, "Roots"; For Queeny's optimism about Randox, see Letter from Edgar Queeny to "Jim," September 4, 1956, series 14, box 16, folder: Queeny, E. M. (1953–56); *Monsanto Corporate Research Community Mirror*, Monsanto in-house publication, February 1991, series 8, box 9, folder: Research Community Mirror, 1989–93, MCR; "CDAA and CDEC," document published by K. E. Maxwell, Monsanto Chemical Company, Santa Clara, California, n.d., posted to University of California Agriculture and Natural Resources website, https://ucanr.edu/repository/fileaccess.cfm?article=163323&p=EDMNSR; Letter from Deloris Graham to Robert Taylor, Re: EPA Registration No. 524-312, Randox, Caswell #284, April 29, 1980, https://www3.epa.gov/pesticides/chem_search/cleared_reviews/csr_PC-019301_29-Apr-80_004.pdf; US EPA pesticide label for Vegadex technical, https://www3.epa.gov/pesticides/chem_search/ppls/000524-00309-19740829.pdf. For more on the toxicity of allidochlor (name for chloroacetamide compound in Randox) and sulfullate (name for thiocarbamate compound in Vegadex), see "Allidochlor," International Union of Pure and Applied Chemistry website, https://sitem.herts.ac.uk/aeru/iupac/Reports/1192.htm#:~:text=Allidochlor%20is%20herbicide%20that%20is,known%20skin%20and%20eye%20irritant; "Sulfullate," International Union of Pure and Applied Chemistry website, https://sitem.herts.ac.uk/aeru/iupac/Reports/2589.htm.

39. Philip C. Hamm, "Discovery, Development, and Current Status of the Chloroacetamide Herbicides," *Weed Science* 22, no. 6 (November 1974), 542; Paul K. Conklin, *A Revolution Down on the Farm: The Transformation of American Agriculture Since 1929* (Lexington: University of Kentucky Press, 2008), 87; Michelle Mart, *Pesticides, a Love Story* (Lawrence: University of Kansas Press, 2018), 13.

40. "News from Monsanto Chemical Company," company news release, September 13, 1959, series 2, box 6, folder: USA (St. Louis, Mo. [Creve Coeur]), MCR.

41. A 1951 company study had shown that "the center of Monsanto general office population" was now located in the western suburbs. Five years later that center "had moved about 1 ½ miles further west." See "News from Monsanto Chemical Company," company news release, October 13, 1957, series 2, box 6, folder: USA (St. Louis, Mo. [Creve Coeur]), MCR; Walter Johnson, *The Broken Heart of America: St. Louis and the Violent History of the United States* (New York: Basic Books, 2020), 225, 312, 315, 319–20, 328.

42. "News from Monsanto Chemical Company," company news release, 1959, series 2, box 6, folder: USA (St. Louis, Mo. [Creve Coeur]), MCR; For more on environmental thought and the suburban landscape, see Adam Rome, *Bulldozer in the Countryside: Suburban Sprawl and the Rise of American Environmentalism* (Cambridge, UK: Cambridge University Press, 2005); See also Paul Robbins, *Lawn People: How Grasses, Weeds, and Chemicals Make Us Who We Are* (Philadelphia: Temple University Press, 2012); Ted Steinberg, *American Green: The Obsessive Quest for the Perfect Lawn* (New York: Norton, 2006).

43. Transcript of Direct Examination of James Ray Boggess in *James R. Boggess, et. al., v. Monsanto Company*, 6189.

44. Transcript of Direct Examination of James Ray Boggess in *James R. Boggess, et. al., v. Monsanto Company*, 6190–91.

45. "Thirteen Centuries Experience Leaves Plant," *Nitrometer* 4, no. 7 (Nitro, WV, Monsanto Company newspaper), January 1984, 1, TCP files.

46. Transcript of Direct Examination of James Ray Boggess in *James R. Boggess, et. al., v. Monsanto Company*, 6387; *Jeffers v. Monsanto*, Civil Action No. 81–2239, Docket No. 225, Plaintiff's Response to Defendant's Motion for Summary Judgment—Statute of Limitations, 55, 71,TCP files.

47. Plaintiff's Response to Monsanto's Motion for Directed Verdict and Brief in Support Thereof, filed by Stuart Calwell in the case of *James R. Boggess, et. al., v. Monsanto Company*, November 9, 1984, 62–73, TCP files. A 1955 Monsanto memorandum cited in this brief said that company officials could share Suskind's finding with select West Virginia public health officials, but it stated that these reports should not "become part of the public records." See Plaintiffs Response to Monsanto's Motion for Directed Verdict, 70; Confidential memorandum from Robert E. Soden, Monsanto chemist, to J. R. Durland, Nitro plant manager, December 7, 1955, Document 2006–10–23 (45), TCP files; "How Industry Handled the Dioxin Question," *St. Louis Post-Dispatch*, November 14, 1983, 32. In 1956, Dr. Suskind said that "the concentration of 2,4,5-T dust in the atmosphere" in a certain area of the Nitro facility "was almost twice that found at the same location in 1953." Focusing on another building associated with 2,4,5-T manufacture, he noted that operations "were found to be exactly as they were in April, 1953." See "Report on Environmental Survey Carried Out in Building 34 of Monsanto Chemical Company at Nitro, West Virginia," unpublished report, February 21, 1956, 1, 4, TCP files.

48. "How Industry Handled the Dioxin Question," *St. Louis Post-Dispatch*, November 14, 1983, 32, 34; *Jeffers v. Monsanto*, Civil Action No. 81–2239, Docket No. 225, Plaintiff's Response to Defendant's Motion for Summary Judgment—Statute of Limitations, 60, TCP files; Plaintiff's Response to Directed Verdict, n.d., *James R Boggess v. Monsanto*, 100, TCP Files; Thomas Whiteside, *The Pendulum and the Toxic Cloud: The Course of Dioxin Contamination* (New Haven: Yale University Press, 1979), 149.

49. R. R. Suskind and V. S. Hertzberg, "Human Health Effects of 2,4,5-T and Its Toxic Contaminants," *Journal of the American Medical Association* 251, no. 8 (May 11, 1984), 2372; Robin, *The World According to Monsanto*, 37; Thomas Whiteside, *The Pendulum and the Toxic Cloud*, 150; "How Industry Handled the Dioxin Question," *St. Louis Post-Dispatch*, November 14, 1983, 32.

50. "Many Workers at Nitro Plant are Sick, But Who's To Blame?," *St. Louis Post-Dispatch*, November 14, 1983, 33; "How Industry Handled the Dioxin Question," *St. Louis Post-Dispatch*, November 14, 1983, 34; Van Strum, *A Bitter Fog*, 13; Plaintiff's Response to Defendant's Motion for Summary Judgment—Statute of Limitations in *Jeffers v. Monsanto*, Civil Action No. 81–2239, Docket No. 225, 6, 67, TCP files.

51. Addendum to 2,4,5-T Agreement, January 2, 1957, A/O Material from Nitro General Files, Item #61; Local Union No. 12610, District 50, U.M.W.A.,

and Monsanto Chemical Company, Nitro, West Virginia, Agreement for the Operation of 2,4,5-T, September 29, 1964, Confidential Contract and Grievance History of Nitro Plant, prepared by Dwight E. Harding, TCP files.

52. Addendum to 2,4,5-T Agreement, January 2, 1957; Local Union No. 12610, District 50, UMWA, and Monsanto Chemical Company, Nitro, West Virginia, Agreement for the Operation of 2,4,5-T, October 16, 1956, TCP Files.

CHAPTER 5: "So You See, I Am Prepared to Argue on Either Side"

1. Peter H. Schuck, *Agent Orange on Trial: Mass Toxic Disasters in the Courts* (Cambridge, MA: Belknap Press of Harvard University Press, 1987), 16; Jeanne Mager Stellman et al., "The Extent and Patterns of Usage of Agent Orange and Other Herbicides in Vietnam," *Nature* 422 (April 17, 2003): 681. This study remains one of the most comprehensive summary assessments of weaponized herbicide use during the Vietnam War. For details of these early spraying campaigns, see Thomas Whiteside, *Defoliation: What Are Our Herbicides Doing to Us?* (New York: Ballantine Books, 1970), 7–8. Excellent studies that focus specifically on the environmental effects of Operation Hades in Vietnam include John Lewallen, *Ecology of Devastation: Indochina* (Baltimore: Penguin, 1971). Lewallen traveled to Vietnam during the war and was able to see firsthand the effects of Agent Orange aerial campaigns. See also Arthur H. Westing, *Herbicides in War: The Long-term Ecological and Human Consequences* (Stockholm: Stockholm International Peace Research Institute, 1984). More recent works include David Zierler, *The Invention of Ecocide: Agent Orange, Vietnam, and the Scientists Who Changed the Way We Think About the Environment* (Athens: University of Georgia Press, 2011); Alvin L. Young, *The History, Use, Disposition, and Environmental Fate of Agent Orange* (New York: Springer, 2009); David Andrew Biggs, *Footprints of War: Militarized Landscapes in Vietnam* (Seattle: University of Washington Press, 2018), 138–40, 170–76, 189–94. For more on Operation Ranch Hand from the perspective of a US pilot who executed herbicidal campaigns, see Frederick Cecil, *Herbicidal Warfare: The Ranch Hand Project in Vietnam* (New York: Praeger, 1986).

2. For a summary of the sequence of events that led Kennedy toward herbicidal warfare, see David Zierler, *Invention of Ecocide*, 48–66, 81–82; Edwin A. Martin, *Agent Orange: History, Science, and the Politics of Uncertainty* (Amherst: University of Massachusetts Press, 2012), 20–26, 42; US Institute of Medicine (US) Committee to Review the Health Effects in Vietnam Veterans of Exposure to Herbicides, *Veterans and Agent Orange: Health Effects of Herbicides Used in Vietnam* (Washington, DC: National Academies Press, 1994), 89.

3. Zierler, *Invention of Ecocide*, 60–61, 78–79.

4. Zierler, *Invention of Ecocide*, 60–61, 79–88.

5. Zierler, *Invention of Ecocide*, 60–61; For more on the natural environment

as military ally, see Richard Tucker and Ed Russell, *Natural Enemy, Natural Ally: Toward an Environmental History of Warfare* (Corvallis: Oregon State University Press, 2004).

6. Incidentally, Dow was the second largest manufacturer of Agent Orange by volume, with 28.6 percent of the market. Peter H. Schuck, *Agent Orange on Trial*, 87, 156.

7. "Monsanto Dissects Pesticide Criticism," *New York Times*, September 22, 1962, 28; "'Silent Spring' Is Now Noisy Summer," *New York Times*, July 22, 1962, 87; "Rachel Carson Dies of Cancer; 'Silent Spring' Author Was 56," *New York Times*, April 15, 1964, 1; Michelle Mart, *Pesticides, a Love Story: America's Enduring Embrace of Dangerous Chemicals* (Lawrence: University Press of Kansas, 2015), 60.

8. "Reviving Rachel," *Globe and Mail* (Toronto, Ont.), December 6, 1997, D19; "Rachel Carson," *New York Times*, April 16, 1964, 36; "Rachel Carson Dies of Cancer," *New York Times*; Justus C. Ward, "The Functions of the Federal Insecticide, Fungicide, and Rodenticide Act," *American Journal of Public Health Nations Health 55*, no. 7 (July 1965), 27–31; *Use of Pesticides*, A report of the President's Science Advisory Committee (Washington, DC: White House, May 15, 1963).

9. Ward, "The Functions of the Federal Insecticide, Fungicide, and Rodenticide Act"; Richard N. L. Andrews, *Managing the Environment, Managing Ourselves: A History of American Environmental Policy*, 3rd ed. (New Haven: Yale University Press, 2020), 216, 243; Frederick Rowe Davis, *Banned: A History of Pesticides and the Science of Technology* (New Haven; London: Yale University Press, 2014), 120; Michelle Mart, *Pesticides, a Love Story*, 36–38.

10. Peter H. Schuck, *Agent Orange on Trial*, 16; Thomas Whiteside, *Defoliation*, 30; Carol Van Strum, *A Bitter Fog: Herbicides and Human Rights* (San Francisco: Sierra Club Books, 1983), 12; Stellman et al., "The Extent and Patterns of Usage of Agent Orange and Other Herbicides in Vietnam," 682, 684.

11. Stellman et al., "The Extent and Patterns of Usage of Agent Orange and Other Herbicides in Vietnam," 682, 685; This 12 million figure may include Agent Orange II, a formulation used later in the war that had a slightly different concentration of active ingredients, but still a 50/50 mixture split between 2,4-D and 2,4,5-T. Charles Bailey and Le Ke Son, *From Enemies to Partners: Vietnam, the U.S. and Agent Orange* (Chicago: G. Anton Publishing, 2018), 156, 411 [Kindle edition].

12. Nguyễn Thị Hồng quoted in Brenda M. Boyle and Jeehyun Lim, eds., *Looking Back on the Vietnam War: Twenty-First Century Perspectives* (New Brunswick, NJ: Rutgers University Press, 2016), 3737, 3750–51, 3760 [Kindle edition].

13. Lewallen, *Ecology of Devastation*, 60, 71.

14. Lewallen, *Ecology of Devastation*, 69–70; Westing, *Herbicides in War*, 18–22.

15. Westing, *Herbicides in War*, 18–22.

16. Lewallen, *Ecology of Devastation*, 59.

17. "What the Government Knew, and What It Did," *St. Louis Post-Dispatch*,

November 14, 1983, 37; Schuck, *Agent Orange on Trial*, 19, 78; David R. Plummer, "Vietnam Veterans and Agent Orange" (master's thesis, California State University Dominguez Hills, 2000), 66; Van Strum, *A Bitter Fog*, 13, 68–69; Whiteside, *Defoliation*, 61; Lewallen, *Ecology of Devastation*, 68, 114–15; Thomas Whiteside editorial, *New Yorker*, August 6, 1971, 54; Zierler, *Invention of Ecocide*, 123; Michelle Mart, *Pesticides, a Love Story*, 98.

18. Plaintiff's Response to Defendant's Motion for Summary Judgment— Statute of Limitations in *Jeffers v. Monsanto*, Civil Action No. 81–2239, Docket No. 225, 73–74, The Calwell Practice law office files, Charleston, West Virginia; "Monsanto Documents on Dioxin Released," *St. Louis Post-Dispatch*, June 8, 1984, 1, 4; "How Industry Handled the Dioxin Question," *St. Louis Post-Dispatch*, November 14, 1983, 80; "1965 Memos Show Dow's Anxiety on Dioxin," *New York Times*, April 19, 1983, A1.

19. Whiteside, *Defoliation*, 2, 11; Schuck, *Agent Orange on Trial*, 19; Zierler, *Invention of Ecocide*, 101.

20. Schuck, *Agent Orange on Trial*, 17; Van Strum, *A Bitter Fog*, 108.

21. Schuck, *Agent Orange on Trial*, 19, 22; Thomas Whiteside, *The Withering Rain: America's Herbicidal Folly* (New York: E. P. Dutton, 1971), 11–12, 41; Whiteside, *Defoliation*, 17, 21, 23; Lewallen, *Ecology of Devastation*, 115; Thomas Whiteside, *The Pendulum and the Toxic Cloud: The Course of Dioxin Contamination* (New Haven: Yale University Press, 1979), 30; "Thomas Whiteside, 79, Dies; Writer Exposed Agent Orange," *New York Times*, October 12, 1977, 44; Thomas Whiteside editorial, *The New Yorker*, August 6, 1971, 54; Institute of Medicine (US) Committee to Review the Health Effects in Vietnam Veterans of Exposure to Herbicides, *Veterans and Agent Orange*, 22; Zierler, *Invention of Ecocide*, 124; Most uses for 2,4,5-T were restricted by 1979, though the government continued to allow spraying on rangelands and fields, Strum, *A Bitter Fog*, 167; David Biggs, "Following Dioxin's Drift: Agent Orange Stories and the Challenge of Metabolic History," *International Review of Environmental History* 4, no. 1 (2018), 8.

22. "Monsanto Organic Chemicals Division, Summer Orientation Program," May 5, 1966, series 1, box 5, folder: Organic Chemical Divisions (History) (Folder 1); Monsanto Company 1966 Annual Report, frontmatter. The year 1962 was the first year that Monsanto reached $1 billion in annual sales. See "Monsanto At a Glance," *Monsanto Magazine* 3 (1995), 32, series 10, box 6, folder: Monsanto Company History (Timeline/Milestones); Leonard A. Paris, "Monsanto: The First 75 Years," corporate public relations publication, 1976, series 10, box 5, folder: Monsanto Company History (Historical Accounts [Misc.]); "Prelude to a Start-Up: A Century of Changes," *Monsanto Magazine* 2 (1997), series 10, box 6, folder: Monsanto Company History (Timeline/Milestones), Monsanto Company Records, Washington University in St. Louis, Julian Edison Department of Special Collections, St. Louis, Missouri [hereinafter MCR]; Monsanto Company 1964 Annual Report, 5; Monsanto Company 1963 Annual Report, 14.

23. Letter from Edgar Queeny to Crawford H. Greenewalt, November 4, 1960, series 14, box 16, folder: Queeny, E. M. (1959–60); Letter from Edgar

Queeny to Skeets and Tom, October 1, 1962, series 14, box 16, folder: Queeny, E. M. (1961–63); Letter from Edgar Queeny to A. K. Chapman, June 26, 1963; Letter from Edgar Queeny to August Belmont, August 13, 1963, series 14, box 16, folder Queeny E. M. (1963–64), MCR.

24. Letter from Edgar M. Queeny to Rev. Bishop and Mrs. William Scarlett, Castine, Maine, February 8, 1963, series 14, box 16, folder: Queeny E. (1963–64), MCR.

25. Letter from Edgar Queeny to Jack Clink, March 7, 1961, series 14, box 16, folder: Queeny, E. M. (1961–63), MCR.

26. Edgar Queeny *New York Times* obituary, July 7, 1968, box 257, folder: Queeny, Edgar M., 1949–68, Robert Winship Woodruff Papers, Stuart A. Rose Manuscript, Archives, and Rare Book Library, Emory University.

27. "Top Officers Are Shifted by Monsanto Chemical," *New York Times*, March 24, 1960, 47; "Who's News: Sommer Elected Monsanto Executive Vice President," June 16, 1959, 20; Paris, "Monsanto: The First 75 Years"; Interview with Bob Rumer, February 11, 2007, series 10, box 5, folder: Monsanto Company History (Historical Accounts [Phillion, Lee]), MCR.

28. "Prelude to a Start-Up: A Century of Changes," *Monsanto Magazine*; Dan Forrestal, *Faith, Hope, and $5,000: The Story of Monsanto* (New York: Simon & Schuster, 1977), 201.

CHAPTER 6: "Sell the Hell out of Them as Long as We Can"

1. Report of a New Chemical Hazard, December 15, 1966, Poison Papers archive, managed by the Bioscience Research Project and the Center for Media and Democracy and based on papers collected by environmentalist and journalist Carol Van Strum, https://www.poisonpapers.org/the-poison-papers/ [hereinafter PPA]; Letter from Henry Strand, Rising and Strand, Sweden, to David Wood, Monsanto Europe, Brussels, Belgium, November 28, 1966, PPA; Gerald Markowitz and David Rosner, "Monsanto, PCBs, and the Creation of a 'World-Wide Ecological Problem,'" *Journal of Public Health Policy* 39 (2018): 482. Markowitz and Rosner adapted this article from a "Report to the Court," which they wrote in 2010 for law firms fighting on behalf of clients claiming injury due to PCB exposure. Markowitz and Rosner draw on many of the same sources featured in this chapter, as does Ellen Griffith Spears in *Baptized in PCBs: Race, Pollution, and Justice in an All-American Town* (Chapel Hill: University of North Carolina Press, 2014). Anyone interested in the history of PCBs should consult their works. For a broader look at the history of chemical companies' efforts to hide damaging information about toxic products, see Gerald Markowitz and David Rosner, *Deceit and Denial: The Deadly Politics of Industrial Pollution* (Berkeley: University of California Press, 2001).

2. Recommended Procedures for the Disposal of PCB-Containing Wastes (Industrial Facilities), February 27, 1976; 1972(?) US Government Printing Office (GPO) Report on PCBs, PPA; Spears, *Baptized in PCBs*, 161;

Markowitz and Rosner, "Monsanto, PCBs, and the Creation of a 'World-Wide Ecological Problem,'" 473. There were approximately fifteen global manufacturers of PCBs, but none produced these compounds on the scale that Monsanto did. For example, Bayer, the second largest producer of PCBs in the twentieth century, posted sales that amounted to less than 23 percent of Monsanto's total PCB output between 1930 and 1977. Most companies sold less than 56,000 metric tons of the chemical, compared to Monsanto's 707,788 metric tons. See International Agency for Research on Cancer, *Polychlorinated Biphenyls and Polybrominated Biphenyls* 107, IARC Monographs on the Evaluation of Carcinogenic Risks to Humans (2016), 72.

3. Memorandum from R. Emmet Kelly to D. Wood, February 10, 1967, Chemical Industry Archives, a project of the Environmental Working Group, https://web.archive.org/web/20170117202824/http://chemicalindustryarchives.org/search/ [hereinafter CIAEWG]. When this book went to press in 2021, this archive was no longer operational online.

4. Letter from R. Emmet Kelly to J. W. Barrett, September 20, 1955, CIAEWG; Spears, *Baptized in PCBs*, 89; For more on what Monsanto knew before 1966, see Markowitz and Rosner, "Monsanto, PCBs, and the Creation of a 'World-Wide Ecological Catastrophe,'" 476–82; Heinz Martin, *Polymers, Patents, Profits: A Classic Case Study for Patent Infighting* (Weinheim: Wiley-VCH, 2007), 56; Letter from Elmer P. Wheeler, Assistant Director, Medical Department, Monsanto, to Illegible, Administrator, Industrial Hygiene, Westinghouse Electric Corporation, October 23, 1959, PPA. Describing the toxicity of PCBs, scientists Sally White and Linda Birnbaum at the National Institute of Environmental Health Sciences explained, "There are multiple, overlapping structural classes of PCBs, but PCBs are inherently found as mixtures, and never exist in the absence of dioxin-like PCBs in the ambient environment. Similarly, TCDD [dioxin] and PCBs are seldom found in the absence of one another. Irrespective of this, the majority of individual PCBs possess their own intrinsic toxicities, and furthermore can interact with dioxins and other PCB congeners additively, synergistically, and/or antagonistically, imparting high variability to the activity of the observed mixtures." See Sally S. White and Linda S. Birnbaum, "An Overview of the Effects of Dioxins and Dioxin-like Compounds on Vertebrates, as Documented in Human and Ecological Epidemiology," *Journal of Environmental Science and Health, Part C: Environmental Carcinogenesis and Ecotoxicology Reviews* 27, no. 4 (October 2009): 197–211.

5. Spears, *Baptized in PCBs*, 130.

6. Memorandum from E. P. Wheeler, St. Louis, October 21, 1968; Memorandum from Elmer Wheeler, Monsanto Manager of Environmental Health, March 3, 1969; Statement of Robert Risebrough before the Committee on Commerce, US Senate, August 4, 1971, PPA; "State Says Some Striped Bass and Salmon Pose a Toxic Peril," *New York Times*, August 8, 1975; Spears, *Baptized in PCBs*, 137–39; Markowitz and Rosner, "Monsanto, PCBs, and the Creation of a 'World-Wide Ecological Problem,'" 488.

7. Report of Aroclor 'Ad Hoc' Committee (Second Draft), October 15, 1969; Memorandum from W. R. Richard, Monsanto Research Center, to Elmer

P. Wheeler, Re: Defense of Aroclor F. Fluids, September 9, 1969, PPA; Markowitz and Rosner, "Monsanto, PCBs, and the Creation of a 'World-Wide Ecological Problem,'" 496–97.

8. On environmental activism in the 1960s and the roots of the Earth Day movement, see Adam Rome, *The Genius of Earth Day: How a 1970 Teach-In Unexpectedly Made the First Green Generation* (New York: Hill & Wang, 2013); Robert Gottlieb, *Forcing the Spring: The Transformation of the American Environmental Movement* (Washington, DC: Island Press, 2005); For a detailed history of the evolution of US environmental policy, see Richard N. L. Andrews, *Managing the Environment, Managing Ourselves: A History of American Environmental Policy*, 3rd ed. (New Haven: Yale University Press, 2020); Sally F. Fairfax and Edmund Russell, eds., *Guide to U.S. Environmental Policy* (Los Angeles: SAGE, 2014).

9. Memorandum from W. R. Richard to E. P. Wheeler, May 26, 1969, CIAEWG; Spears, *Baptized in PCBs*, 160; Memorandum from W. R. Richard, Monsanto Research Center, to Elmer P. Wheeler, Re: Defense of Aroclor F. Fluids, September 9, 1969; Letter from Howard S. Bergen, Director, Functional Fluids, to Fred H. Dierker, Executive Officer, State of California-Resources Agency, San Francisco Bay Region, March 27, 1969, PPA.

10. Spears, *Baptized in PCBs*, 155; Report and Comments on Meeting on Chlorinated Biphenyls in the Environment at Industrial Biotest Laboratories, Chicago, March 21, 1969, by Robert Metcalf, U. of Illinois, CIAEWG; Memorandum from W. R. Richard, Monsanto Research Center, to Elmer P. Wheeler, Re: Defense of Aroclor F. Fluids, September 9, 1969; Report of Aroclor 'Ad Hoc' Committee (Second Draft), October 15, 1969, PPA. IBT later came under fire in the 1970s and early 1980s when the EPA discovered that that laboratory had falsified findings in toxicological reports on various chemicals. Michelle Mart, *Pesticides, A Love Story: America's Enduring Embrace of Dangerous Chemicals* (Lawrence: University Press of Kansas, 2015), 155.

11. Report of Aroclor 'Ad Hoc' Committee (Second Draft), October 15, 1969, PPA; Spears *Baptized in PCBs*, 142–43; Markowitz and Rosner, "Monsanto, PCBs, and the Creation of a 'World-Wide Ecological Problem,'" 465.

12. Report of Aroclor 'Ad Hoc' Committee (Second Draft), October 15, 1969, PPA; Markowitz and Rosner, "Monsanto, PCBs, and the Creation of a 'World-Wide Ecological Problem,'" 492.

13. Quoted in Attorney General of Washington Press Release, "AG Ferguson Makes Washington First State to Sue Monsanto Over PCB Damages, Cleanup Cost," December 8, 2016, http://www.atg.wa.gov/news/news -releases/ag-ferguson-makes-washington-first-state-sue-monsanto-over -pcb-damages-cleanup; See also Confidential Report of Aroclor 'Ad Hoc' Committee, October 2, 1969; Undated PCB Presentation to Corporate Development Committee, listed as January 1, 1970, in the Chemical Industry Archives, CIAEWG; "Monsanto Releases PCB Production Figures to Department of Commerce," Monsanto Industrial Chemicals Company News Release, November 30, 1971, PPA; Spears, *Baptized in PCBs*, 129; Markowitz and Rosner, "Monsanto, PCBs, and the Creation of a 'World-Wide Ecological Problem,'" 503–4.

14. Memorandum from W. R. Richard, Monsanto Research Center to Elmer Wheeler, Re: Defense of Aroclor F. Fluids, September 9, 1969; The PCB Pollution Problem, St. Louis Meeting with General Electric, January 21 and 22, 1970, PPA; Letter from Elmer P. Wheeler, Medical Department, Monsanto, Re: Status of Aroclor Toxicological Studies, January 29, 1970, CIAEWG; Spears, *Baptized in PCBs*, 156; Markowitz and Rosner, "Monsanto, PCBs, and the Creation of a 'World-Wide Ecological Problem,'" 494.

15. Outline: PCB Environmental Pollution Abatement Plan, November 10, 1969, PPA.

16. The EPA assessed whether less chlorinated PCBs were in fact less toxic to human health and concluded in 1979 that "all PCB mixtures then in use, including the less chlorinated ones (e.g., Aroclor 1016), are capable of inducing severe toxic effects at low levels in mammals and aquatic organisms." See Environmental Protection Agency, Support Document/Voluntary Environmental Impact Statement for Polychlorinated Biphenyls (PCBs) Manufacturing, Processing, Distribution in Commerce, and Use Ban Regulation (Section 6(e) of TSCA), Prepared by Office of Toxic Substances, April 1979, PPA; Memorandum from N. T. Johnson, Monsanto, St. Louis, to Various, Re: Pollution Letter, February 16, 1970, CIAEWG; Spears, *Baptized in PCBs*, 143; Markowitz and Rosner, "Monsanto, PCBs, and the Creation of a 'World-Wide Ecological Problem,'" 500.

17. The PCB Pollution Problem, St. Louis Meeting with General Electric, January 21 and 22, 1970, PPA; Outline, PCB Environmental Pollution Abatement Plan, Rough Draft, November 10, 1969, CIAEWG.

18. The PCB Pollution Problem, St. Louis Meeting with General Electric, January 21 and 22, 1970.

19. Monsanto Press Release, Monsanto Replies to Charge That PCB Threatens Environment, April 10, 1970, PPA.

20. 1972(?) US Government Printing Office (GPO) Report on PCBs, PPA.

21. "Monsanto's PCB Problem," W. B. Papageorge, Presented at ANSI Committee C-107 Meeting, September 14, 1971; Memorandum from John Mason, Monsanto, General Office—St. Louis, Subject: PCB's, Report on Meeting with Congressman Wm. F. Ryan (Dem.) 20th District, New York, Washington DC, July 8, 1970, July 13, 1970, PPA.

22. Letter from R. Emmet Kelly to W. B. Papageorge, March 30, 1970, PPA; Markowitz and Rosner, "Monsanto, PCBs, and the Creation of a 'World-Wide Ecological Problem,'" 502.

23. Spears, *Baptized in PCBs*, 138; Letter forwarding Testimony of William Papageorge Re: Proposed Toxic Pollutant Effluent Standards for Aldrin-Dieldrin, et al—FWPCA (307)—Docket No. 1, addressed to Betty J. Billings, Hearing Clerk, Environmental Protection Agency, March 12, 1974, PPA.

24. Memorandum from W. B. Papageorge to E. P. Wheeler, et al., Re: PCB Environmental Problem August Status Report, September 8, 1970, CIAEWG; Letter from W. E. Shalk, Director of Sales, Plasticizers, to Customers, June 1, 1970, PPA; Memorandum from W. B. Papageorge, Re: PCB Environmental Problem September Status Report, October 6, 1970, CIAEWG; Spears, *Baptized in PCBs*, 152.

25. Outline: PCB Environmental Pollution Abatement Plan, November 10, 1969 (Rough Draft), PPA; Spears, *Baptized in PCBs*, 129.

26. Spears, *Baptized in PCBs*, 151; Memorandum from W. B. Papageorge, Re: PCB Environmental Problem September Status Report, October 6, 1970, CIAEWG.

27. Undated PCB Presentation to Corporate Development Committee, listed as January 1, 1970, in the Chemical Industry Archives, CIAEWG; Report from Working Group on Disposal of Toxic and Hazardous Waste, June 1973, PPA.

28. "Monsanto's PCB Problem," W. B. Papageorge, Presented at ANSI Committee C-107 Meeting, September 14, 1971, PPA; Council on Environmental Quality, *Toxic Substances* (Washington, DC: GPO, 1971), 13–14.

29. Special Undertaking by Purchasers of Polychlorinated Biphenyls, Signed by Monsanto and GE, January 21, 1972, CIAEWG; Markowitz and Rosner, "Monsanto, PCBs, and the Creation of a 'World-Wide Ecological Problem,'" 518–19.

30. Letter forwarding Testimony of William Papageorge Re: Proposed Toxic Pollutant Effluent Standards for Aldrin-Dieldrin, et al—FWPCA (307)—Docket No. 1, addressed to Betty J. Billings, Hearing Clerk, Environmental Protection Agency, March 12, 1974, PPA.

31. Minutes of Meeting on Proposed PCB Effluent Standards, Monsanto Company, St. Louis, MO, February 28, 1974, PPA.

32. Letter from Dr. E. L. Simons, Manager, Environmental Protection Operation, General Electric, to Dr. C. High Thompson, Chairman, Hazardous and Toxic Substance, Regulation Task Force, Office of Water Program Operations, EPA, November 21, 1973, PPA.

33. Memorandum from H. S. Bergen, B2SL, Monsanto, to W. B. Papageorge, et al., Re: Polychlorinated Biphenyl Effluent Standards, March 8, 1974, PPA; Letter forwarding Testimony of William Papageorge Re: Proposed Toxic Pollutant Effluent Standards for Aldrin-Dieldrin, et al—FWPCA (307)—Docket No. 1, addressed to Betty J. Billings, Hearing Clerk, Environmental Protection Agency, March 12, 1974, PPA.

34. Letter from W. B. Papageorge to Dan A. Albert, Westinghouse Electric Corporation, March 18, 1975, PPA; Markowitz and Rosner, "Monsanto, PCBs, and the Creation of a 'World-Wide Ecological Problem,'" 521.

35. Letter from W. B. Papageorge to Dan A. Albert, Westinghouse Electric Corporation, March 18, 1975, PPA.

36. Letter from W. B. Papageorge to Dan A. Albert, Westinghouse Electric Corporation, March 18, 1975, PPA.

37. Letter from Robert A. Emmett, Chief, Legal Branch, Water Enforcement Division, EPA, to William B. Papageorge, June 5, 1975, PPA.

38. Letter from Robert A. Emmett, Chief, Legal Branch, Water Enforcement Division, EPA, to William B. Papageorge, June 5, 1975, PPA.

39. Memo from Floyd A. Bean, Monsanto Marketing Manager, to All Marketing, Re: Inerteen Information, August 29, 1975, PPA.

40. Spears, *Baptized in PCBs*, 165; EPA Administrator Russell Train's Press Conference Regarding the Toxic Substances Control Act, December 1976,

https://www.youtube.com/watch?v=qA5sx6nsQw8; ANSI Committee C107 Meeting Minutes on Use and Disposal of Askarel and Askarel-Soaked Materials, January 12 and January 13, 1976, PPA.

41. Environmental Protection Agency, Support Document/Voluntary Environmental Impact Statement for Polychlorinated Biphenyls (PCBs) Manufacturing, Processing, Distribution in Commerce, and Use Ban Regulation (Section 6(c) of TSCA), Prepared by Office of Toxic Substances, April 1979, PPA.

42. PCB Preparedness Q & A, From D. Bishop and D. Wood, St. Louis, to J. Carr, Brussels, September 29, 1976, CIAEWG; "Through Six Reigns: The Story of Our Company from 1867 to 1953," *The Autoclave* 5, no. 3 (June 1953), 12.

43. Memorandum from Pierre R. Wilkins, New York, Monsanto, to Mr. Earle H. Harbison, Re: Report by PCB Study Group, December 10, 1975, CIAEWG.

44. In 1975, Russell Train wrote to Monsanto to say that "the economic impacts" of a PCB ban "would have to be considered." Letter from Russell Train, EPA, to Mr. Hanley, Monsanto, December 22, 1975, CIAEWG; Environmental Protection Agency, Support Document/Voluntary Environmental Impact Statement for Polychlorinated Biphenyls (PCBs) Manufacturing, Processing, Distribution in Commerce, and Use Ban Regulation (Section 6(e) of TSCA), Prepared by Office of Toxic Substances, April 1979, PPA. For a list of the companies that petitioned the EPA, see pages 107–16 of this document.

45. Environmental Protection Agency, Support Document/Voluntary Environmental Impact Statement for Polychlorinated Biphenyls (PCBs) Manufacturing, Processing, Distribution in Commerce, and Use Ban Regulation (Section 6(e) of TSCA), Prepared by Office of Toxic Substances, April 1979, PPA.

46. *Environmental Defense Fund v. Environmental Protection Agency*, 636 F. 2d 1267 (D. C. Cir., 1980).

47. Letter from Thomas E. Kotoske, lawyer representing Indiana University, to Roger Strelow, Vice President, Corporate Environmental Program, General Electric Company, December 12, 1988, PPA.

48. Dana Loomis et al., "Cancer Mortality Among Electric Utility Workers Exposed to Polychlorinated Biphenyls," *Occupational and Environmental Medicine* 54 (1997): 720.

49. Thomas Sinks et al., "Mortality Among Workers Exposed to Polychlorinated Biphenyls," *American Journal of Epidemiology* 136, no. 4 (1992): 389–98, PPA; The EPA's PCB Transformer Database can be found here: https://www.epa.gov/pcbs/registering-transformers-containing-polychlorinated-biphenyls-pcbs.

50. Recommended Procedures for the Disposal of PCB-Containing Wastes (Industrial Facilities), February 27, 1976, PPA; "State Says Some Striped Bass and Salmon Pose a Toxic Peril," *New York Times*, August 8, 1975, 1; "GE Nears End of Hudson River Cleanup," *Wall Street Journal*, November 11, 2015, https://www.wsj.com/articles/ge-nears-end-of-hudson-river-cleanup

-1447290049#:~:text=As%20the%20dredging%20phase%20comes%20
to%20a%20close,%20GE%20faces.

CHAPTER 7: "Strategic Exit"

1. This account is based on the author's field reporting in June 2016; on Budweiser barley in this area, see "Teetotalling Mormons Grow Barley for
   Beer," *Baltimore Sun*, September 8, 2010, https://www.baltimoresun.com/
   bs-mtblog-2010–09–teetotalling_mormons_grow_barl-story.html (All
   URLs accessed November 26, 2018). The author would like to thank *Agricultural History* for allowing me to reuse material here and in Chapter 9
   from a journal article it published in 2019. "Roundup from the Ground Up:
   A Supply-Side Story of the World's Most Widely Used Herbicide," *Agricultural History* 91, no. 1 (Winter 2019): 102–38.
2. "Idaho Rejects Request to Limit Mercury," *Spokesman-Review*, April
   26, 2008, https://www.spokesman.com/stories/2008/apr/26/idaho-rejects
   -request-to-limit-mercury/; "Monsanto Released Dangerous Chemicals at
   Idaho Plant," *St. Louis Post-Dispatch*, March 26, 2015, https://www.stltoday
   .com/business/local/monsanto-released-dangerous-chemicals-at-idaho
   -plant/article_d08e6444-fc62-5347-a30d-295d3f62a8bb.html; "Monsanto
   Wants Tighter Mercury Rules in Idaho," *Times-News* (Twin Falls, ID), July
   24, 2009. I confirmed the existence of Monsanto's mercury problem with an
   anonymous source. Air Pollutant Report for Monsanto P4 Production Plant,
   EPA Enforcement and Compliance History Online, https://echo.epa.gov/air
   -pollutant-report?fid=110000743982; Hubert Kay, "Monsanto and the American Idea," draft copy, June 1958, "Part II, Chapter 5, How Monsanto Grew,"
   E-4, E-39, folder: Monsanto and the America Idea (Hubert Kay) 1st Draft
   (Forrestal's Copy), box 1, series 6, Monsanto Company Records, Washington
   University in St. Louis, Julian Edison Department of Special Collections, St.
   Louis, Missouri [hereinafter MCR]; Mitch Hart (former Monsanto Company
   mining engineer), interview by the author, Soda Springs, Idaho, July 14, 2016.
3. Monsanto Chemical Company 1952 Annual Report, 17, 54.
4. Interview with John E. Franz conducted by James J. Bohning, St. Louis,
   Missouri, November 29, 1994, Chemical Heritage Foundation Oral History
   Program, 1–3, 10.
5. Interview with John E. Franz, 2, 5.
6. Interview with John E. Franz, 2.
7. Interview with John E. Franz, 11.
8. Interview with John E. Franz, 11, 14–16.
9. Interview with John E. Franz, 24.
10. In 1964, Monsanto noted that detergents represented "the largest market for
    Monsanto phosphates." Monsanto Company 1964 Annual Report, 3; Mon-
    santo Chemical Company 1952 Annual Report, 17; Hubert Kay, "Monsanto
    and the American Idea," draft copy, June 1958, "Part II, Chapter 5, How
    Monsanto Grew," E-46, series 6, box 1, folder: Monsanto and the America
    Idea (Hubert Kay) 1st Draft (Forrestal's Copy), MCR; Dan Forrestal, *Faith*,

Hope, & $5,000: The Story of Monsanto (New York: Simon & Schuster, 1977), 106–7, 136; "Monsanto Plans Expansion of its Phosphorous Facilities," *Wall Street Journal*, August 2, 1954, 5; "Phosphate Shuffling Nearing End Game," *Chemical and Engineering News* 77, no. 12 (March 22, 1999), 17–20; Monsanto Chemical Company News Release, May 24, 1957, series 04, box 1, folder: all (Laundry detergent); Lever House News Release, "Coin-Op Vending Box Containing Active 'all' Detergent Selected for 5,000-Year Time Capsule at World's Fair," n.d., series 04, box 1, folder: all (Laundry detergent), MCR.

11. The answer was NTA: a new chemical builder that could replace the phosphate-based STPP. Lever Brothers, which purchased Monsanto's "all" brand in 1957, made the switch to the new chemical supplied by Monsanto in 1970, but by December of that year the Office of the Surgeon General raised new concerns about NTA as a potential carcinogen. The agency urged caution in adopting NTA, and for years a debate raged about the suitability of the chemical for large-scale use in the detergent industry. By the 1980s, however, Monsanto finally attained federal sanction for NTA, officially ending its dependence on STPP. "Lever Brothers to Use More NTA in Products and Less Phosphates," *Wall Street Journal*, May 28, 1970, 10; House Committee on Interstate and Foreign Commerce, Subcommittee on Oversight and Investigations, *EPA's Action Concerning Nitrilotriacetic Acid (NTA)*, 96th Cong., 2nd Sess., June 26, 1980; For more on the history of phosphate detergent and eutrophication, see Chris Knud-Hansen, "Historical Perspective of the Phosphate Detergent Conflict," Conflict Research Consortium Working Paper 94–54, February 1994, 3; National Academy of Sciences, "Eutrophication: Causes, Consequences, and Correctives," report from proceedings of symposium held at the University of Wisconsin, Madison, June 11–15, 1967 (Washington, DC: NAS, 1969), 6; "Lake Erie Choked by Eutrophication," *Hartford Courant*, February 12, 1968, 8A; "Phosphates Help Little Firms Slip Past Soap Giants," *Los Angeles Times*, November 15, 1970, 11; "Detergents Held Pollution Factor," *New York Times*, December 15, 1969, 1; Terence Kehoe, "Merchants of Pollution?: The Soap and Detergent Industry and the Fight to Restore Great Lakes Water Quality, 1965–1972," *Environmental History Review* 16, no. 3 (Autumn 1992), 31; "Report Urges Phosphate Ban," *Atlanta Journal Constitution*, April 10, 1970, 19A.

12. "Anti-Phosphate Move Bodes Ill for Plants," *Idaho State Journal* (Pocatello, ID), December 24, 1970, 11.

13. Interview with John E. Franz, 24.

14. Interview with John E. Franz, 18–21, 24, 26, 28; Monsanto Company 1972 Annual Report, 12.

15. "Q&A—Monsanto in Idaho," transcript of interview with Monsanto's Randy Vranes, President of the Idaho Mining Association, published on the Idaho Mining Association website, February 20, 2014, http://mineidaho .com/2014/02/20/713/. Website no longer available. Text in author's possession.

16. "Roundup," Monsanto Company website, https://web.archive.org/ web/20190320150119/http://www.monsantoglobal.com/global/au/

products/pages/roundup.aspx; Interview with John E. Franz, 33; EPA Fact Sheet, "Regulatory Status of 2–4–5T," (October 1978), 1; "Backgrounder: History of Monsanto's Glyphosate Herbicides," Monsanto publication posted online, June 2005, https://web.archive.org/web/20190619171544/ https://monsanto.com/app/uploads/2017/06/back_history.pdf. For an excellent history of 2,4-D and 2,4,5-T, see Carol Van Strum, *A Bitter Fog: Herbicides and Human Rights* (San Francisco: Sierra Club Books, 1983); Thomas Whiteside, *Defoliation: What Are Our Herbicides Doing to Us?* (New York: Ballantine Books, 1970); Thomas Whiteside, *The Withering Rain: America's Herbicidal Folly* (New York: E. P. Dutton, 1971); Monsanto did not know that glyphosate was an EPSP synthase inhibitor until the 1980s when German scientists made that discovery. For many years, in other words, the company was really in the dark when it came to understanding exactly how glyphosate worked. Daniel Charles, *The Lords of the Harvest* (Cambridge, MA: Perseus Publishing, 2001), 62.

17. For scientific studies related to glyphosate, microbiomes, and EPSP synthase inhibition, see Qixing Mao et al., "The Ramazzini Institute 13-Week Pilot Study on Glyphosate and Roundup administered at Human-Equivalent Dose to Sprague Dawley Rats: Effects on the Microbiome," *Environmental Health* 17 (2018): 1–12; Monsanto challenged the validity of this study, saying the researchers were part of an "activist group" trying to ban glyphosate. See "Glyphosate Shown to Disrupt Microbiome 'At Safe Levels,' Study Claims," *Guardian*, May 16, 2018, https:// www.theguardian.com/environment/2018/may/16/glyphosate-shown-to -disrupt-microbiome-at-safe-levels-study-claims. Yassine Airbali et al., "Glyphosate-Based Herbicide Exposure Affects Gut Microbiota, Anxiety and Depression-like Behaviors in Mice," *Neurotoxicology and Teratology* 67 (2018): 44–49; Veronica L. Lozano et al., "Sex-Dependent Impact of Roundup on the Rat Gut Microbiome," *Toxicology Reports* 5 (2018): 96–107; Erick V. S. Motta et al. "Glyphosate Perturbs the Gut Microbiota of Honey Bees," *Proceedings of the National Academy of Sciences of the United States of America* 115, no. 41 (2018): 10305–10; Nicolas Blot et al., "Glyphosate, but Not Its Metabolite AMPA, Alters the Honeybee Gut Microbiota," *PLoS One* 14, no. 4 (2019): e01215366; Lola Rueda-Ruzafa et al., "Gut Microbiota and Neurological Effects of Glyphosate," *NeuroToxicology* 75 (2019): 1–8. Evolutionary biologist Jonathan Eisen at the University of California, Davis, has compiled a database of studies looking into glyphosate and effects on soil, human, and animal microbiomes. See https://www.zotero.org/groups/341914/glyphoate_ microbiota_and_microbiomes/items/H37I47CR/library. On soil bacteria and glyphosate, see, for example, L. H. S. Zobiole, "Glyphosate Affects Micro-Organisms in Rhizospheres of Glyphosate-Resistant Soybeans," *Journal of Applied Microbiology* 110, no. 1 (January 2011): 118–27. For a study calling for more research into how glyphosate may affect the soil biome, see J. R. Powell and C. J. Swanton, "A Critique of Studies Evaluating Glyphosate Effects on Diseases Associated with *Fusarium* spp.," *Weed Research* 48, no. 4 (August 2008): 307–18. A study in the prestigious

journal *Nature* also noted the lack of research on this topic: "Surprisingly despite the enormous use of glyphosate-based herbicides around the world and the active research around glyphosate products, little is known of their potential effects on non-target soil organisms." Nevertheless, the study came to this conclusion: "When glyphosate degradation is effective and Roundup is used within recommended limits, the effect of weed control with Roundup likely has minor and transient effects on the structure and functioning of food webs in agricultural soils." See Marleena Hagner, "Effects on Glyphosate-Based Herbicide on Soil Animal Trophic Groups and Associated Ecosystem Functioning in a Northern Agricultural Field," *Scientific Reports* 9 (2019): 2, 9.

18. Monsanto Company 1972 Annual Report, 12; Monsanto Company 1975 Annual Report, 2. On the history of the Pesticide Control Act of 1972, see Richard N. L. Andrews, *Managing the Environment, Managing Ourselves: A History of American Environmental Policy*, 3rd ed. (New Haven: Yale University Press, 2020), 243–44, 262; "Farmers Risk Fine for Herbicide Use," *Messenger-Inquirer* (Owensboro, KY), July 20, 1977, 1A, 8A; On the history of this early registration, see Carey Gillam, *Whitewashed: The Story of a Weed Killer, Cancer, and the Corruption of Science* (Washington, DC: Island Press, 2017), 25–36.

19. "Monsanto Goes Outside Industry in Choosing New President-Chief," *Wall Street Journal*, October 27, 1972, 25; "The Outsider at Monsanto," *New York Times*, November 19, 1972, F9; "New Monsanto President," *St. Louis Post-Dispatch*, October 26, 1972, 9C; Forrestal, *Faith, Hope &5,000*, 227–29; Monsanto Company 1972 Annual Report, 1.

20. Monsanto 1975 Annual Report, 4–5, 23; Monsanto Company 1976 Annual Report, 67–68; Interview with William B. Daume (former head of Executive Compensation Committee) conducted by James E. McKee Jr., November 17, 1981; Interview with Winthrop R. Corey conducted by James E. McKee Jr., December 30, 1981, series 10, box 5, folder: Monsanto Company History (Oral History Project, 1981, Folder 2), MCR; "N. C. Wyeth, Inventor, Dies at 78; Developed the Plastic Bottle," *New York Times*, July 7, 1990, 12; For the story of Coca-Cola's response to this plastic bottle debate, see "Monsanto to Expand Plastic Bottle Output, Has Accord to Supply Coca-Cola Bottlers," *Wall Street Journal*, October 15, 1973, 13; "A Market Thirst, Never Quenched," *New York Times*, April 9, 1978, F1; "Technology: The Dispute Over Plastic Bottles," *New York Times*, April 13, 1977, 79; Letter from Paul Austin to Woodruff, February 15, 1977, box 16, folder 5, Robert Winship Woodruff Papers, Stuart A. Rose Manuscript, Archives, and Rare Book Library, Emory University. Building on the work of DuPont, Goodyear actually produced some of the first PET plastic bottles for Pepsi, Coca-Cola's arch-rival, in the late 1970s. See "Goodyear Commits to Plastic Recycling," *The Akron Beacon*, December 23, 1990, D12.

21. "Johnsongrass Cure Looks Promising During Tests," *El Paso Times*, February 25, 1973, 8-D; "No More Mr. Nutgrass," *Honolulu Advertiser*, March 23, 1975; "Wet and Weedy," *Guardian*, November 23, 1974, 12; Carey Gillam, *Whitewashed*, 28; "Roundup Herbicide Promising," *Red Deer*

Advocate (Red Deer, Alberta, Canada), May 28, 1976, 12; Daniel Charles, *Lords of the Harvest*, 62.

22. Paul K. Conklin, *A Revolution Down on the Farm: The Transformation of American Agriculture Since 1929* (Lexington: University Press of Kentucky, 2009), 131; 1977 Monsanto advertisement featured in Mathieu Asselin, *Monsanto: A Photographic Investigation* (Arles: Actes Sud, 2019), n.p.; Tore Olsson, *Agrarian Crossings: Reformers and the Remaking of the US and Mexican Countryside* (Princeton: Princeton University Press, 2017), 155; Nick Cullather, *The Hungry World: America's Cold War Battle Against Poverty in Asia* (Cambridge, MA: Harvard University Press, 2013), 68, 70, 78, 134, 231, 261, 266–71; John Perkins, *Geopolitics and the Green Revolution: Wheat, Genes, and the Cold War* (New York: Oxford University Press, 1997), vi, 187, 258–59; Michelle Mart, *Pesticides, a Love Story*, 84–85.

23. "Johnsongrass Cure Looks Promising During Tests," *El Paso Times*, February 25, 1973, 8-D; "No More Mr. Nutgrass," *Honolulu Advertiser*, March 23, 1975; "Monsanto Faces A Possible Rival Line of Herbicide," *Wall Street Journal*, April 8, 1982, 7.

24. Monsanto Company 1975 Annual Report, 1-2.

25. "Interest Growing in Phosphate-rich Region," *Idaho State Journal* (Pocatello, ID), February 3, 1975, 9; "Development of Phosphate Mines Could Have Huge Economic Impact," *Idaho State Journal*, December 15, 1974, C-2.

26. US Department of the Interior, Geological Survey, Bureau of Land Management, US Department of Agriculture, and Forest Service, Draft Environmental Impact Statement, "Development of Phosphate Resources in Southeastern Idaho," Volume 1 (1976), 1–34, 1–349, 1–373, 1–464; Letter from Donald Dubois, EPA Regional Manager, to Vincent McKelvey, US Geological Survey, July 23, 1976, in *Development of Phosphate Resources in Southeastern Idaho*, 3 (1976), 35.

27. EPA, Office of Radiation Programs, Las Vegas Facility, *Idaho Radionuclide Study* (April 1990), 1; EPA. *Idaho Radionuclide Exposure Study— Literature Review* (October 1987), 3.1-5.2, https://www.osti.gov/servlets/purl/5811680.

28. North Wind, Inc., private contractor commissioned by the Idaho Department of Environmental Quality, Pocatello Regional Office Mining Program, "Ballard, Enoch Valley, and Henry (P4) Mines Community Involvement Plan," (2012), 2.

29. For an environmental history that details the EPA's early deployment of the Superfund Act, see David Brooks, *Restoring the Shining Waters: Superfund Success at Milltown, Montana* (Norman: University of Oklahoma Press, 2015). On Love Canal, see Elizabeth Blum, *Love Canal Revisited: Race, Class, and Gender in Environmental Activism* (Lawrence: University of Kansas Press, 2008).

30. Letter from David A. Becker, Regional Project Manager, Ecology and Environment, Inc., Seattle, Washington, to Kent Lott, Monsanto Company, March 4, 1987; Letter to John Osborn, USEPA, Region X from Jeffrey Whidden, Ecology & Environment, Inc., April 10, 1987, Re: Trip Report,

Monsanto Chemical Company, Soda Springs, Idaho; Letter from Jeffrey Whidden, Ecology and Environment, to John Osborn, EPA, Region X, September 9, 1987; On May 5, 1989, the EPA proposed placing the Soda Springs plant on the National Priorities List (NPL), which would make it eligible for Superfund cleanup protocols. The site officially attained Superfund status a little over a year later, on August 30, 1990. EPA Record of Decision (Cleanup Plan) for Monsanto site, with attachments including administrative record index, April 30, 1997, FOIA request number EPA-R10–2015–007749. Filed by author.

CHAPTER 8: "They Can Have My House; I Just Need Thirty Days to Get Out"

1. Account based on author visit to Stuart Calwell's law office, Charleston, West Virginia, October 11, 2016.
2. Stuart Calwell (attorney), interview by the author, Charleston, West Virginia, October 11, 2016.
3. Calwell, interview, October 11, 2016.
4. Calwell, interview, October 11, 2016.
5. Calwell, interview, October 11, 2016.
6. Calwell, interview, October 11, 2016.
7. Calwell, interview, October 11, 2016. On his decision to leave California, Calwell said, "Aside from the glow of romance that descends on some experience like that, the harsh reality is, when you wake up on a stinking abalone boat before dawn . . . and you reach down into the engine compartment to pull your wetsuit out and it's slimy with diesel residue and you slip into that and you jump overboard at first light and you're cold and wet all the time—that is not as romantic as one might think."
8. "Dioxin Leaves Mark on a City Called Nitro, But Extent of Impact Is at Issue," *New York Times*, August 4, 1983; Calwell, interview, October 11, 2016.
9. "Workers Sue Monsanto, Claiming Chemical Company Ignored Health Risk," *Baltimore Sun*, July 8, 1984, 11A; Calwell, interview, October 11, 2016.
10. Calwell, interview, October 11, 2016.
11. Plaintiffs' Amended Complaint, 56, April 13, 1981, Docket No. 4, *Dorothy Adkins et al. v. Monsanto Company*, Civil Action 81–2098, US District Court for the Southern District of West Virginia, National Archives at Philadelphia, Philadelphia, Pennsylvania [hereinafter *Adkins* case file, Philadelphia NARA]; "Monsanto, Workers Settle Poisoning Case," *St. Louis Post-Dispatch*, June 9, 1988, 8D; Monsanto Company 1981 Annual Report, 1.
12. Monsanto Company 1980 Annual Report, 26; The following analysis of Monsanto's oil dependency first appeared in an article I published with *Enterprise & Society* in 2018. See Bartow J. Elmore, "The Commercial Ecology of Scavenger Capitalism: Monsanto, Fossil Fuels, and the Remaking of a Chemical Giant," *Enterprise & Society* 19, no. 1 (March 2018):

153–78; Richard J. Mahoney, *A Commitment to Greatness* (St. Louis: Monsanto Company, 1988), 9–10; Fred Aftalion, *A History of the International Chemical Industry: From the 'Early Days' to 2000* (Philadelphia: Chemical Heritage Foundation, 2002), 262–64, 320; Alfred D. Chandler Jr., *Shaping the Industrial Century: The Remarkable Story of the Evolution of the Modern Chemical and Pharmaceutical Industries* (Cambridge, MA: Harvard University Press, 2005), 146–52.

13. Chandler Jr., *Shaping the Industrial Century*, 10.

14. Remarks to Shareholders by Howard A. Schneiderman, Monsanto Shareholders Meeting, April 23, 1982, 2, series 14, box 6, folder: Hanley, John W. (Remarks [Shareholders Mtg., 1982–83]), Monsanto Company Records, Washington University in St. Louis, Julian Edison Department of Special Collections, St. Louis, Missouri [hereinafter MCR]; Daniel Charles, *The Lords of the Harvest* (Cambridge, MA: Perseus Publishing, 2001), 12.

15. Monsanto Company 1982 Annual Report, 7; Monsanto Company 1980 Annual Report, 26; "Dramatic Climb by Oil Prices Gives U.S. Chemical Industry Executives a Headache," *The Sun* (Baltimore), December 2, 1979, K9.

16. Bartow J. Elmore, "The Commercial Ecology of Scavenger Capitalism," 16; Richard J. Mahoney, *Commitment to Greatness*, 10; Richard J. Mahoney, *In My Opinion: Writings on Public Policy* (St. Louis: Murray Weidenbaum Center on the Economy, Government, and Public Policy at Washington University in St. Louis, 2003), 8, 19–20.

17. Monsanto EMC Report on Biotech, April 23, 1984, series 1, box 1, folder: Biotechnology (Association of Reserves, 1984–89), MCR; Mahoney, *Commitment to Greatness*, 20; Jack Ralph Kloppenburg Jr., *First the Seed: The Political Economy of Plant Biotechnology, 1492–2000* (1988; repr., Madison: University of Wisconsin Press, 2004), 209.

18. Chandler, *Shaping the Industrial Century*, 30, 59; Aftalion, *A History of the International Chemical Industry*, 320, 322, 324, 332; Charles, *Lords of the Harvest*, 6; Thomas K. McCraw and William R. Childs, eds., *American Business Since 1920*, 3rd ed. (John Wiley & Sons, 2018), 166; Mark Lynas, *Seeds of Science: Why We Got It So Wrong on GMOs* (London: Bloomsbury Sigma, 2020), 91.

19. Carol Van Strum, *A Bitter Fog: Herbicides and Human Rights* (San Francisco: Sierra Club Books, 1983); Thomas Whiteside, "A Reporter at Large: Contaminated," *New Yorker* (September 1978), 34–81; "Panetta Introduces Agent Orange Bill," *Santa Cruz Sentinel* (Santa Cruz, CA), March 10, 1983, 8; Thomas Whiteside, *The Pendulum and the Toxic Cloud: The Course of Dioxin Contamination* (New Haven: Yale University Press, 1979); Thomas Whiteside, "The Pendulum and the Toxic Cloud," *New Yorker* (July 1977), 55; Whiteside published two earlier treatises on dioxin in *The Withering Rain: America's Herbicidal Folly* (New York: E. P. Dutton, 1971) and *Defoliation: What Are Our Herbicides Doing to Us?* (New York: Ballantine Books, 1970); Marie-Monique Robin, *The World According to Monsanto* (New York: New Press, 2010), 48.

20. Plaintiff's Response to Defendant's Motion for Summary Judgment—Statute

of Limitations in *Jeffers v. Monsanto*, Civil Action No. 81–2239, Docket No. 225, 75–76, The Calwell Practice law office files, Charleston, West Virginia [hereinafter TCP files].

21. Plaintiff's Response to Defendant's Motion for Summary Judgment—Statute of Limitations in *Jeffers v. Monsanto*, 75–76; Letter from James R. Boggess to Senator Robert Byrd, March 17, 1986, Freedom of Information Act (FOIA) request EPA-R3–2016–005049. Filed by author.

22. During the trial, this 1977 document became known as the "Wonders" document because Monsanto's Manager of Personnel for Specialty Chemicals Division Compensation, John S. Wonders, authored it. Memorandum from John S. Wonders, Employee/Union Requests Employee Health Data, November 16, 1977, Exhibit A attached to Plaintiff's Complaint, Docket No. 1, *Adkins* case file, Philadelphia NARA; Plaintiff's Response to Defendant's Motion for Summary Judgment—Statute of Limitations in *Jeffers v. Monsanto*, Civil Action No. 81–2239, Docket No. 225, 78, TCP files; "How Industry Handled the Dioxin Question," *St. Louis Post-Dispatch*, November 14, 1983, 32.

23. Robin, *The World According to Monsanto*, 46; *Jeffers v. Monsanto*, Civil Action No. 81–2239, Docket No. 225, Plaintiff's Response to Defendant's Motion for Summary Judgment—Statute of Limitations, 83, TCP files.

24. J. A. Zack and R. R. Suskind, "The mortality experience of workers exposed to tetrachlorodibenzodioxin in a trichlorophenol process accident," *Journal of Occupational Medicine* 22, no. 1 (January 1980): 11–14; R. R. Suskind and V. S. Hertzberg, "Human Health Effects of 2,4,5-T and Its Toxic Contaminants," *Journal of the American Medical Association* 251, no. 18 (1984): 2372–80; Plaintiff's Response to Defendant's Motion for Summary Judgment—Statute of Limitations, 77, TCP files; Letter from Monsanto Company to Frank Cagnetti, Executive Director, Medical Center Fund of Cincinnati, July 13, 1979, Docket No. 2228, *Adkins* case files, Philadelphia NARA.

25. Linda Birnbaum, interview by the author, October 26, 2016; Marilyn A. Fingerhut et al., "Cancer Mortality in Workers Exposed to 2,3,7,8-Tetrachlorodibenzo-P-Dioxin," *New England Journal of Medicine* 324, no. 4 (1991): 212; Manolis Kogevinas et al., "Cancer Mortality in Workers Exposed to Phenoxy Herbicides, Chlorophenols, and Dioxins: An Expanded and Updated International Cohort," *American Journal of Epidemiology* 145, no. 12 (June 15, 1997): 1061–75; Sally S. White and Linda S. Birnbaum, "An Overview of the Effect of Dioxin and Dioxin-like Compounds on Vertebrates, as Documented in Human and Ecological Epidemiology," *Journal of Environmental Science and Health, Part C: Environmental Carcinogenesis and Ecotoxicology Reviews* 27, no. 4 (October 2009): 197–211; Clapp quoted in "EPA Wants to Delay Cleanup of Dioxin in Kanawha River," *Charleston Gazette*, May 21, 2000; US Department of Health and Human Services, *14th Report on Carcinogens* (Washington, DC: Department of Health and Human Services, 2016), https://ntp.niehs.nih.gov/pubhealth/roc/index-1.html; "Study Concludes Dioxin Definitive Cause of Cancer: Report will Be Used to Determine Future Regulations," *Charleston Gazette*, January 23, 2001, P3A. For the most comprehensive synthesis of current health studies related to dioxin exposure, see the National Academy of Sciences, *Veterans and Agent Orange:*

Update 2014 (Washington, DC: National Academy of Sciences Press, 2016). This report has been updated biennially since the mid-1990s.

26. *Jeffers v. Monsanto*, Civil Action No. 81–2239, Docket No. 225, Plaintiff's Response to Defendant's Motion for Summary Judgment—Statute of Limitations, 77, TCP files; Judith A. Zack and William R. Gaffey, "A Mortality Study of Workers Employed at the Monsanto Company Plant in Nitro, West Virginia," in *Human and Environmental Risks of Chlorinated Dioxins and Related Compounds*, eds. Richard E. Tucker, Alvin L. Young, and Allan P. Gray (Boston: Springer US, 1983), 575–91; Alastair Hay and Ellen Silbergeld, "Assessing the risk of dioxin exposure," *Nature* 315 (May 9, 1985), 102.

27. *James Mandolidis v. Elkins Industries, Inc.*, 246 S. E. 2d 907 (1978).

28. *James Mandolidis v. Elkins Industries, Inc.* On misspelling, see "Employer Liability In West Virginia: Compensation Beyond The Law," *Washington & Lee Law Review* 151 36, no. 1 (Winter 1979), 152.

29. Calwell, interview, October 11, 2016.

30. Transcript of Proceedings before the Honorable John T. Copenhaver Jr. filed on January 23, 1984, 25–27, Docket No. 2228, *Adkins* case files, Philadelphia NARA; Stuart Calwell, interview by the author, September 15, 2017.

31. Transcript of Proceedings before the Honorable John T. Copenhaver Jr. filed on January 23, 1984, 31, 34–35, Docket No. 2228, *Adkins* case files, Philadelphia NARA.

32. "Mule-Headed," *Charleston Gazette*, November 28, 1985; Calwell, interview, September 15, 2017; "Jury Awards $58 Million to 47 Railroad Workers Exposed to Dioxin," *New York Times*, August 27, 1982; "Disaster Defense," *Wall Street Journal*, June 7, 1957, 1; "Monsanto Suits Fester in W. Va.," *Pittsburgh Press*, April 13, 1986, 1B.

33. "Monsanto Denies It Favored Profit Over Safety," *St. Louis Post-Dispatch*, June 26, 1984, 1A, 9A. Weather Underground reported that the temperature averaged 70 degrees that day with no precipitation: https://www.wunderground.com/history/airport/KCRW/1984/6/25/DailyHistory.html?req_city=&req_state=&req_statename=&reqdb.zip=&reqdb.magic=&reqdb.wmo=&MR=1.

34. Peter H. Schuck, *Agent Orange on Trial: Mass Toxic Disasters in the Courts* (Cambridge, MA: Belknap Press of Harvard University Press, 1987), 87, 112, 166; Jack B. Weinstein, "Preliminary Reflections on Administration of Complex Litigations," *Cardozo Law Review De-Novo* 1 (2009), 8.

35. "Monsanto Denies It Favored Profit Over Safety," *St. Louis Post-Dispatch*, June 26, 1984, 1A, 9A; "Dioxin Trial Opens Against Monsanto," *New York Times*, June 26, 1984, A10. In a telling reflection of the times, the same day the *Dispatch* reported this story, it also discussed Reagan's deregulatory policies that resulted in reduced stringency in EPA implementation of the Clean Air Act; "Monsanto Trial Starts; 'Chemical Soup' Cited," *The Tennessean* (Nashville, TN), June 26, 1984, 1-B. Journalist William Freivogel, then in his thirties, covered the trial for the *St. Louis Post-Dispatch* in the 1980s. He spoke with me about the experience in an interview conducted on September 26, 2016.

36. "Monsanto Denies It Favored Profit Over Safety," *St. Louis Post-Dispatch*, June 26, 1984, 1A, 9A; "Monsanto Trial Starts; 'Chemical Soup' Cited," *The Tennessean* (Nashville, TN), June 26, 1984, 1-B; "The Canadian Connection," *Asbury Park Press* (Asbury, NY), June 13, 1983, C8.

37. "Monsanto Charted Ills' Costs Against Profits, Attorney Says," *Charleston Gazette*, June 26, 1984, 7A.

38. "Monsanto Co. Defends Its Health Record," *St. Louis Post-Dispatch*, June 27, 1984, 12A; "Monsanto Trial Starts; 'Chemical Soup' Cited," *The Tennessean*.

39. "Monsanto Co. Defends Its Health Record," *St. Louis Post-Dispatch*.

40. "Workers Not Told of All Risks, Monsanto's Doctor Says," *St. Louis Post-Dispatch*, July 3, 1984, 17A; "Workers Sue Monsanto, Claiming Chemical Company Ignored Health Risk," *Baltimore Sun*, July 8, 1984, 11A.

41. Written Communication Between the Court and the Jury During Deliberations, filed May 7, 1985, Docket No. 2911, *Adkins* case files, Philadelphia NARA; "Monsanto Suits Fester in W. Va.," *Pittsburgh Press*, April 13, 1986, 1B.

42. Motion by Plaintiffs for Stay on Motion for New Trial and For Injunction Pending Appeal, May 28, 1985, 2, Docket No. 2915, *Adkins* case files, Philadelphia NARA.

43. Judgment, May 17, 1985, Docket No. 2912; Bill of Costs by Defendant, July 26, 1985, Docket No. 2928; Memorandum Order, December 30, 1985, Docket No. 3235; Order, June 11, 1986, Docket No. 3247, Order, January 24, 1986, Docket No. 3241, *Adkins* case files, Philadelphia NARA.

44. "Monsanto Suits Fester in W. Va.," *Pittsburgh Press*, April 13, 1986, 1B.

45. "Monsanto Suits Fester in W. Va.," *Pittsburgh Press*.

46. "Monsanto Suits Fester in W. Va.," *Pittsburgh Press*.

47. Calwell, interview, October 11, 2016.

48. "Monsanto is Upheld in a Suit on Dioxin Brought by Workers," *New York Times*, August 29, 1987, 9; "Monsanto, Workers Settle Poisoning Case," *St. Louis Post-Dispatch*, June 9, 1988, 8D.

49. "Weary Monsanto Case Jury Voices Relief Upon Dismissal," *Charleston Gazette*, May 8, 1985, 1A, 6A; Summary Guide to Notebooks Containing Jury Instructions, Jury Charge, and Cases and Secondary Authorities, Docket No. 2867, *Adkins* case files, Philadelphia NARA.

50. "Only Medical Records Studied By Witness," *Charleston Gazette*, January 18, 1985, 8A; "Judge Again Refuses EPA Map As Evidence," *Charleston Gazette*, March 14, 1985, 11A.

51. "Former Juror Questions Monsanto's Actions," *Charleston Gazette*, March 29, 1986, 9A.

52. "Former Juror Questions Monsanto's Actions," *Charleston Gazette*.

CHAPTER 9: **"Trespassing to Get to Our Own Property"**

1. "Making a Difference at Monsanto," an article based on an interview with Dr. Ernest G. Jaworski, *Monsanto Corporate Research Community Mirror* (February, 1991), 1–4, series 8, box 9, folder: MCR-Research Community

Mirror, 1989–93, Monsanto Company Records, Washington University in St. Louis, Julian Edison Department of Special Collections, St. Louis, Missouri [hereinafter MCR]; For more on Jaworski's biography, see Daniel Charles, *The Lords of the Harvest* (Cambridge, MA: Perseus Publishing, 2001), 8.

2. Yvonne Cripps, "A Legal Perspective on the Control of the Technology of Genetic Engineering," *Modern Law Review* 44, no. 4 (July 1981): 369–70; Jack Ralph Kloppenburg Jr., *First the Seed: The Political Economy of Plant Biotechnology, 1492–2000* (1988; repr., Madison: University of Wisconsin Press, 2004), 252; Marie-Monique Robin, *The World According to Monsanto* (New York: New Press, 2010), 133–35; Errol C. Friedberg, *Biography of Paul Berg: The Recombinant DNA Controversy Revisited* (Singapore: World Scientific, 2014), 138–40; "Origins of Recombinant DNA," Interview with Janet E. Mertz conducted by Stephanie Chen, April 5, 2013, https://dukespace.lib.duke.edu/dspace/bitstream/handle/10161/11704/2013%2005%20April%20Janet%20Mertz%20Interview%20REDACTED.pdf?sequence=1&isAllowed=y.

3. Robin, *The World According to Monsanto*, 133–35; Kloppenburg Jr., *First the Seed*, 196–97; Fred Aftalion, *A History of the International Chemical Industry: From the 'Early Days' to 2000* (Philadelphia: Chemical Heritage Foundation, 2002), 342; On the early financial growth of Genentech and other biotech startups, see Charles, *Lords of the Harvest*, 11–12; "People," *Los Angeles Times*, March 26, 1987, F3; Nichola Kalaitzandonakes, "Mycogen: Building a Seed Company for the Twenty-first Century," *Review of Agricultural Economics* 19, no. 2 (Winter 1997), 456; Thomas K. McCraw and William R. Childs, eds., *American Business Since 1920*, 3rd ed. (John Wiley & Sons, 2018), 169.

4. "Making a Difference at Monsanto"; National Academy of Sciences (NAS), *Genetically Engineered Crops: Experiences and Prospects* (Washington, DC: National Academies Press, 2016), 67–73; On Schell and Chilton and their discoveries regarding *Agrobacterium tumefaciens*, see Charles, *Lords of the Harvest*, 3–6, 10, 14–17, 21–23; On the development of the gene gun, see Charles, *Lords of the Harvest*, 74–91; Mark Lynas, *Seeds of Science: Why We Got It So Wrong on GMOs* (London: Bloomsbury Sigma, 2018, 2020), 59–70.

5. Mark Lynas, *Seeds of Science*, 95–96.

6. Charles, *Lords of the Harvest*, 41–49.

7. Charles, *Lords of the Harvest*, 62.

8. Charles, *Lords of the Harvest*, 65–68; L. Comai et al., "Expression in Plants of Mutant *aroA* Gene from *Salmonella typhimurium* Confers Tolerance to Glyphosate," *Nature* 317 (1985): 741–44.

9. Charles, *Lords of the Harvest*, 67–69; Robin, *The World According to Monsanto*, 141; For more details on the bacteria-derived EPSP synthase gene and the Luling discovery, see Jerry M. Green and Michael D. K. Owen, "Herbicide-Resistant Crops: Utilities and Limitations for Herbicide-Resistant Weed Management," *Journal of Agricultural and Food Chemistry* 59, no. 11 (2011), 5825.

10. Senate Committee on Environment and Public Works, Subcommittee on

Nuclear Regulation, *Phosphate Slag Risk*, 101st Cong., 2nd Sess., August 21, 1990, 47; "Soda Springs Temporarily Bans Use of Slag on Roads," *Deseret News* (Salt Lake City, UT), June 25, 1990; "Betting the Farm on Biotech," *New York Times*, June 10, 1990, 36; Joseph M. Hans Jr. et al., *Above Ground Gamma Ray Logging for Locating Structures and Areas Containing Elevated Levels of Uranium Decay Chain Radionuclides*, prepared by the Office of Radiation Programs of the EPA (April 1978), 1, 17; Tom Gesell, interview by the author, Pocatello, Idaho, July 14, 2016; "State Wants Ban on Phosphate Slag," *Idaho State Journal*, September 27, 1976, 2; EPA, "Radiological Surveys of Idaho Phosphate Ore Processing—The Thermal Process Plant," report produced by EPA's Office of Radiation Programs (November 1977), 2.

11. For a detailed description of this aerial study, see EPA, Office of Radiation Programs, Las Vegas Facility, *Idaho Radionuclide Study* (April 1990), I, 2–8; H. A. Barry, Project Scientist, Nuclear Radiation Department, Department of Energy Remote Sensing Department operated for the US Department of Energy by EG&G Energy Measurements, Inc., "An Aerial Radiological Survey of Pocatello and Soda Springs, Idaho, and Surrounding Area" (February 1987). An EPA official in Region 10 provided the author with this report by Barry; Gesell, interview; Senate Committee on Environment and Public Works, Subcommittee on Nuclear Regulation, *Phosphate Slag Risk*, 101st Cong., 2nd Sess., August 21, 1990, 47.

12. "Soda Springs Temporarily Bans Use of Slag on Roads," *Deseret News* (Salt Lake City, UT), June 25, 1990; "Betting the Farm on Biotech," *New York Times*, June 10, 1990, 36.

13. "Republican State Senators," *Caribou County Sun*, June 22, 1989, FOIA request number EPA-R10–2015–007749. Filed by author.

14. Matthew Cheramie (former Soda Springs resident), email correspondence with the author, January 18, 2020. The author would like to thank Mr. Cheramie for offering insights about his time living in Soda Springs.

15. Cheramie email correspondence; Deberah Hansen obituary, https://www .findagrave.com/memorial/26544930/debera-hansen.

16. It is unclear exactly when or if EPA officials made such a statement regarding retaliation, but Senator James A. McClure seemed to recall in his congressional testimony that such "threats" of "stronger regulations" had been made "nearly three months" prior by an "EPA field officer . . . quoted in an Idaho news article" who "reportedly said that since the communities did not react to earlier statements about slag as EPA wished, the agency decided on a stronger strategy of persuasion." Senate Committee on Environment and Public Works, Subcommittee on Nuclear Regulation, *Phosphate Slag Risk*, 101st Cong., 2nd Sess., August 21, 1990, 2, 8; Steven Douglas Symms and Larry Grupp, *The Citizen's Guide to Fighting Government* (Ottawa, IL: Jameson Books, 1994).

17. Senate Committee on Environment and Public Works, Subcommittee on Nuclear Regulation, *Phosphate Slag Risk*, 49. This fear of declining property values following Superfund designation was not unfounded. Environmental historian Kent Curtis explained that in the copper-smelting city of Anaconda, Montana, homeowners could "not even sell their houses for a

loss" after large regions of the town were declared Superfund areas in 1983. Kent Curtis, "Greening Anaconda: EPA, ARCO, and the Politics of Space in Postindustrial Montana," *Beyond the Ruins: The Meaning of Deindustrialization*, eds. Jefferson Cowie and Joseph Heathcott (Ithaca: ILR Press, 2003), 101.

18. Letter from Raymond C. Loehr, Oddvar F. Nygaard, and James E. Martin, EPA Science Advisory Board, to William R. Reilly, January 21, 1991; Letter from Dana A. Rasmussen, Regional EPA administrator, to Senator Larry Craig, January 24, 1992; "Phase II Remedial Investigation Report - Volume I," Golder Associates, November 21, 1995, 1–13, FOIA request number EPA-R10–2015–007749.

19. Letter from Raymond C. Loehr, Oddvar F. Nygaard, and James E. Martin, EPA Science Advisory Board, to William R. Reilly.

20. Environmental Protection Agency (EPA), Graded Decision Guidelines for Phosphorous Slag, https://web.archive.org/web/20131111125111/http://yosemite.epa.gov/r10/cleanup.nsf/ID+slag/Graded+Decision+Guidelines+for+Phosphorus+Slag; Environmental Protection Agency (EPA), informational website for the southeast Idaho Phosphorus Slag Program, "What if Your Building Has Slag in It," https://web.archive.org/web/20170512180830/https://yosemite.epa.gov/R10/CLEANUP.NSF/webpage/what+if+your+building+has+slag+in+it?OpenDocument; Southeastern Idaho Public Health website on phosphate slag, http://www.sdhdidaho.org/comhealth/slag.php.

21. Environmental Protection Agency (EPA), Graded Decision Guidelines for Phosphorous Slag; Southeastern Idaho Public Health website on phosphate slag. For more radiation comparisons, see the United States Nuclear Regulatory Commission, "Doses in Our Daily Lives," https://www.nrc.gov/about-nrc/radiation/around-us/doses-daily-lives.html.

22. Environmental Protection Agency (EPA), Graded Decision Guidelines for Phosphorous Slag; Southeastern Idaho Public Health website on phosphate slag; Environmental Protection Agency (EPA), "What if Your Building Has Slag in It."

23. Southeastern Idaho Public Health website on phosphate slag; Environmental Protection Agency (EPA), "What if Your Building Has Slag in It."

24. Letter from Dana S. Rasmussen to Senator Larry Craig, January 24, 1992, FOIA request number EPA-R10–2015–007749; Letter from Raymond C. Loehr et al., Science Advisory Board, to William K. Reilly, EPA, January 21, 1991, FOIA request number EPA-R10–2015–007749; Richard N. L. Andrews, *Managing the Environment, Managing Ourselves: A History of American Environmental Policy*, 3rd ed. (New Haven: Yale University Press, 2020), 267.

25. Letter from Raymond C. Loehr et al., Science Advisory Board, to William K. Reilly, EPA; Community Relations Plan for Monsanto Chemical Corporation, Caribou County, Idaho, written by EPA, December 17, 1991, FOIA request number EPA-R10–2015–007749.

26. Community Relations Plan for Monsanto Chemical Corporation, Caribou County, Idaho; "Republican State Senators Tour Plants," *Caribou County Sun*, June 22, 1988, FOIA request number EPA-R10–2015–007749. For

comparisons to other Superfund communities, the case of Picher, Oklahoma, is instructive. Geographer David Robertson studied this lead and zinc mining town, which EPA placed on the Superfund list in 1983, and he showed that "Picherites recognized the area's environmental threats and in the 1990s formed a number of grassroots advocacy groups whose mission focused" on publicly exposing the pollution problems in their community. In one survey, 80 percent of the townspeople said they would take part in a proposed buyout program initiated by the EPA that would help relocate families to new homes. Thus, here was a case, unlike Soda Springs, where a mining community actively worked to increase awareness about the environmental issues in their town and where many community members were willing to consider radical remediation initiatives. Admittedly, there are many differences between Picher and Soda Springs (including the fact that links between health problems and lead contamination were clearer in Picher than were connections between environmental pollution and health problems in Soda Springs). But it is important to point out that mining firms in Picher had long ago abandoned this Oklahoma community, closing their zinc and lead mines in the 1960s. In Soda Springs, the polluter was still operating and was a critical driver of the local economy. There, many citizens clearly expressed concern about adversely impacting such a big employer. David Robertson, *Hard as the Rock Itself: Place and Identity in the American Mining Town* (Boulder: University Press of Colorado, 2006), 161, 165.

27. "Phase II Remedial Investigation Report - Volume I," Golder Associates, November 21, 1995; Monsanto Superfund Site Public Meeting transcript for meeting held on August 13, 1996, Soda Springs High School Auditorium, Soda Springs, Idaho, FOIA request number EPA-R10–2015–007749; EPA, Office of Radiation Programs, Las Vegas Facility, *Idaho Radionuclide Study*, 1; Golder Associates, an environmental consulting firm hired by Monsanto, reported in 1995 that radionuclide emissions (including polonium-210) coming from kilns at the Soda Springs plant produced ".6 curies" of radiation a year, which was below the National Emission Standards for Hazardous Air Pollutants standard for elemental phosphorus plants of 2 curies per year set by the EPA in the 1990s. Nevertheless, polonium-210 could be found in soil and stream-sediment samples around the plant. See "Phase II Remedial Investigation Report - Volume I," Golder Associates, November 21, 1995, 1-14, 4-16, 4-19, and Table 2-3, Table 2-4, Table 4-13, Table 5-1, and Figure 4-18.

28. Robert Gunnell, interview by the author, Provo, Utah, February 28, 2019.

29. Monsanto Superfund Site Public Meeting transcript for meeting held August 13, 1996; Letter from Robert D. Gunnell to Tim Brincefield, EPA, September 3, 1996; Letter from Charlotte Gunnell to Tim Brincefield, EPA, September 8, 1996, FOIA request number EPA-R10–2015–007749; "Phase II Remedial Investigation Report - Volume I," Golder Associates, Table 4-17, Figure 4-13; *Fourth Fifth-Year Review Report for Monsanto Chemical Co. (Soda Springs Phosphorous Plant), Superfund Site, Caribou County, Idaho* (September 12, 2018), 8–9, https://semspub.epa.gov/work/10/100113049.pdf.

30. Letter from Robert D. Gunnell to Tim Brincefield, EPA, September 3, 1996, FOIA request number EPA-R10–2015–007749.

31. Letter from Charlotte Gunnell to Tim Brincefield, EPA, September 8, 1996, FOIA request number EPA-R10–2015–007749. Robert's brother also wrote a letter supporting his family. See letter from E. Leroy Gunnell to Tim Brincefield, September 9, 1996, FOIA request number EPA-R10–2015–007749.

32. Letter from Ron and Carolyn Lau to Tim Brincefield, EPA, September 26, 1996; Memorandum from Timothy Brincefield to Monsanto Site File and Administrative Record, April 30, 1997, FOIA request number EPA-R10–2015–007749.

33. Phone Conversations Considered/Relied Upon During the Selection of Remedy, 4/30/97, Summary from Earlier Handwritten notes taken by the Remedial Project Manager (RPM), FOIA request number EPA-R10–2015–007749; Gunnell, interview.

34. Phone Conversations Considered/Relied Upon During the Selection of Remedy; Letter from Robert L. Geddes, Monsanto, to Kathleen Stryker, EPA, June 17, 1997, FOIA request number EPA-R10–2015–007749.

35. Letter from Hyland P. James, Monsanto, to Randy Smith, EPA, March 7, 1996, FOIA request number EPA-R10 2015–007749.

36. EPA, *Fourth Fifth-Year Review Report for Monsanto Chemical Co.*, 11, 21–22.

37. EPA, *Third Five-Year Review Report for Monsanto Chemical Co. (Soda Springs Phosphorous Plant)*, September 2013, vii, 28, 41, https://www3.epa .gov/region10/pdf/sites/monsanto/monsanto_3rd_FYR_091013.pdf.

38. US Government Accountability Office (GAO), *Phosphate Mining: Oversight Has Strengthened, but Financial Assurances and Coordination Still Need Improvement* (May 2012), 9; Idaho Mining Association, Selenium Committee, *Final—Summer 2001 Area-Wide Investigation Data Summary*, Southeast Idaho Phosphate Resource Area Selenium Project, prepared by Montgomery Watson Harza (July 2002), 1–2; Elizabeth Niven, "Reaching Out to the Community in Idaho," Monsanto company blog, October 19, 2012, http://monsantoblog.com/2012/10/19/reaching-out-to-the -community-in-idaho/; North Wind, Inc., private contractor commissioned by the Idaho Department of Environmental Quality, Pocatello Regional Office Mining Program, "Ballard, Enoch Valley, and Henry (P4) Mines Community Involvement Plan," (2012), 3; EPA News Release, "Idaho mining company agrees to pay $1.4 million penalty to settle alleged clean water act violations," April 20, 2011, http://yosemite.epa.gov/opa/admpress.nsf/ eeffe922a687433c85257359003f5340/8bf6aa09197a30f5852578780079625 2!OpenDocument.

39. Shoshone-Bannock Tribal Council, interview by the author, Fort Hall Indian Reservation, Idaho, July 14, 2016.

40. "Monsanto's Phosphate Operation," *Chemical & Engineering News* 87, no. 33 (August 17, 2009); "APNewsBreak: EPA Says Monsanto Mine Violates Law," *San Diego Union-Tribune*, June 25, 2009, http://www .sandiegouniontribune.com/sdut-us-monsanto-mine-violations-062509– 2009jun25–story.html.

41. On asymmetries of power in pollution remediation, see, for example, Robert Bullard, *Dumping in Dixie: Race, Class, and Environmental Quality*

(Boulder: Westview Press, 1990), which documents how communities of color have disproportionately born the burdens of dealing with toxic waste and have often been unable to convince government authorities not to site hazardous operations near their home communities. On the deep history of environmental racism in the United States, see Carl A. Zimring, *Clean and White: A History of Environmental Racism in the United States* (New York: New York University Press, 2017). Soda Springs' population was predominately white (over 95%), but minority communities living nearby were still affected by the decisions EPA made. Representatives of the Shoshone-Bannock Tribes clearly expressed frustration that their voice did not seem to carry the same weight with EPA as did those of other constituencies.

CHAPTER 10: "The Only Weed Control You Need"

1. Stephen O. Duke, "The History and Current Status of Glyphosate," *Pest Management Science* 74, no. 5 (May 2018): 1030.
2. Marie-Monique Robin penned this critical quip referring to Monsanto in her 2008 book, *The World According to Monsanto* (New York: New Press, 2010), iii; 1985 Searle Annual Report, 1–9; Daniel Charles, *The Lords of the Harvest* (Cambridge, MA: Perseus Publishing, 2001), 114.
3. Robin, *The World According to Monsanto*, 189; "Monsanto Visionary in a Cubicle; Could His Company's Special Culture Survive A Merger," *New York Times*, March 3, 1999, C1, C2.
4. "Monsanto Visionary in a Cubicle," *New York Times*; "Monsanto's Bet: There's Gold in Going Green," *Fortune*, April 14, 1997, 166, http://archive .fortune.com/magazines/fortune/fortune_archive/1997/04/14/224981/ index.htm; I experienced this swift response time to the author in 2014 when I reached out to Mr. Shapiro. Email from Bob Shapiro, February 18, 2014. Journalist Marie-Monique Robin details a similar experience in *The World According to Monsanto*.
5. Robin, *The World According to Monsanto*, 306; Carl Franken, "Monsanto Breaks the Mold," *Tomorrow* magazine (May/June, 1996), 62, series 14, box 26, folder: Shapiro, R. (Speeches) 1996, Monsanto Company Records, Washington University in St. Louis, Julian Edison Department of Special Collections, St. Louis, Missouri [hereinafter MCR]; Joan Magretta, "Growth Through Global Sustainability: An Interview with Monsanto's CEO, Robert Shapiro," *Harvard Business Review* (January–February 1997): 82; Charles, *Lords of the Harvest*, 269.
6. Magretta, "Growth Through Global Sustainability," 82.
7. 1993 Monsanto Company Annual Report, 40-41.
8. 1992 Monsanto Company Annual Report, frontmatter; 1993 Monsanto Company Annual Report, 3, 11.
9. Monsanto Company 1993 Annual Report, 32. For an excellent history of how productivity became the watchword of the feed-the-world movement, despite the fact that productivity was never really the root problem of famine in many parts of the world, see Nick Cullather, *The Hungry World:*

America's Cold War Battle Against Poverty in Asia (Cambridge, MA: Harvard University Press, 2010). For additional analysis of the Cold War foreign policy imperatives that drove policy makers to invest heavily in increasing agricultural productivity in America, see Shane Hamilton, *Supermarket USA: Food and Power in the Cold War Farms Race* (New Haven: Yale University Press, 2018).

10. Monsanto Company 1993 Annual Report, 8, 14, 17; Monsanto Company 1994 Annual Report, 8; Monsanto BST Public Affairs Plan (September 1986), 1, series 1, box 2, folder: Biotechnology (Public Affairs Plan), MCR. The author would like to thank Matthew Bonner, who found this document while conducting research for his undergraduate honors thesis at The Ohio State University. See Matthew Bonner, "From the Boardroom to the Courtroom: The Monsanto Corporate Influence and Liabilities," (undergraduate honors thesis, The Ohio State University, Department of Management and Human Resources, May 2020), 38. Charles, *Lords of the Harvest*, 94–95.

11. Monsanto Company 1994 Annual Report, 1, 6, 8; "Betting the Farm on Biotech," *New York Times*, June 10, 1990, 36.

12. Charles, *Lords of the Harvest*, 72, 126–48, 151.

13. *Diamond v. Chakrabarty*, 447 US 303 (1980); Committee on a National Strategy for Biotechnology in Agriculture, National Research Council, "Agricultural Biotechnology: Strategies for National Competitiveness," published by the National Academies of Science (1987), 57; Jack Ralph Kloppenburg Jr., *First the Seed: The Political Economy of Plant Biotechnology, 1492–2000* (1988; repr., Madison: University of Wisconsin Press, 2004), 261–63.

14. Kloppenburg Jr., *First the Seed*, 132–51, 262–63.

15. Kloppenburg Jr., *First the Seed*, 263–65.

16. For more on David Kingsbury and his views on biotechnology, see "Kingsbury on NSF, Biotech Regulation," *The Scientist*, March 23, 1987, https://www.the-scientist.com/news/kingsbury-on-nsf-biotech-regulation-63924; "Scientist and Rule-Maker: Dr. David T. Kingsbury," *New York Times*, July 21, 1986, A10; Fact Sheet: Proposal for a Coordinated Framework for Regulation of Biotechnology, John A. Svahn (Jack) Files, box 13531, folder: Biotechnology (2), Ronald Reagan Presidential Library, Simi Valley, CA.

17. "Redesigning Nature: Hard Lessons Learned; Biotechnology Food: From the Lab to Debacle," *New York Times*, January 25, 2001, A1; Robin, *The World According to Monsanto*, 142–43; Charles, *Lords of the Harvest*, 25–29.

18. George H. W. Bush's 1987 Visit to Monsanto, https://www.youtube.com/watch?v=EeS6usKzMTE; Robin, *The World According to Monsanto*, 143–44.

19. President's Council on Competitiveness, Report on National Biotechnology Policy, June 28, 1990, NLGB Control No. 40029; Stephen Hopgood, *American Foreign Environmental Policy and the Power of the State* (New York: Oxford University Press, 1998), 141; Memorandum from Vice President Dan Quayle to President George H. W. Bush, July 24, 1990, NLGB Control No. 2106; Michael Boskin, Council of Economic Advisers, Briefing for the Council on Competitiveness, Fostering the Competitiveness of the U.S.

Biotechnology Industry, n.d., NLGB Control No. 8436, George H. W. Bush Library, College Station, Texas ([hereinafter GHWBL].

20. Larry Lindsey, domestic economic policy adviser to George H. W. Bush, Memorandum for the Biotechnology Working Group of the Council of Competitiveness, March 9, 1990, NLGB Control No. 2105, GHWBL; Rebecca Goldburg et al., *Biotechnology's Bitter Harvest: Herbicide-Tolerant Crops and the Threat to Sustainable Agriculture*, a report of the Biotechnology Working Group (March 1990), 1, 5, https://blog.ucsusa.org/wp-content/uploads/2012/05/Biotechnologys-Bitter-Harvest.pdf; Charles, *Lords of the Harvest*, 92–97.

21. Lindsey, Memorandum.

22. Food and Drug Administration, "Food For Human Consumption and Animal Drugs, Feeds, and Related Products: Foods Derived from New Plant Varieties; Policy Statement, 22984," *Federal Register* 57, no. 104 (May 29, 1992), 22984, https://www.fda.gov/regulatory-information/search-fda-guidance-documents/statement-policy-foods-derived-new-plant-varieties; Robin, *The World According to Monsanto*, 144–46.

23. "Making Decisions By Consensus Suits Shapiro's Management Style," *St. Louis Post-Dispatch*, December 10, 1996, A12, series 10, box 07, folder: Organization (1998 Merger), MCR; "Company News; A Promotion to President at Monsanto," *New York Times*, December 8, 1992, D6.

24. For the history of Roth IRA and 401k plans, see John Cassidy, *Dot.con: The Greatest Story Ever Sold* (New York: Penguin, 2003), 29–31, 118. For an excellent graph detailing the explosion of middle-class equity in the stock market in the 1980s, see Julia C. Ott, *When Wall Street Met Main Street: The Quest for an Investors' Democracy* (Cambridge, MA: Harvard University Press, 2014), 101.

25. Monsanto Company 1995 Annual Report, 4, 8; Monsanto Company 1997 Annual Report, 6. Bob Shapiro made reference to Monsanto as the "Microsoft of genetic engineering" in an internal email communication to employees in 1997. Email to Charlotte J. Kuhn, Re: Monsanto Tomorrow—LSC, May 14, 1997, series 10, box 07, folder: Organization (1997 Solutia Separation), MCR; Robb Fraley, vice president of agricultural research, also promoted this analogy to Microsoft. See Charles, *Lords of the Harvest*, 110.

26. Charles, *Lords of the Harvest*, 149–50; Monsanto Company 1996 Annual Report, 3.

27. Charles, *Lords of the Harvest*, 164–67, 239.

28. Patrick J. Tranel and Terry R. Wright, "Review: Resistance of Weeds to ALS-inhibiting Herbicides: What Have We Learned?" *Weed Science* 50 (November–December 2002): 700–712; Stephen O. Duke and Stephen B. Powles, "Mini-Review: Glyphosate: A Once-in-a-Century Herbicide," *Pest Management Science* 64 (2008): 321. For a graph showing the composition of herbicide use in the United States prior to the Roundup Ready revolution, see Scott M. Swinton and Braeden Van Deynze, "Hoes to Herbicides: The Economics of Evolving Weed Management," *European Journal of Development Research* 29, no. 3 (2017): 565; Mark Loux (weed scientist at The Ohio

State University), interview by the author, September 5, 2018; Mark Loux, interview by the author, November 14, 2018.

29. Loux, interview, September 5, 2018; Loux, interview November 14, 2018.

30. Monsanto Agricultural Group, "Roundup Ready™ Soybean Checklist," *Midwest MAGnifier* 3, no. 4 (June 1996), 1 (emphasis in the original); Monsanto advertisement for Roundup Ready Cotton (April/May 1998), attached to author email correspondence with Mark Loux, September 4, 2018; Monsanto Agricultural Group, "Growers Give New Soybeans an A+," *Midwest MAGnifier* 4, no. 1 (March 1997), 1; Monsanto Agricultural Group, "Clean Fields and High Yields with Roundup Ready Soybeans," *Midwest MAGnifier* 4, no. 2 (May 1997), 1–2, all in author's possession; Sylvie Bonny, "Genetically Modified Herbicide-*Tolerant* Crops, Weeds, and Herbicides: Overview and Impact," *Environmental Management* 57 (2016): 39.

31. Laura D. Bradshaw et al., "Perspectives on Glyphosate Resistance," *Weed Technology* 11, no. 1 (January–March 1997): 189, 196.

32. Christopher Preston et al., "A Decade of Glyphosate-Resistant *Lolium* around the World: Mechanisms, Genes, Fitness, and Agronomic Management," *Weed Science* (2009): 435; Stephen B. Powles et al., "Evolved Resistance to Glyphosate in Rigid Ryegrass (*Lolium rigidum*) in Australia," *Weed Science* 46 (1998): 604–7; Bonny, "Genetically Modified Herbicide-Tolerant Crops, Weeds, and Herbicides," 39.

33. Roundup Ready Canola, FarmCentral.com [a Monsanto-sponsored website], https://web.archive.org/web/19981202105106/http://www.farmcentral.com/s/rr/s3rrzzzz.html; Monsanto, *1996 Environmental Annual Review*, 3, https://web.archive.org/web/19970714063108/http://www.monsanto.com:80/monpub/environment/monsantoear96/96earall.pdf; Monsanto Chairman Robert B. Shapiro, Remarks to Society of Environmental Journalists, October 28, 1995, https://web.archive.org/web/19961111115403/http://www.monsanto.com:80/MonPub/InTheNews/Speeches/951028Shapiro_Robert.html; Magretta, "Growth Through Global Sustainability," 82.

34. Charles, *Lords of the Harvest*, 14, 60, 109, 193.

35. Monsanto Company 1996 Annual Report, frontmatter; Charles, *Lords of the Harvest*, 60–61, 159, 195; "Monsanto to Buy Seed Concern for as Much as $1.02 Billion," *Wall Street Journal*, January 7, 1997, https://www.wsj.com/articles/SB852557052805035500; "Monsanto in a Big Seed Deal Whose Price Raises Eyebrows," *New York Times*, January 7, 1997, D8.

36. Charles, *Lords of the Harvest*, 120–23, 160.

37. Charles, *Lords of the Harvest*, 121–22, 161.

38. Charles, *Lords of the Harvest*, 123–24; 152–55.

39. Charles, *Lords of the Harvest*, 187, 210; Duke, "The History and Current Status of Glyphosate," 1030.

40. Email to Charlotte J. Kuhn, Re: Monsanto Tomorrow—Monday, December 9, 1996, Email to Charlotte J. Kuhn, Re: Monsanto Tomorrow—Thursday, December 17, 1996, Email to Charlotte J. Kuhn, Re: Monsanto Tomorrow—Wednesday, December 18, 1996, series 10, box 07, folder: Organization (1997 Solutia Separation); "It's Official: Monsanto to Divide Into 2 Firms," *St. Louis Post-Dispatch*, December 10, 1996, A1, series 10,

box 07, folder: Organization (1998 Merger); "Monsanto's Shapiro Repeats As Top-Paid Executive," *St. Louis Post-Dispatch*, July 19, 1998, E1, series 10, box 7, folder: Organization (1998 Merger), MCR; "Solutia's Ultimatum is Called Bid for Attention," *St. Louis Post-Dispatch*, December 7, 2003, G1, Rich Sauget Sr. scrapbook made available to the author in Sauget, Illinois; "Bankruptcy Stalks Solutia if Bondholders Won't Relent," *St. Louis Post-Dispatch*, Rich Sauget Sr. scrapbook; "Merger Between Monsanto and Pharmacia," *St. Louis Post-Dispatch*, December 21, 1999, A8.

41. Monsanto Company 1997 Annual Report, 24.
42. "It's Official: Monsanto to Divide Into 2 Firms," *St. Louis Post-Dispatch*.

CHAPTER 11: **"I Have to Cry for Them"**

1. Affidavit of Robert G. Kaley II, Ph.D., Director of Environmental Affairs, Monsanto, in *Walter Owens et al. v. Monsanto Company*, August 14, 1996, M01569, M01571–M01572, Chemical Industry Archives, a project of the Environmental Working Group https://web.archive.org/web/20170117202824/http://chemicalindustryarchives.org/search/ [hereinafter CIAEWG]; Monsanto Property Purchase Program, Document sent to Anniston homeowners, 1995, Poison Papers archive, managed by the Bioscience Research Project and the Center for Media and Democracy and based on papers collected by environmentalist and journalist Carol Van Strum, https://www.poisonpapers.org/the-poison-papers/ [hereinafter PPA]; Ellen Griffith Spears, *Baptized in PCBs: Race, Pollution, and Justice in an All-American Town* (Chapel Hill: University of North Carolina Press, 2014), 215. In 1995, the Alabama Department of Public Health determined that 5,296 people lived within one mile of the Monsanto site, and that 44 percent of that population was Black. See Health Consultation, Monsanto Company, Anniston, Calhoun County, Alabama, CERCLIS No. ALD004019048, prepared by Alabama Department of Public Health, 1995, CIAEWG.
2. David Baker admitted his ignorance about the PCB issue in his early life in an interview with historian Ellen Griffith Spears. See Spears, *Baptized in PCBs*, 215. See also Dennis Love, *My City Was Gone: One American Town's Secret, It's Angry Band of Locals, and a $700 Million Day in Court* (New York: Harper Perennial, 2006), 159. This section of the book would not have been possible without the work of Ellen Griffith Spears and journalist Dennis Love, both of whom completed interviews with David Baker and conducted in-depth investigations into the Anniston story. For more details on Monsanto's effect on this town, see their excellent works.
3. Love, *My City Was Gone*, 23; Spears, *Baptized in PCBs*, 96–100.
4. CBS's *60 Minutes* produced an episode about Anniston called "Toxic Town" in which reporter Steve Kroft interviewed citizens who reported extremely high levels of PCBs in their blood. See "Toxic Town," *60 Minutes*, broadcast on November 7, 2002. David Baker revealed to journalist Dennis Love that his PCB blood level was 341 parts per billion (ppb), some 227 times

the normal concentration of 1.5 ppb found in an average American's blood; Health Consultation, Monsanto Company, Anniston, Calhoun County, Alabama, CERCLIS No. ALD004019048, prepared by Alabama Department of Public Health, 1995, CIAEWG; ADEM, Draft Study Proposal— Choccolocco Creek Watershed, November 1993, CIAEWG; Love, *My City Was Gone*, 209, 227; Spears, *Baptized in PCBs*, 5, 203, 229.

5. Love, *My City Was Gone*, 206; Spears, *Baptized in PCBs*, 204, 215–16, 235, 238.

6. Love, *My City Was Gone*, 31, 145–52, 157.

7. Spears, *Baptized in PCBs*, 214.

8. Quoted in Spears, *Baptized in PCBs*, 214–15; Love, *My City Was Gone*, 158–59.

9. Love, *My City Was Gone*, 163.

10. Letter from David Baker of Community Against Pollution, Anniston, Alabama, to Richard D. Green, Director, Waste Management Division, Region IV, EPA, Re: Anniston PCB contamination, February 18, 1999, CIAEWG.

11. Report of Aroclor 'Ad Hoc' Committee (Second Draft), October 15, 1969, PPA; Spears, *Baptized in PCBs*, 229, 241.

12. "St. Louis CEOs Pocketed $430 Million Last Year," *St. Louis Post-Dispatch*, July 19, 1998, A1; "Monsanto's Shapiro Repeats As Top-Paid Executive," *St. Louis Post-Dispatch*, July 19, 1998, E1, series 10, box 7, folder: Organization (1998 Merger), Monsanto Archives; "Merger Between Monsanto and Pharmacia," *St. Louis Post-Dispatch*, December 21, 1999, A8.

13. "Merger Between Monsanto and Pharmacia," *St. Louis Post-Dispatch*, December 21, 1999, A8; Daniel Charles, *The Lords of the Harvest* (Cambridge, MA: Perseus Publishing, 2001), 235.

14. "Timeline: The EU's Unofficial GMO Moratorium," *Financial Times*, February 7, 2006, https://www.ft.com/content/624a88c6–97db-11da-816b -0000779e2340; USDA Economic Research Service, "Impacts of Adopting

15. Genetically Engineered Crops in the U.S.—Preliminary Results," July 20, 1999, 1–2, folder: Agriculture Biotech/USDA [United States Department of Agriculture], binder 3, Global Environmental Affairs-Bowles, Ian, Clinton Presidential Records.

16. Charles, *Lords of the Harvest*, 216–217.

17. Charles, *Lords of the Harvest*, 218–21, 258; Dan Charles, "Top Five Myths of Genetically Modified Seeds, Busted," *NPR*, October 18, 2012, https:// www.npr.org/sections/thesalt/2012/10/18/163034053/top-five-myths -of-genetically-modified-seeds-busted; "Royal Society: GM Food Hazard Claim is 'Flawed,'" *Nature* 399 (May 20 1999): 188; "Killer Potatoes: Where's the Data," *Nature Biotechnology* 17 (1999): 207; Stanley Ewen and Arpad Pusztai, "Effects of Diets Containing Genetically Modified Potatoes Expressing *Galanthus nivalis* Lectin on Rat Small Intestine," *Lancet* 354 (1999): 1354–55; Marie-Monique Robin, *The World According to Monsanto* (New York: New Press, 2010), 179–83.

18. Charles, *Lords of the Harvest*, 222, 238.

19. Charles, *Lords of the Harvest*, 243–44; John E. Losey et al., "Transgenic Pollen Harms Monarch Larvae," *Nature* 399 (1999): 214; Patricia

Anderson et al., "Effects on Fitness and Behavior of Monarch Butterfly Larvae Exposed to a Combination of CRY1AB-Expressing Corn Anthers and Pollen," *Environmental Entomology* 34, no. 4 (August 2005): 944–52; Tom Clarke, "Monarch Safe from Bt," *Nature* news website, September 12, 2001; John M. Pleasants and Karen S. Oberhauser, "Milkweed Loss in Agricultural Fields Because of Herbicide Use: Effect on the Monarch Butterfly Population," *Insect Conservation and Diversity* 6, no. 2 (March 2013): 135–44.

20. Charles, *Lords of the Harvest*, 240.
21. Charles, *Lords of the Harvest*, 229, 257.
22. Spears, *Baptized in PCBs*, 285–89.
23. "Monsanto: Pharmacia & Upjohn, Monsanto Will Merge," *St. Louis Post-Dispatch*, December 20, 1999, A8; Spears, *Baptized in PCBs*, 11, 240, 270.
24. Spears, *Baptized in PCBs*, 235, 249–53, 263.
25. Love, *My City Was Gone*, 288; Robin, *The World According to Monsanto*, 27; Senate Subcommittee of the Committee on Appropriations, *PCB Contamination in Anniston, Alabama*, 107 Cong., 2nd Sess., April, 19, 2002, 39.
26. Spears, *Baptized in PCBs*, 259–60; "Proposed Settlement in PCB Case Denounced," *Washington Post*, March 24, 2002, A6.
27. Senate Subcommittee of the Committee on Appropriations, *PCB Contamination in Anniston, Alabama*, 37.
28. Senate Subcommittee of the Committee on Appropriations, *PCB Contamination in Anniston, Alabama*, 37–38.
29. Spears, *Baptized in PCBs*, 16, 235, 262–63.
30. Spears, *Baptized in PCBs*, 16, 270, 275. Solutia came out of bankruptcy in 2008. Monsanto continued to share responsibility with the firm for legacy environmental liabilities. "Saving Solutia," *Pensacola News Journal*, March 16, 2008, 1B; "Everyone Gets Something in Complex Case," *St. Louis Post-Dispatch*, September 27, 2007, C4.
31. Spears, *Baptized in PCBs*, 270; Monsanto Company 2003 10-K form filed with SEC, 17; Monsanto Company 2004 Annual Report, 2; Monsanto Company 2007 Annual Report, 2.
32. "St. Louis CEOs Pocketed $430 Million Last Year," *St. Louis Post-Dispatch*, July 19, 1998, A1; "Monsanto's Shapiro Repeats As Top-Paid Executive," *St. Louis Post-Dispatch*, July 19, 1998, E1, series 10, box 7, folder: Organization (1998 Merger), MCR; "Merger Between Monsanto and Pharmacia," *St. Louis Post-Dispatch*, December 21, 1999, A8; "Timeline: The EU's Unofficial GMO Moratorium," *Financial Times*, February 7, 2006, https://www.ft.com/content/624a88c6–97db-11da-816b-0000779e2340; "Reed Considered Past Pay of Board Nominees," *Los Angeles Times*, November 6, 2003, C4; "Lawsuit Gives Rise to Dark Theories in Solutia Spinoff," *St. Louis Post-Dispatch*, August 25, 2006, C1; Love, *My City Was Gone*, 302; "Money Grab," *Forbes*, November 15, 2004, https://www.forbes.com/forbes/2004/1115/162.html#594bdacd196a.
33. Spears, *Baptized in PCBs*, 272–73; "Money Grab," *Forbes*, November 15, 2004, https://www.forbes.com/forbes/2004/1115/162.html#594bdacd196a.
34. Spears, *Baptized in PCBs*, 292–93; The EPA still maintains an "Anniston PCB

Site (Monsanto Co.)" Superfund site here: https://cumulis.epa.gov/supercpad/
SiteProfiles/index.cfm?fuseaction=second.docdata&id=0400123/.

CHAPTER 12: "Oh Shit, the Margins Were Very, Very, Very Good"

1. The following account is based on author conversations with a former Monsanto employee who preferred to remain anonymous.
2. Opinion in the *National Family Farm et al.* v. *U. S. Environmental Protection Agency* case, filed June 3, 2020, at 5, https://www.courthousenews.com/wp-content/uploads/2020/06/Dicamba.pdf.
3. Fred Pond, interview by the author, September 20, 2018.
4. Pond, interview, September 20, 2018.
5. Philip H. Howard, "Visualizing Consolidation in the Global Seed Industry: 1995–2008," *Sustainability* 1, no. 4 (2009), 1274–75; "Monsanto Buying Leader in Fruit and Vegetable Seeds," *New York Times*, January 25, 2005, C7; "Monsanto's Family Tree: Monsanto Acquisitions and Collaborations, 1965–2008," *Seed Today* (Second Quarter 2008), provided to the author by John Armstrong, secretary/manager of the Ohio Seed Improvement Association; For a graphic detailing Monsanto's position in the global seed industry, see "2017 Family Tree," published by AgWeb, https://www.agweb.com/assets/1/6/2017%20Seed%20Family%20Tree2.pdf.
6. For soybean and corn seed costs, see USDA Economic Research Service, "Commodity Costs and Returns," https://www.ers.usda.gov/data-products/commodity-costs-and-returns/commodity-costs-and-returns/#Recent%20Cost%20and%20Returns; "As Crop Prices Fall, Farmers Focus on Seeds," Wall Street Journal, October 16, 2016, https://www.wsj.com/articles/as-crop-prices-fall-farmers-focus-on-seeds-1476669901; Description of the TUA system on FarmCentral.com, https://web.archive.org/web/20021104140425/http://www.farmcentral.com:80/s/rrs/s4rsstzzz.htm; Pond, interview, September 20, 2018.
7. Pond, interview, September 20, 2018.
8. Pond, interview, September 20, 2018; Fred Pond, email correspondence with the author, February 18, 2018.
9. Fred Pond, interview by the author, October 5, 2018; Pond, email correspondence, February 18, 2018.
10. Pond, interview, October 5, 2018.
11. Paul K. Conkin, *Revolution Down on the Farm: The Transformation of American Agriculture Since 1929* (Lexington: University of Kentucky Press, 2009), 30; David B. Danbom, *Born in the Country: A History of Rural America* (Baltimore: Johns Hopkins University Press, 1995), 229; USDA National Agricultural Statistics Service, *Crop Production Historical Track Records* (April 2018), 31, 163–64, https://www.nass.usda.gov/Publications/Todays_Reports/reports/croptr18.pdf; USDA Economic Research Service, "Historical Costs and Returns: Soybeans"; USDA Economic Research Service, "Historical Costs and Returns: Corn," https://www.ers.usda

.gov/data-products/commodity-costs-and-returns/commodity-costs-and -returns/#Historical%20Costs%20and%20Returns:%20Corn.

12. Pond, interview, September 20, 2018.
13. Pond, interview, September 20, 2018; Pond, email correspondence, February 18, 2001; Mark Lynas, *Seeds of Science: Why We Got It so Wrong on GMOs* (London: Bloomsbury Sigma, 2020),104.
14. Pond, interview, September 20, 2018.
15. Marc Vanacht, interview by the author, November 26, 2018: Marc Vanacht LinkedIn page, https://www.linkedin.com/in/vanacht-marc-8b101362; *Farmers Edge Inc., and Farmers Edge (US) v. Farmmobile LLC*, Case No. 8:16-CV-00191-JFB-SMB in the United States District Court for the District of Nebraska, Order on Final Pretrial Conference, Addenda 2, https://www.govinfo.gov/content/pkg/USCOURTS-ned-8_16-cv-00191/pdf/USCOURTS-ned-8_16-cv-00191-8.pdf.
16. Vanacht, interview.
17. Vanacht, interview.
18. Vanacht, interview; Monsanto Company 2000 Annual Report, 8, 21, 24; "The Power of Roundup; A Weed Killer Is A Block For Monsanto to Build On," *New York Times*, August 2, 2001, C1; Carey Gillam, *Whitewashed: The Story of a Weed Killer, Cancer, and the Corruption of Science* (Washington, DC: Island Press, 2017), 46; On Roundup's complicated patent history, see "Last Roundup: Monsanto Herbicide's Patents Expiring," *St. Louis Post-Dispatch*, September 30, 1991, 9BP.
19. Vanacht, interview.
20. Vanacht, interview.
21. Pond, interview, September 20, 2018.
22. Stephen O. Duke, "The History and Current Status of Glyphosate," *Pest Management Science* 74, no. 5 (May 2018), 1027–30. Duke provided the data on glyphosate-resistant crops used in this study to the author in email correspondence on December 17, 2020. Sylvie Bonny, "Genetically Modified Herbicide-Tolerant Crops, Weeds and Herbicides: Overview and Impact," *Environmental Management* 57 (2016): 36.
23. Michael Livingston et al., *The Economics of Glyphosate Resistance Management in Corn and Soybean Production*, Economic Research Service, Economic Research Report Number 184 (April 2015), 1; USDA scientists explained in a 2014 study that "herbicide-tolerant crops have enabled the substitution of glyphosate for more toxic and persistent herbicides." See Jorge Fernandez-Conejo et al., *Genetically Engineered Crops in the United States*, Economic Research Service, Economic Research Report Number 162 (February 2014); Swinton and Van Deynze, "Hoes to Herbicides," 565.
24. Justin G. Gardner et al., "Genetically Modified Crops and Household Labor Savings in US Crop Production," *AgBioForum* 12 (2009): 310; National Academy of Sciences (NAS), *Genetically Engineered Crops: Experiences and Prospects* (Washington, DC: National Academies Press, 2016), 267; Fernandez-Conejo et al., *Genetically Engineered Crops in the United States*, 22.

25. Pond, interview, September 20, 2018.

26. The author would like to thank Ohio State weed scientist Mark Loux, who toured Ohio farms discussing weed-resistance issues with the author.

27. Special thanks to Ohio State alum Elliot Ping, who investigated this history of corporate financing in fall 2020. She filed a Freedom of Information Act request with The Ohio State University's Public Records Office. To access the data referenced here, see The Ohio State University Public Records Office FOIA Request 21-0509: Financial requests related to donations, research funding, and monies given by Monsanto to the University within the Office of Sponsored Programs and Endowments from 2000-2020; "Public Research, Private Gain: Corporate Influence Over University Agricultural Research," Food and Water Watch report (April 2012), 1–5, https://www.foodandwaterwatch.org/sites/default/files/Public%20 Research%20Private%20Gain%20Report%20April%202012.pdf; Letter from Thomas P. Hardy, Chief Records Officer at University of Illinois at Urbana-Champaign, to Monica Eng, Chicago Public Radio, forwarding response to FOIA No. 16–132, March 4, 2016; Monica Eng, "Why Didn't an Illinois Professor Have to Disclose GMO Funding?" WBEZ Chicago public radio website, March 15, 2016, https://www.wbez.org/stories/why-didnt-an -illinois-professor-have-to-disclose-gmo-funding/eb99bdd2–683d-4108– 9528–de1375c3e9fb; For more on public universities' lack of transparency when it comes to private donations, see Molly McCluskey, "Public Universities Get an Education in Private Industry," The Atlantic, April 3, 2017, https://www.theatlantic.com/education/archive/2017/04/public-universities -get-an-education-in-private-industry/521379/. On the corporate mind-set that drove early extension work and land-grant research as early as the 1920s, see Deborah Fitzgerald, Every Farm a Factory: The Industrial Ideal in American Agriculture (New Haven: Yale University Press, 2003), 8. For a work that shows the complicated relationship between businesses and agricultural improvement associations even earlier (1830s and 1840s), see Emily Pawley, The Nature of the Future: Agriculture, Science, and Capitalism in the Antebellum North (Chicago: University of Chicago Press, 2020).

28. Mark J. VanGessel, "Rapid Publication: Glyphosate-Resistant Horseweed from Delaware," Weed Science 49 (2001): 703–5; Mark Loux, interview by the author, September 5, 2018.

29. Mark Loux, interview by the author, September 5, 2018.

30. Loux, interview, May 1, 2017; Loux, interview, September 5, 2018.

31. Monsanto 2005 Technology Use Guide, 1, https://web.archive.org/ web/20060316093655/https://monsanto.com/monsanto/us_ag/content/ stewardship/tug/tug2005.pdf; Marie-Monique Robin, The World According to Monsanto (New York: New Press, 2010), 208–9; "Seeds of Discord: Monsanto's Gene Police Raise Alarm on Farmers' Rights, Rural Tradition," Washington Post, February 3, 1999, A1; Center for Food Safety, Monsanto vs. U.S. Farmers (2005), 25, 31, https://www.centerforfoodsafety.org/files/ cfsmonsantovsfarmerreport11305.pdf; Monsanto, 1996 Environmental Annual Review, 3, https://web.archive.org/web/19970714063108/http://www .monsanto.com:80/monpub/environment/monsantoear96/96earall.pdf.

32. "Seed Makers' Suits Show Hostility," *Arkansas Democrat-Gazette*, May 18, 2003, quoted here https://www.grain.org/article/entries/2051-syngenta -also-cracks-down-on-seed-saving; Center for Food Safety, *Monsanto vs. U.S. Farmers* (2005), 33, 45; "Monsanto Wins Patent Case on Plant Genes," *New York Times*, May 22, 2004, C1.

33. "Seeds of Discord: Monsanto's Gene Police Raise Alarm on Farmers' Rights, Rural Tradition," *Washington Post*, February 3, 1999, A1. For more on Schmeiser's story, see Robin, *The World According to Monsanto*, 213–16. Dan Charles, author of *The Lords of the Harvest*, has noted, "As far as I can tell, Monsanto has never sued anybody over trace amounts of GMOs that were introduced into fields simply through cross-pollination." See Dan Charles, "Top Five Myths of Genetically Modified Seeds, Busted," *NPR*, October 18, 2012, https://www.npr.org/sections/thesalt/2012/10/18/163034053/top-five -myths-of-genetically-modified-seeds-busted; Daniel Charles, *The Lords of the Harvest* (Cambridge, MA: Perseus Publishing, 2001), 188–89.

34. *Monsanto Canada Inc v. Schmeiser*, 1 S. C. R. 902, 2004 SCC 34; "Monsanto Wins Patent Case on Plant Genes," *New York Times*, May 22, 2004, C1; "*Monsanto Canada Inc., v. Schmeiser*," *Berkeley Technology Law Journal* 20, no. 1 (January 2005): 179.

35. "Seeds of Discord: Monsanto's Gene Police Raise Alarm on Farmers' Rights, Rural Tradition," *Washington Post*, February 3, 1999, A1; Robin, *The World According to Monsanto*, 208; Charles, *Lords of the Harvest*, 187; Author call to 1–800–ROUNDUP, January 15, 2018.

36. For a complete list of glyphosate-resistant weeds in the United States and around the world, see the International Survey of Herbicide Resistant Weeds, Weeds Resistant to the Herbicide Glyphosate, WeedScience.org, http://www.weedscience.org/Summary/ResistbyActive.aspx; "After years of research," Monsanto told farmers in a 2001 company brochure, "Roundup UltraMAX alone is the best choice for most weed control programs." Two years later, the company said that "the development of weed resistance to glyphosate is less likely than active ingredients in many other herbicides due to glyphosate's unique mode of action." Quoted in Sylvie Bonny, "Genetically Modified Herbicide-Tolerant Crops, Weeds and Herbicides: Overview and Impact," *Environmental Management* 57 (2016): 39; Livingston et al., *The Economics of Glyphosate Resistance Management in Corn and Soybean Production*, 21–22; A 2017 USDA study found that "glyphosate resistance had a significant impact on weed control costs and corn yields of US farmers in 2005 and 2010." Seth J. Wechsler, Jonathan R. McFadden, and David J. Smith, "What Do Farmers' Weed Control Decisions Imply About Glyphosate Resistance? Evidence from Surveys of US Corn Fields," *Pest Management Science* 74, no. 5 (2018): 1143.

37. Loux, interview, May 1, 2017.

38. Bonny, "Genetically Modified Herbicide-Tolerant Crops, Weeds and Herbicides," 43; Stephen O. Duke and Stephen B. Powles, "Mini-Review: Glyphosate: A Once-In-A-Century Herbicide," *Pest Management Science* 64 (2008): 319.

39. "Hugh Grant Is Elected President and Chief Executive Officer of Monsanto

Company," Monsanto News Release, May 29, 2003; "Verfaillie Resigns as CEO of Monsanto, Company Says It Was 'Mutual,'" *St. Louis Business Journal*, December 18, 2002; "Chief of Monsanto Resigns After String of Poor Results," *New York Times*, December 19, 2002, C2; "Monsanto Struggles Even as It Dominates," *New York Times*, May 31, 2003, C1.

40. "Monsanto's CEO Isn't Deterred by Rival's Rebuffs," *Wall Street Journal*, June 23, 2015, https://www.wsj.com/articles/monsantos-ceo-isnt-deterred-by-rivals-rebuffs-1435026949; "Life After the Monsanto Sale," *Wall Street Journal*, June 21, 2018, https://www.wsj.com/articles/life-after-the-monsanto-sale-1529586001; "Planting the Seeds of Growth," *Barron's*, August 10, 2013, https://www.barrons.com/articles/SB50001424052748703759004578 650384123880880; "Can Monsanto Save the Planet," *Fortune*, June 6, 2016, https://fortune.com/longform/monsanto-fortune-500-gmo-foods/.

41. "Planting the Seeds of Growth," *Barron's*, August 10, 2013; "Monsanto's Hugh Grant Has Altered Course When Needed," *Wall Street Journal*, June 23, 2015, https://www.wsj.com/articles/monsantos-hugh-grant-has-altered-course-when-needed-1435026641.

42. "Monsanto's CEO Isn't Deterred by Rival's Rebuffs," *Wall Street Journal*, June 23, 2015; "Planting the Seeds of Growth," *Barron's*, August 10, 2013.

43. "The Power of Roundup; A Weed Killer is a Block for Monsanto to Build On," *New York Times*, August 2, 2001, C1.

44. Interview with Hugh Grant for PBS documentary *Harvest of Fear*, December 2000, https://www.pbs.org/wgbh/harvest/interviews/grant.html.

45. Livingston et al., *The Economics of Glyphosate Resistance Management in Corn and Soybean Production*, 1; "Doubts About a Promised Bounty," *New York Times*, October 30, 2016, A1; Bonny, "Genetically Modified Herbicide-Tolerant Crops, Weeds and Herbicides," 36; Swinton and Van Deynze, "Hoes to Herbicides," 572; National Academy of Sciences (NAS), *Genetically Engineered Crops: Experiences and Prospects* (Washington, DC: National Academies Press, 2016), 135.

46. Gale E. Peterson, "The Discovery and Development of 2,4-D," *Agricultural History* 41, no. 3 (July 1967): 245; James R. Troyer, "In the Beginning: The Multiple Discovery of the First Hormone Herbicides," *Weed Science* 49, no. 2 (March–April, 2001): 290–97; Duke and Powles, "Mini-Review: Glyphosate," 319; Swinton and Van Deynze, "Hoes to Herbicides," 563, 571; In making this assessment, Swinton and Van Deynze relied on a USDA study conducted by Jorge Fernandez-Cornejo et al., "Conservation Tillage, Herbicide Use, and Genetically Engineered Crops in the United States: The Case of Soybeans," *AgBioForum* 15, no. 3 (2012): 235. French National Institute for International Research scholar Sylvie Bonny noted that the use of some herbicides, "notably 2,4-D, might have some harmful health or environmental effects"; Bonny, "Genetically Modified Herbicide-Tolerant Crops, Weeds and Herbicides," 32; William S. Pease et al., "Pesticide Use in California: Strategies for Reducing Environmental Health Impacts," An Environmental Health Program Report, Center for Occupational and Environmental Health, School of Public Health, University of California, Berkeley (1996), 29, 63; Lois Levitan, "An Overview of Pesticide Assessment

Systems (a.k.a. 'Pesticide Risk Indicators') based on Indexing or Ranking Pesticides by Environmental Impact," Background Paper Prepared for the Organisation of Economic Cooperation and Development (OECD), Workshop on Pesticide Risk Indicators, April 21–23, 1997, Copenhagen, Denmark, 5, 9, http://citeseerx.ist.psu.edu/viewdoc/download?doi=10.1.1.195 .3449&rep=rep1&type=pdf; Loux, interview, May 1, 2017.

47. F. L. Timmons, "A History of Weed Control in the United States and Canada," *Weed Science* 53 (2005), 754; Douglas J. Doohan and Roger A. Downer, "Reducing 2,4-D and Dicamba Drift Risk to Fruits, Vegetables and Landscape Plants," The Ohio State University Extension Fact Sheet, published January 21, 2016, https://ohioline.osu.edu/factsheet/hyg-6105; Bob Hartzler, "A Historical Perspective on Dicamba," Iowa State University Extension and Outreach PowerPoint presentation, December 19, 2017, https://crops .extension.iastate.edu/blog/bob-hartzler/historical-perspective-dicamba.

48. Hartzler, "A Historical Perspective on Dicamba"; Ellery Knake, "Weed Control in Corn and Soybeans," 140–41, conference presentation published in Summaries of Presentations, January 26 & 27, 1972, Twenty-Fourth Illinois Custom Spray Operators Training School, Urbana, Illinois; Monsanto Company 2005 Annual Report, 15.

49. Text message from Dan Jenkins to Michael Dykes, Vice President of Government Affairs, November 6, 2014; Text message from Dan Jenkins to Ty Vaughn, Monsanto's Director of Regulatory Affairs, August 27, 2015, marked "Confidential—Produced Subject to Protective Order," files released by the law firm of Baum Hedlund Aristei & Goldman [hereinafter Monsanto Papers] and made available at the firm's website, http://baumhedlundlaw .com/pdf/monsanto-documents/55–Text-Messages=Detailing=Monsantos -Collusion-with-EPA.pdf; Dan Jenkins LinkedIn profile, https://www .linkedin.com/in/dan-jenkins-7858286; Michael Dykes LinkedIn profile, https://www.linkedin.com/in/michael-dykes-ba60587.

50. On ghost writing, see, for example, Dr. David Saltmiras "custodial file," "Glyphosate activities," August 4, 2015, Monsanto Papers. On EPA's 2016 review of glyphosate, see EPA Office of Pesticide Programs, "Glyphosate Issue Paper: Evaluation of Carcinogenic Potential," 140, September 12, 2016, https://www.epa.gov/sites/production/files/2016–09/documents/glyphosate _issue_paper_evaluation_of_carcincogenic_potential.pdf.

51. Text messages involving Dan Jenkins and other Monsanto officials, February 11, 2003, to March 10, 2016, Document No.: MONGLY03293245; Email from Dan Jenkins to William F. Heydens, April 28, 2015; Text message from Mary Manibusan to Eric Sachs, June 21, 2015; Text message from Eric Sachs to Mary Manibusan, June 21, 2015, Monsanto Papers; Maria Dinzeo, "Roundup Cancer Trial: Emails Show Monsanto Cozy With Feds," *Courthouse News Service*, April 15, 2019, https://www.courthousenews .com/roundup-cancer-trial-emails-show-monsanto-cozy-with-feds/.

52. "Scant Oversight, Corporate Secrecy Preceded U.S. Weed Killer Crisis," *Reuters*, August 9, 2017, https://www.reuters.com/article/us-usa-pesticides -dicamba-insight/sc%E2%80%A6ght-corporate-secrecy-preceded-u-s -weed-killer-crisis-idUSKBN1AP0DN.

53. For Monsanto's argument regarding "misuse" of older dicamba formula-tions, see Defendant's Reply Brief: In Support of Motion to Dismiss, *Steven W. Landers v. Monsanto Company*, Case No. 1:17–cv-20–SNJ, US Eastern District Court, Southeastern Division, http://fingfx.thomsonreuters.com/gfx/rngs/MONSANTO-DICAMBA/010051MK3NZ/data/monsanto.pdf; For an excellent timeline of the dicamba-drift problem in 2015 and 2016, see "Special Report: The Decisions Behind Monsanto's Weed-Killer Crisis," *Reuters*, November 9, 2017, https://www.reuters.com/article/us-monsanto-dicamba-specialreport/special-report-the-decisions-behind-monsantos-weed-killer-crisis-idUSKBN1D91PZ.

54. EPA, "Dicamba/Auxin Formulations: An Update on Label Changes in Response to Reported Incidents," PowerPoint presentation at the Pesti-cide Program Dialogue Committee Meeting, November 1, 2017; "Dicamba Scofflaws A Worry for Board," *Arkansas Democrat-Gazette*, July 31, 2019, https://www.arkansasonline.com/news/2019/jul/31/dicamba-scofflaws-a-worry-for-board-201/.

55. Douglas J. Doohan and Roger A. Downer, "Reducing 2,4-D and Dicamba Drift Risk to Fruits, Vegetables and Landscape Plants," The Ohio State University Extension Service Fact Sheet, January 21, 2016; Memorandum and Order, *Bader Farms, Inc. v. Monsanto*, US District Court, Eastern District of Missouri, Southeastern Division, Case No. 1:16-CV-299-SNLJ, https://cases.justia.com/federal/district-courts/missouri/moedce/1:201 6cv00299/150890/50/0.pdf?ts=1491905010; Bev Randles (attorney repre-senting Bader Farms), interview by the author, January 10, 2018; Bev Ran-dles (attorney representing Bader Farms), email correspondence with the author, May 14, 2019.

56. Mark Loux, interview by the author, September 5, 2018.

57. "EPA Scientists' Dicamba Input Went Unheeded," *Arkansas Democrat-Gazette*, November 21, 2018, https://www.arkansasonline.com/news/2018/nov/21/epa-scientists-dicamba-input-went-unhee/.

58. "Seeds, Weeds and Divided Farmers" *New York Times*, September 21, 2017, B1; "Monsanto Attacks Scientists After Studies Show Trouble For Weed-killer Dicamba," *NPR*, October 26, 2017, https://www.npr.org/sections/thesalt/2017/10/26/559733837/monsanto-and-the-weed-scientists-not-a-love-story; "EPA Scientists' Dicamba Input Went Unheeded," *Arkan-sas Democrat-Gazette*, November 21, 2018, https://www.arkansasonline.com/news/2018/nov/21/epa-scientists-dicamba-input-went-unhee/; "EPA Announces Changes to Dicamba Registration," EPA News Release, Octo-ber 31, 2018, https://www.epa.gov/newsreleases/epa-announces-changes-dicamba-registration.

59. Jason Parker et al., "Symposium Introduction," The Ohio State University Agricultural Risk Analysis Program Symposium Proceedings, The New 2,4-D and Dicamba-Tolerant Crops: October 31 to November 1, 2011, http://fingfx.thomsonreuters.com/gfx/rngs/MONSANTO-DICAMBA /010051MK3NZ/data/riskanalysis.pdf; "Special Report: The Decisions Behind Monsanto's Weed-Killer Crisis," *Reuters*, November 9, 2017, https://www.reuters.com/article/us-monsanto-dicamba-specialreport/special

-report-the-decisions-behind-monsantos-weed-killer-crisis-idUSKBN 1D91PZ.

60. Robert B. Shapiro, Remarks to Society of Environmental Journalists, October 28, 1995, https://web.archive.org/web/19961111115403/http://www .monsanto.com:80/MonPub/InTheNews/Speeches/951028Shapiro_Robert .html; Charles, *Lords of the Harvest*, 222; "Monsanto: Genetics in Farming Raises Controversy in India," *St. Louis Post-Dispatch*, November 22, 1998, A12.

61. Stephen O. Duke et al., "Glyphosate Effects on Plant Mineral Nutrition, Crop Rhizosphere, Microbiota, and Plant Disease in Glyphosate-Resistant Crops," *Journal of Agricultural and Food Chemistry* 60 (2012): 10390; The National Academy of Sciences, citing a Duke study published three years later, referenced a graph showing yield-per-acre changes for various commodity crops, saying "there is no obvious change in the slope of cotton and maize, which have the BT and HR traits, or for soybean, which has only the HR trait." See National Academy of Sciences (NAS), *Genetically Engineered Crops*, 102; The author discussed Duke's findings and confirmed this statement on yield in an interview: Stephen O. Duke (USDA researcher), interview by the author, September 19, 2018; Marti Crouch described Stephen O. Duke as "Mr. Roundup": Marti Crouch, interview by the author, December 10, 2015.

62. Fernandez-Cornejo et al., *Genetically Engineered Crops in the United States*, 12, 16.

63. F. J. Areal et al., "Economic and Agronomic Impact of Commercialized GM Crops: A Meta-Analysis," *Journal of Agricultural Science* 151 (2013): 7; Giani Mariza Bärwald Bohm, "Glyphosate Effects on Yield, Nitrogen Fixation, and Seed Quality in Glyphosate-Resistant Soybean," *Crop Science* 54 (July–August 2014), 1737.

64. "Doubts About a Promised Bounty," *New York Times*, October 30, 2016, A1. For the data used in this article, see the Food and Agricultural Organization of the United Nations FAOSTAT website here, http://www.fao. org/faostat/en/#home. The author would like to thank *New York Times* reporter Danny Hakim for corresponding with me about this article. The Alliance for Science, an organization funded by the Gates Foundation that promotes GE technology at Cornell University, argued that the *New York Times* 2016 analysis was misleading and that it did not represent the full findings of the 2016 NAS study. The organization pointed to a 2017 study, Elisa Pellegrino, "Impact of Genetically-Engineered Maize on Agronomic, Environmental and Toxicological Traits; A Meta-Analysis of 21 Years of Field Data," *Scientific Reports* 8 (2018): 1–12, which found evidence of yield advantages for GE corn crops over non-GE corn crops. However, after reviewing this study, Fred Gould, the chair of the 2016 NAS study, said the following: "Our large scale [NAS] analysis was to look at overall yields of corn, cotton and soybeans to see if the rate of yield increase over time was impacted by genetic engineering—and we found no effect. This study does not refute that large scale finding." Fred Gould, correspondence with the author, November 27, 2020.

65. National Academy of Sciences (NAS), *Genetically Engineered Crops: Experiences and Prospects*, 14, 102, 154; Fred Gould (chair of the 2016 NAS study), interview by the author, August 24, 2018.

66. National Academy of Sciences (NAS), *Genetically Engineered Crops: Experiences and Prospects*, 14, 102, 154; Duke, interview; Allison Snow (Distinguished Professor Emerita in Arts and Sciences, Department of Evolution, Ecology, and Organismal Biology and reviewer for the 2016 NAS study), interview by the author, September 26, 2018.

67. Gould, interview; Duke and Powles, "Mini-Review: Glyphosate," 322.

68. "USDA Coexistence Fact Sheets: Soybeans," USDA Office of Communications publication (February 2015), https://www.usda.gov/sites/default/files/documents/coexistence-soybeans-factsheet.pdf; USDA Coexistence Fact Sheets: Corn," USDA Office of Communications publication (February 2015), https://www.usda.gov/sites/default/files/documents/coexistence-corn-factsheet.pdf; Jonathan Foley, "A Five-Step Plan to Feed the World," *National Geographic Magazine* (May 2014), https://www.nationalgeographic.com/foodfeatures/feeding-9-billion/.

69. Roundup Ready Xtend Crop System advertisement, NorthStar Genetics 2016/2017, Product Guide, 18, https://issuu.com/intrepid604/docs/nsg_ca_product_guide_v2_20160714jf_; Monsanto Company, "New Roundup Ready® Xtend Crop System To Extend Weed Control and Maximize Yield," PR Newswire website, March 1, 2012, https://www.prnewswire.com/news-releases/new-roundup-ready-xtend-crop-system-to-extend-weed-control-and-maximize-yield-141022713.html.

70. Duke, "The History and Current Status of Glyphosate," 1030.

CHAPTER 13: **"They Are Selling Us a Problem We Don't Have"**

1. João Paulo Capobianco (former executive secretary of Ministério do Meio Ambiente ([MMA]), interview by the author, São Paulo, Brazil, May 3, 2019; On the history of Luiz Inácio Lula da Silva's political rise, see Thomas E. Skidmore, *Brazil: Five Centuries of Change*, 2nd ed. (New York: Oxford University Press, 2010), 203–4, 215–25, 229–32.

2. Marie-Monique Robin, *The World According to Monsanto* (New York: New Press, 2010), 276; "Monsanto recorrerá da proibição a transgênico," *O Estado de S. Paulo*, August 18, 1999, A14; Felipe Amin Filomeno, *Monsanto and Intellectual Property in South America* (Basingstoke, UK: Palgrave Macmillan, 2014), 30–31, 54–55. The author would like to thank Felipe Amin Filomeno for his assistance in investigating the history of Monsanto's seed operations in South America.

3. On Monsanto's early interest in Brazil in the 1960s, see "International Operations, Long Range Plan, 1965–1969," Monsanto company document marked confidential (June 1965), 66, series 1, box 4, folder: International Division (History) (Folder 2), Monsanto Company Records, Washington University in St. Louis, Julian Edison Department of Special Collections, St. Louis, Missouri. Monsanto Company 2001 Annual Report, 13, 32;

Monsanto Company 2004 Annual Report, 10; Jan Peter Nap et al., "The Release of Genetically Modified Crops into the Environment," *Plant Journal* 33 (2003): 1–2; Clara Craviotti, "Which Territorial Embeddedness? Territorial Relationships of Recently Internationalized Firms of the Soybean Chain," in *Soy, Globalization, and Environmental Politics in South America*, eds. Gustavo de L. T. Oliveira and Susanna B. Hecht (New York: Routledge, 2018), 85; There were legal reasons why Brazil was considered such an important place for Monsanto to expand its seed business. In 1996 the Brazilian government had passed the Law of Industrial Property and in 1997 offered enhanced intellectual property protections for agricultural products through the Law of Protection of the Cultivars. The first piece of legislation specifically offered protections for biotechnological innovations while the later offered protection to plant breeders that developed unique varieties. If Monsanto could gain approval for Roundup Ready seed sales in Brazil, it could profit from patent protections it did not enjoy in Argentina. See Filomeno, *Monsanto and Intellectual Property in South America*, 90.

4. Filomeno, *Monsanto and Intellectual Property in South America*, 87–88; Capobianco, interview; "Rocky Outlook for Genetically Engineered Crops," *New York Times*, December 20, 1999, C8.

5. Capobianco, interview; "A MP da soja transgênico," *O Estado de S. Paulo*, March 31, 2003, A3; Filomeno, *Monsanto and Intellectual Property in South America*, 90.

6. Capobianco, interview.

7. 2006 Pledge Report, 39; Christine M. Du Bois and Ivan Sergio Freire de Sousa, "Genetically Engineered Soy," in *The World of Soy*, eds. Christine M. Du Bois, Chee-Beng Tan, and Sidney Mintz (Urbana: University of Illinois Press, 2008), 84; Craviotti, "Which Territorial Embeddedness?," 85; Herbert S. Klein and Francisco Vidal Luna, *Feeding the World: Brazil's Transformation into a Modern Agricultural Economy* (New York: Cambridge University Press, 2019), 173; Robin, *The World According to Monsanto*, 277.

8. Capobianco, interview.

9. Gustavo de L. T. Oliveira and Susanna B. Hecht, "Sacred Groves, Sacrifice Zones and Soy Production: Globalization, Intensification and Neo-nature in South America," in *Soy, Globalization, and Environmental Politics in South America*, eds. Gustavo de L. T. Oliveira and Susanna B. Hecht (New York: Routledge, 2018), 5–6.

10. "We decided that, near term, our biggest block should be the Americas, where the majority of crops are grown," Hugh Grant told reporters in 2004. "Piracy on the High Plains," *Forbes*, April 12, 2004, https://www.forbes .com/forbes/2004/0412/135.html#787f03a21eb4; "Monsanto's CEO Isn't Deterred by Rival's Rebuffs," *Wall Street Journal*, June 23, 2015, https:// www.wsj.com/articles/monsantos-ceo-isnt-deterred-by-rivals-rebuffs -1435026949; "Monsanto Vai Investir US$ 630 milhões até 2001," *O Estado de S. Paulo*, November 21, 1998, B11; Robin, *The World According to Monsanto*, 278; Direct Testimony of Kevin P. Lawrence Before the Public

Utilities Commission of the State of Idaho, in the Matter of the Application of Rocky Mountain Power for Approval of Changes to Its Electric Service Rules, Case No. PAC-E-10-07, October 14, 2010, in author's possession.

11. The government reduced this tariff dramatically in 2008 and again in 2009. Filomeno, *Monsanto and Intellectual Property in South America*, 76–77; "Mais da metade dos recursos do Finor foi para uma só empresa," *O Estado de S. Paulo*, June 24, 2000, A4; "Prefeito visita fábrica da Monsanto em São José," Prefeitura São José Dos Campos website, July 25, 2015, https://servicos2.sjc.sp.gov.br/noticias/noticia.aspx?noticia_id=21327.

12. Stephen O. Duke, "The History and Current Status of Glyphosate," *Pest Management Science* (2018): 1029; Sylvia Bonny, "Genetically Modified Herbicide-Tolerant Crops, Weeds, and Herbicides: Overview and Impact," *Environmental Management* (2016): 34; "Brazil Boasts World's Second Largest Genetically Modified Crop Area: ISAAA," *Reuters*, June 27, 2018, https://www.reuters.com/article/us-brazil-gmo/brazil-boasts-worlds-second-largest-genetically-modified-crop-area-isaaa-idUSKBN1JN1KW.

13. Klein Luna, *Feeding the World*, 243, 312; Philip F. Warnken, *The Development and Growth of the Soybean Industry in Brazil* (Ames: Iowa State University Press, 1999), 28.

14. Warnken, *Development and Growth of the Soybean Industry in Brazil*, 32–33, 47, 50.

15. Klein and Luna, *Feeding the World*, 39, 159, 161, 165; Filomeno, *Monsanto and Intellectual Property in South America*, 70; Warnken, *Development and Growth of the Soybean Industry in Brazil*, 42–43, 46–48, 148; Ivan Sergio Freire de Sousa and Rita de Cássia Milagres Texeira Vieria, "Soybeans and Soyfoods in Brazil, with Notes on Argentina: Sketch of an Expanding World Commodity," in *The World of Soy*, eds. Christine M. Du Bois, Chee-Beng Tan, and Sidney Mintz (Singapore: National University of Singapore Press, 2008), 239; Thomas E. Skidmore, *Brazil: Five Centuries of Change*, 2nd ed. (New York: Oxford University Press, 2010), 170–71; Francisco Jose Becker Reifschneider (secretary of intelligence and strategic relations, Embrapa), interview by the author, Embrapa headquarters, Brasilia, Brazil, May 7, 2019. None of this would have been possible were it not for dictator Getúlio Vargas's 1940s "March to the West," a government initiative that focused on developing roadways and other key infrastructure in the Cerrado and beyond. There were also international players that played a big role, including the Japanese government, which offered funding to Brazil to help Japanese immigrants settle in the Cerrado in the 1970s and 1980s. On the "March to the West," see Ludivine Eloy et al., "On the Margins of Soy Farms: Traditional Populations and Selective Environmental Policies in the Brazilian Cerrado," in *Soy, Globalization, and Environmental Politics in South America*, eds. Gustavo de L. T. Oliveira and Susanna B. Hecht (New York: Routledge, 2018), 248; Gustavo de L. T. Oliviera, "The Geopolitics of Brazilian Soybeans," in *Soy, Globalization, and Environmental Politics in South America*, eds. Gustavo de L. T. Oliveira and Susanna B. Hecht (New York: Routledge, 2018), 101.

16. Warnken, *Development and Growth of the Soybean Industry in Brazil*, 42, 46, 48, 133. Scholars Hecht and Oliviera point out that increases in

productivity for soybean farmers was less dramatic than for corn growers, largely because soybean farmers started out practicing more "modern," market-oriented farming techniques, deploying advanced mechanized equipment and chemical fertilizers, whereas many corn farmers made the switch from subsistence production to commodity farming, which resulted in a much bigger change in farming practices. See Klein and Luna, *Feeding the World*, 117.

17. "Concentração e Lógica de Mercado," *O Estado de S. Paulo*, July 22, 1998, A3; Robin, *The World According to Monsanto*, 278; Filomeno, *Monsanto and Intellectual Property in South America*, 81.

18. "Concentração e Lógica de Mercado," *O Estado de S. Paulo*, July 22, 1998, A3; Professor Ricardo Shirota, (Economics Administration and Sociology Department at ESALQ–University of Sao Paulo), interview by the author, Piracicaba, Brazil, April 29, 2019; Rachael D. Garrett and Lisa L. Rausch, "Green for Gold: Social and Ecological Tradeoffs Influencing the Sustainability of the Brazilian Soy Industry," In *Soy, Globalization, and Environmental Politics in South America*, eds. Gustavo de L. T. Oliveira and Susanna B. Hecht (New York: Routledge, 2018), 221–22

19. See, for example, discussion of Brazilian farmer on page 10 of the Monsanto Company 2017 Annual Report; "Monsanto Seeks Big Increase in Crop Yields," *New York Times*, June 5, 2008, C3.

20. Mato Grosso Institute of Agricultural Economics (IMEA), "Custo de Produção de Soja, Safra 2017/2018–Matto Grosso, Outubro/2016," https://web.archive.org/web/20180619052941/http://www.imea.com.br/upload/publicacoes/arquivos/CPSoja_oUTUBRO_16.pdf. See detail on bottom of the seventh page, "Productividade Média Ponderada Estimada para Convencional 60 sacas/ha . . . Productividade Média Ponderada Estimada para Transgênico: 55 sacas/ha." As this book went to press, IMEA published data for 2020, which showed that in some parts of Mato Grosso, GE crops outyielded conventional crops, but in the "oeste" region of the state, conventional and GE farmers reported the exact same productivity on their farms, "64.78 sacas/ha." See IMEA, "Custo de Produção, Soja Convencional, Região Oeste, Mato Grosso" and "Custo de Produção, Soja GMO, Região Oeste, Mato Grosso," https://www.imea.com.br/imea-site/relatorios-mercado-detalhe?c=4&s=3.

21. Farmer, interview by the author, Rolândia, Brazil, April 26, 2019; Mauro Tribulato (farmer and proprietor of Silo Panamá grain storage facility), interview by the author, Rolândia, Brazil, April 26, 2019; Dr. Fernando Storniolo (Adegas, Pesquisador in Plantas Daninhas) and Dr. Antonio Eduardo Pípolo (Fitotecnica), interview by the author, Londrina, Brazil, April 26, 2019. The author would like to thank Felipe Fadel Sartori, PhD crop scientist at ESALQ–University of Sao Paulo, for arranging many of these interviews and serving as a guide when the author visited Brazil in 2019; Warnken, *Development and Growth of the Soybean Industry in Brazil*, 35; Evgenia (Jenia) Ustinova (agricultural attaché, US Department of Agriculture, Foreign Agricultural Service), interview by the author, Brasilia, Brazil, May 6, 2019; Amalia Leguizamón, "Disappearing Nature? Agribusiness, Biotechnology and Distance in Argentine Soybean

Production," in *Soy, Globalization, and Environmental Politics in South America*, eds. Gustavo de L. T. Oliveira and Susanna B. Hecht (New York: Routledge, 2018), 71, 72; Gustavo de L. T. Oliveira and Susanna B. Hecht, "Introduction: Sacred Groves, Sacrifice Zones and Soy Production: Globalization, Intensification and Neo-Nature in South America," in *Soy, Globalization, and Environmental Politics in South America*, eds. Gustavo de L. T. Oliveira and Susanna B. Hecht (New York: Routledge, 2018), 12, 16.

22. Mato Grosso Institute of Agricultural Economics (IMEA), "Custo de Produção de Soja, Safra 2016/2017–Matto Grosso, Outubro/2015," http://www .imea.com.br/upload/publicacoes/arquivos/R410_CPSoja_10_2015.pdf. http://www.imea.com.br/upload/publicacoes/arquivos/R410_CPSoja_10 _2015.pdf

23. Site is no longer active.

24. Antonio L. Cerdeira et al., "Review of Potential Environmental Impacts of Transgenic Glyphosate-Resistant Soybean in Brazil," *Journal of Environmental Science and Health, Part B: Pesticides, Food Contaminants, and Agricultural Wastes* 42, no. 5 (2007), 539.

25. Sebastião Pedro da Silva Neto, (Embrapa Innovation and Technology Department), interview by the author, Embrapa headquarters, Brasilia, Brazil, May 7, 2019.

26. Professor Rafael Pedroso, (ESALQ), interview by the author, Piracicaba, Brazil, Monday, April 29, 2019.

27. Pedroso, interview.

28. "Brazil Grain Growers Wary of Dicamba as Bayer Launches New GM Soy Seed," *Reuters*, September 30, 2019, https://br.reuters.com/article/us-brazil -dicamba-idUSKBN1WF1UY.

29. Klein and Luna, *Feeding the World*, 166; Filomeno, *Monsanto and Intellectual Property in South America*, 101; "Brazil's Mato Grosso Leads Push for GM-Free Soy," *Reuters*, May 11, 2017, https://www.reuters.com/ article/brazil-grains-gmo/brazils-mato-grosso-leads-push-for-gm-free-soy -idUSL1N1IA0KW; "Soja Livre Institute," Embrapa website, https://www .embrapa.br/soja/convencional/sojalivre.

30. "Brazil Grain Growers Wary of Dicamba as Bayer Launches New GM Soy Seed," *Reuters*.

31. "Monsanto Unfazed by Legal Wrangles, Keeps Brazil Dicamba-Tolerant Seed Launch," *Reuters*, March 16, 2018, https://www.reuters.com/article/us -monsanto-gmo/monsanto-unfazed-by-legal-wrangles-keeps-brazil-dicamba -tolerant-seed-launch-idUSKCN1GS2IT; "Monsanto Wins Approval in Brazil for GM Soy Seed Intacta2 Xtend," *Reuters*, March 8, 2018, https://www .reuters.com/article/brazil-grains-monsanto/monsanto-wins-approval-in -brazil-for-gm-soy-seed-intacta2–xtend-idUSE6N1ND02K; "Agroeste Apresenta variedades de soja com tecnologia Intacta 2 Xtend®," Bayer corporate website, December 10, 2020, https://www.bayer.com.br/pt/midia/agroeste -apresenta-variedades-de-soja-com-tecnologia-intacta-2-xtend.

32. This account is based on the author's visit to VAVA's Hanoi office in June 2017. On estimates of Agent Orange victims in Vietnam, see the assessment

of Jeanne Mager Stellman's research team in 2003 in "The Extent and Patterns of Usage of Agent Orange and Other Herbicides in Vietnam," *Nature* 422 (April 17, 2003): 685; Charles Bailey and Le Ke Son, *From Enemies to Partners: Vietnam, the U.S. and Agent Orange* (Chicago: G. Anton Publishing, 2018), 1399, 2757 [Kindle edition].

33. Hatfield Consultants, "Preliminary Assessment of Environmental Impacts Related to Spraying Of Agent Orange Herbicide During the Vietnam War," Full Report (January 1998); "Development of Impact Mitigation Strategies Related to the Use of Agent Orange Herbicide in the Aluoi Valley, Viet Nam," Main Report (April 2000), https://www.hatfieldgroup.com/services/contaminant-monitoring-agent-orange/hatfield-agent-orange-reports-and-presentations/; Bailey and Son, *From Enemies to Partners*, 437, 510–701. For more on Agent Orange contamination at US air bases and cleanup campaigns, see Edwin A. Martini, " 'This is Really Bad Stuff Buried Here,' " in *Proving Ground: Militarized Landscapes, Weapons Testing, and the Environmental Impact of U.S. Bases*, ed. Edwin A. Martini (Seattle: University of Washington Press, 2015), 111–42.

34. Trần Thị Tuyết Hạnh et al., "Environmental Health Risk Assessment of Dioxin in Foods at the Two Most Severe Dioxin Hot Spots in Vietnam," *International Journal of Hygiene and Environmental Health* 218, no. 5 (July 2015): 471. For an earlier Hạnh study on dioxin and human health, see Trần Thị Tuyết Hạnh et al., "Environmental Health Risk Assessment of Dioxin Exposure Through Foods in a Dioxin Hot Spot—Bien Hoa City, Vietnam," *International Journal of Environmental Research and Public Health* 7, no. 5 (May 2010): 2395–2406.

35. See Arnold Schecter et al., "Food As a Source of Dioxin Exposure in the Residents of Bien Hoa City, Vietnam," *Journal of Occupational and Environmental Medicine* 45, no. 8 (August 2003): 781. Schecter edited the touchstone textbook on dioxin and human health. See Arnold Schecter, ed., *Dioxins and Health: Including Other Persistent Organic Pollutants and Endocrine Disruptors* (Hoboken, NJ: John Wiley & Sons, 2012). Schecter testified before Congress in 2008 to push for assistance to Vietnam victims of Agent Orange. See House Hearing of the Subcommittee on Asia, the Pacific, and the Global Environment, Committee on Foreign Affairs, *Our Forgotten Responsibility: What Can We Do To Help Victims of Agent Orange?*, May 15, 2008, 110 Cong., Sess. 2, 72–73. The author is grateful to Dr. Schecter for discussing his involvement in dioxin research: Dr. Arnold Schecter, telephone interview by the author, March 8, 2017.

36. *In re 'Agent Orange' Product Liability Litigation. The Vietnam Association for Victims of Agent Orange/Dioxin et al. v. The Dow Chemical Company; Monsanto Chemical Company, et al.* 373 F. Supp. 2d 7 (2005), 1734.

37. *In re 'Agent Orange' Product Liability Litigation. The Vietnam Association for Victims of Agent Orange/Dioxin et al. v. The Dow Chemical Company; Monsanto Chemical Company et al.*, 373 F. Supp. 2d 7, Amended Memorandum, Order, and Judgement, 15–19, 27–46, https://www.courtlistener.com/opinion/2312256/in-re-agent-orange-product-liability-litigation/; For fascinating insight into Judge Weinstein's thoughts on Agent Orange

litigation he presided over, see Jack B. Weinstein, "Preliminary Reflections on Administration of Complex Litigations," *Cardozo Law Review De-Novo* 1 (2009): 1–19 [inaugural article]. For a summary assessment of this case by legal scholar Takeshi Uesugi, see "Is Agent Orange a Poison?: Vietnamese Agent Orange Litigation and the New Paradigm of Poison," *Japanese Journal of American Studies* No. 24 (2013): 203–22; Bailey and Son, *From Enemies to Partners*, 1734, 1858, 2784.

38. David R. Plummer, "Vietnam Veterans and Agent Orange" (master's thesis, California State University Dominguez Hills, 2000), 68–69; For official Department of Veterans Affairs policy regarding Agent Orange, see "Veterans Exposed to Agent Orange," US Department of Veterans Affairs benefits website, https://www.benefits.va.gov/compensation/claims-postservice -agent_orange.asp.

39. Peter H. Schuck, *Agent Orange on Trial: Mass Toxic Disasters in the Courts* (Cambridge, MA: Belknap Press of Harvard University Press, 1987), 156; Bailey and Son, *From Enemies to Partners*, 1757–68; Department of Veterans Affairs, Office of Public Affairs Media Relations, "Over $2.2 Billion in Retroactive Agent Orange Benefits Paid to 89,000 Vietnam Veterans and Survivors for Presumptive Conditions," Veterans Affairs Press Release, August 31, 2011, https://www.va.gov/opa/pressrel/pressrelease .cfm?id=2154.

40. "What To Do When Cynicism Becomes Your Political Default," *The Hill*, September 12, 2016, http://thehill.com/blogs/pundits-blog/lawmaker -news/295411–what-to-do-when-cynicism-becomes-your-political-default; "Tim Rieser: Senate Aide," *Politico Magazine*, Politico 50, 2015 issue, https://www.politico.com/magazine/politico50/2015/tim-rieser; Bailey and Son, *From Enemies to Partners*, 2172; Tim Rieser, email correspondence with the author, February 5, 2018.

41. Tim Rieser, telephone interview by the author, July 11, 2017; Rieser, email correspondence.

42. Bailey and Son, *From Enemies to Partners*, 231. Charles Bailey offered insights into the Agent Orange cleanup program in an interview with the author on April 5, 2017.

43. Rieser, telephone interview.

44. Rieser, telephone interview.

45. Bailey and Son, *From Enemies to Partners*, 643, 2771, 2738, 2744; Rieser, email correspondence.

46. Chris Abrams (USAID), interview by the author, Hanoi, Vietnam, June 7, 2017; The author visited the Da Nang site in June 2017; USAID, *Vietnam: Environmental Remediation of Dioxin Contamination at Danang Airport*, USAID Progress Report for September 1, 2016, to September 30, 2016; Progress Report for August 1, 2013, to August 31, 2013; Progress Report for December 1, 2013, to December 31, 2013; Progress Report for October 1, 2016, to December 31, 2016, https://www.usaid.gov/vietnam/progress -reports-environmental-remediation-dioxin-contamination-danang-airport.

47. Abrams, interview; Schecter, telephone interview; Bailey Son, *From Enemies to Partners*, 1427; US Congressional Research Service, "US Agent

Orange/Dioxin Assistance," CRS report, January 15, 2021, frontmatter, https://fas.org/sgp/crs/row/R44268.pdf.

48. The five times figure comes from conversations with Tim Rieser and Chris Abrams: Abrams, interview; Rieser, telephone interview. Bailey and Son, *From Enemies to Partners*, 559, 671.

49. Rieser, telephone interview; Rieser, email correspondence.

50. Rieser, telephone interview; Rieser, email correspondence.

51. 2017 Monsanto Company SEC Form 10–K, 16.

52. Dien Luong, "55 Years After Agent Orange Was Used in Vietnam, One of Its Creators is Thriving Here," *Huffington Post*, October 31, 2017. Reporter Dien Luong was kind enough to meet with the author in Vietnam and talk about his investigations into this topic. This section of the chapter would not have been possible without his reporting and assistance.

53. Luong, "55 Years After Agent Orange Was Used in Vietnam." American seed sellers had long deployed this strategy of creating visual spectacles on farms to sell a particular vision of modernity. Especially during the Green Revolution, US foreign-policy makers hoped eye-popping agricultural demonstrations in the developing world would serve as powerful symbols of the superiority of America's economic system. See, for example, Nick Cullather, "Miracles of Modernization: The Green Revolution and the Apotheosis of Technology," *Diplomatic History* 28, no. 2 (April 2004), 227–54.

54. Luong, "55 Years After Agent Orange Was Used in Vietnam"; "Monsanto Subsidiary Launches Genuity Corn Seeds in Vietnam," *Vietnam Investment Review*, October 12, 2015, http://www.vir.com.vn/monsanto-subsidiary -launches-genuity-corn-seeds-in-vietnam.html

55. Luong, "55 Years After Agent Orange Was Used in Vietnam"; USDA Foreign Agricultural Service (FAS) Global Agriculture Information Network (GAIN) Report, "Vietnam Biotechnology Update 2008," July 11, 2008, GAIN Report No. VM8051, 7–8; USDA FAS GAIN Report, "Vietnam Agricultural Biotechnology Annual 2015," July 8, 2015, GAIN Report No. VM5042, 13–15.

56. Luong, "55 Years After Agent Orange Was Used in Vietnam"; Brian Leung, "Vietnam, Agent Orange, and GMOS: An Agent Orange Maker is Being Welcomed Back to Vietnam to Grow Genetically Modified Organisms," *The Diplomat*, November 24, 2014, https://thediplomat.com/2014/11/vietnam -agent-orange-and-gmos/; "Monsanto's Answer to Vietnam's Burgeoning Nutrition Demand," *Vietnam Investment Review*, July 11, 2017, https:// www.vir.com.vn/monsantos-answer-to-vietnams-burgeoning-nutrition -demand-53597.html; "Monsanto and Vietnam University of Agriculture Collaborate to Develop Talents in Agricultural Biotechnology," *Vietnam Investment Review*, December 10, 2017, http://www.vir.com.vn/monsanto -and-vietnam-university-of-agriculture-collaborate-to-develop-talents-in -agricultural-biotechnology.html.

57. USDA FAS GAIN Report, "Vietnam Agricultural Biotechnology Annual 2015," 14–15.

58. For a description of the genes inserted into MON 89034 and NK 603, see the International Service for the Acquisition of Agri-Biotech Applications,

"Event Name: MON8934," http://www.isaaa.org/gmapprovaldatabase/event/default.asp?EventID=95; "Event Name: NK603," http://www.isaaa.org/gmapprovaldatabase/event/default.asp?EventID=86.

59. Luong, "55 Years After Agent Orange Was Used in Vietnam"; Studies published as this book moved to press showed that Vietnamese adopters of GE crops have seen increased revenue streams and yield increases, at least in the short term. See Graham Brookes and Tran Xuan Dinh, "The Impact of Using Genetically Modified (GM) corn/maize in Vietnam: Results of the First Farm-Level Survey" *GM Crops and Food* 12, no. 1 (2020), 71–83.

CONCLUSION: **"Malicious Code"**

1. 2020 Bayer Shareholders' Meeting, webcast available here: https://www.bayer.com/en/annual-stockholders-meeting-2020.aspx.
2. "In 1918 Pandemic, Another Possible Killer: Aspirin," *New York Times*, October 12, 2009, D5.
3. Harry A. Stewart, "It's Dangerous to be Too Good a Loser!," *American Magazine*, May 1925, series 14, box 22, folder: Queeny, John F. (Misc.), Folder 1; Letter from Juanita McCarthy to H. A. Marple, May 2, 1961, series 1, box 6, folder: Monsanto Company History (WWI), Monsanto Company Records, Washington University in St. Louis, Julian Edison Department of Special Collections, St. Louis, Missouri; Walter Sneader, "The Discovery of Aspirin: A Reappraisal," *British Medical Journal* 321 (December 23–30, 2000): 1591–94.
4. "The Quiet Bayer Boss Hit by Germany's First Big No Confidence Vote," *Financial Times*, May 3, 2019, https://www.ft.com/content/abc5ad80-6d71-11e9-80c7-60ee53e6681d.
5. "The Sobering Details Behind the Latest Seed Monopoly Chart," *Civil Eats*, January 11, 2019, https://civileats.com/2019/01/11/the-sobering-details-behind-the-latest-seed-monopoly-chart/#:~:text=History%20shows%20us%20that%20seed,seed%20saving%20and%20research%20purposes.
6. "Bayer Pursued Monsanto Despite Weedkiller Suits and Executive Concerns," *Wall Street Journal*, November 25, 2018, https://www.wsj.com/articles/bayer-pursued-monsanto-despite-weedkiller-suits-and-executives-concern-1543147201; "With Each Roundup Verdict, Bayer's Monsanto Purchase Looks Worse," *Bloomberg Businessweek*, September 19, 2019, https://www.bloomberg.com/news/features/2019-09-19/bayer-s-monsanto-purchase-looks-worse-with-each-roundup-verdict.
7. "Bayer Pursued Monsanto Despite Weedkiller Suits and Executive Concerns," *Wall Street Journal*; "Bayer Transforms to Pure Life Sciences Leader with Winning Ways," *Financial Times*, September 16, 2015, https://www.ft.com/content/9b8b7194-5b9f-11e5-9846-de406ccb37f2.
8. "Bayer Pursued Monsanto Despite Weedkiller Suits and Executive's Concern," *Wall Street Journal*, November 25, 2018; "Monsanto Weedkiller Roundup Was 'Substantial Factor' in Causing Man's Cancer, Jury Says," *New York Times*, March 19, 2019; "From Toxic to Turnaround," *Wall Street Journal*, February 10, 2020, https://www.wsj.com/articles/from

-toxic-to-turnaround-bayers-ceo-fights-to-fix-a-problem-of-his-own
-making-11581332607; "Bayer Execs Face Investor Heat After Rare No
Confidence Vote," *Financial Times*, April 28, 2019, https://www.ft.com/
content/0a6cc01c-69a3-11e9-80c7-60ee53e6681d.

9. "$2 Billion Verdict Against Monsanto Is Third to Find Roundup Caused
Cancer," *New York Times*, May 13, 2019; "Bayer CEO Has 9 Months to
Overcome Shareholders' Extraordinary Rebuke of His Monsanto Deal,"
Fortune, April 30, 2019, https://fortune.com/2019/04/30/bayer-werner
-baumann-shareholder-meeting/; "Will Bayer CEO Lose Another Confidence Vote?" *Fierce Pharma*, April 15, 2020, https://www.fiercepharma
.com/pharma/will-bayer-ceo-lose-another-confidence-vote-2-top-proxy
-advisers-have-different-opinions; "Bayer CEO Seeks Investor Patience
as Cloud of Roundup Claims Persist," *Bloomberg*, April 28, 2020,
https://www.bloomberg.com/news/articles/2020-04-27/bayer-seeks
-investor-forbearance-over-lack-of-closure-on-roundup-k9ipoggd?utm_
source=google&utm_medium=bd&utm_campaign=HP&cmpId=GP.HP;
For the status of various glyphosate bans around the world, see "Where
is Glyphosate Banned?", Baum Hedlund law firm website, https://www.
baumhedlundlaw.com/toxic-tort-law/monsanto-roundup-lawsuit/where-is-
glyphosate-banned-/; Vietnam Extends the Use of Glyphosate Until June
2021, USDA Foreign Agricultural Service GAIN Report VM 2020-0045,
published May 28, 2020, https://apps.fas.usda.gov/newgainapi/api/Report
/DownloadReportByFileName?fileName=Vietnam%20Extends%20the
%20Use%20of%20Glyphosate%20until%20June%202021_Hanoi_
Vietnam_05-14-2020#:~:text=On%20April%2010%2C%202019%2C
%20MARD,Plant%20Protection%20Products%20in%20Vietnam.&text
=Accordingly%2C%20MARD%20suspended%20all%20registrations
,glyphosate%20on%20June%2010%2C%202019.

10. Transcript of Testimony at 2313, *Bader Farms, Inc. v. Monsanto Co., and
BASF Corporation*, MDL No. 1:18-md-2820-SNLJ, Case No. 1:16-CV-299-
SNLJ, US District Court, E. D. Missouri, Southeastern Division (2019);
Stock prices for Bayer came from the company's "Shareholder Information"
webpage,   https://www.bayer.com/en/investors/shareholder-information
?gclid=Cj0KCQiA4L2BBhCvARIsAO0SBdYfnSyW4I-JzvQ3awSmSuKlu_
z1fMCLSqYWLBYBl9jk5FcJj-zHMEQaAvd-EALw_wcB.

11. 2020 Bayer Shareholders' Meeting, webcast available here: https://www
.bayer.com/en/annual-stockholders-meeting-2020.aspx; "From Toxic to
Turnaround," *Wall Street Journal*, February 10, 2020; "Roundup Maker
Agrees to Pay More Than $10 Billion to Settle Thousands of Claims That
the Weedkiller Causes Cancer," *New York Times*, June 25, 2020, B1; "The
Quiet Bayer Boss Hit by Germany's First Big No Confidence Vote," *Financial Times*, May 3, 2019, https://www.ft.com/content/abc5ad80-6d71-
11e9-80c7-60ee53e6681d; On job cutting after the 2018 Monsanto merger,
see "Bayer to Sell Businesses, Cut Jobs After Monsanto Deal," WSAU News,
November 29, 2018, https://wsau.com/news/articles/2018/nov/29/bayer-to
-sell-businesses-cut-jobs-after-monsanto-deal/.

12. Opinion in the *National Family Farm et al. v. U. S. Environmental Protection*

Agency case, filed June 3, 2020, at 5, https://www.courthousenews.com/wp
-content/uploads/2020/06/Dicamba.pdf; Carey Gillam, "Court Orders EPA
Approval of Bayer Dicamba Herbicide Vacated; Says Regulator 'Under-
stated the Risks,'" US Right to Know, June 3, 2020, https://usrtk.org/
pesticides/court-orders-epa-approvals-of-bayer-dicamba-herbicide-vacated
-says-regulator-understated-the-risks/.

13. "EPA Offers Clarity to Farmers in Light of Recent Court Vacatur of
    Dicamba Registrations," EPA News Release, June 8, 2020, https://www
    .epa.gov/newsreleases/epa-offers-clarity-farmers-light-recent-court-vacatur
    -dicamba-registrations; Carey Gillam, "Big Ag Group Argue Court Cannot
    Tell EPA When to Ban Dicamba," US Right to Know, June 17, 2020, https://
    usrtk.org/uncategorized/big-ag-groups-tell-court-it-has-no-authority-to
    -tell-epa-when-to-ban-dicamba/; Mark Loux (Ohio State weed scientist),
    correspondence with the author, June 17, 2020; Email from Michal Freed-
    hof, Acting Assistant Administrator in the EPA's Office of Chemical Safety
    and Pollution Prevention, to EPA employees, March 10, 2021, https://www2
    .dtn.com/ag/assets/EPA-Memorandum-Scientific-Integrity.pdf.

14. "Brazil Grain Growers Wary of Dicamba as Bayer Launches new GM Soy
    Seed," *Reuters*, September 30, 2019, https://www.reuters.com/article/us
    -brazil-dicamba/brazil-grain-growers-wary-of-dicamba-as-bayer-launches
    -new-gm-soy-seed-idUSKBN1WF1UY.

15. "Suit Accusing Monsanto of Polluting Baltimore Waterways Advances,"
    Courthouse News Service, April 1, 2020, https://www.courthousenews
    .com/suit-accusing-monsanto-of-polluting-baltimore-waterways
    -advances/; "Seattle Seeks Millions from Monsanto to Clean Up PCBs from
    Duwamish," *Seattle Times*, January 26, 2016, https://www.seattletimes
    .com/seattle-news/environment/seattle-sues-monsanto-seeking-millions
    -to-clean-up-pcbs-from-duwamish/; "AG Ferguson Makes Washing-
    ton First State to Sue Monsanto Over PCB Damages," press release from
    the Washington attorney general's office, December 8, 2016, http://www
    .atg.wa.gov/print/12385; "Lawsuit: Toxic PCBs Still in Rivers, Plants,
    Air," *Cincinnati Enquirer*, March 5, 2018, https://www.cincinnati.com/
    story/news/2018/03/05/ohio-ag-mike-dewine-sues-monsanto-pay-cleanup
    -pcbs/396836002/; "PCB Maker Says State Lawsuit is 'Without Merit,'"
    New Hampshire Union Leader, October 28, 2020, https://www.unionleader
    .com/news/environment/pcb-maker-says-state-lawsuit-is-without-merit/
    article_5c6647e9–a88b-5ddd-af3e-af963e622550.html; "Monsanto Must
    Pay to Clean Up PCBs, New Suit by D. C. Says," *Bloomberg Law*, May 7,
    2020, https://news.bloomberglaw.com/environment-and-energy/monsanto
    -must-pay-to-clean-up-pcbs-new-suit-by-d-c-says.

16. "Roundup Maker to Pay $10 Billion to Settle Cancer Suits," *New York
    Times*, June 24, 2020, B1; Carey Gillam, "Bayer Inks Deals with Three
    Roundup Cancer Law Firms As Settlement Progresses," September 15, 2020,
    US Right to Know, https://usrtk.org/monsanto-roundup-trial-tracker/after
    -bayer-inks-deals-with-three-roundup-cancer-law-firms-as-settlement
    -progresses/.

17. "United States and Vietnam Strengthen Partnership to Address War

Legacies," USAID Press Release, December 5, 2019, https://www.usaid.gov/vietnam/press-releases/dec-5–2019–united-states-and-vietnam-strengthen-partnership-address-war-legacies; "USAID New Round of Agent Orange Cleanup in Vietnam," *devex* [global development industry media outlet], January 8, 2020, https://www.devex.com/news/usaid-begins-new-round-of-agent-orange-cleanup-in-vietnam-96222.

18. "From the 1980's forward," one judge said of Calwell, "he chose an almost solitary course to make the Defendant's accountable for their actions." "Like Robert the Bruce," he went on, "watching the spider spin his web, he picked himself up each time and went back to the fray." *Zina G. Bibb et al. v. Monsanto Company et al.*, Civil Action No. 04-C-465, Final Order Awarding Attorneys' Fees and Litigation Expenses and Awarding Class Representatives' Incentive Payments, 12, filed in the Circuit Court of Putnam County, West Virginia, January 25, 2013; "A Town Embraces Its Explosive Past," *Wall Street Journal*, March 1, 2012, https://www.wsj.com/articles/SB10001424052970204571404577253610618376708; "Monsanto Plaintiffs at Odds," *Charleston Gazette-Mail*, May 19, 2012, https://www.wvgazettemail.com/news/legal_affairs/monsanto-plaintiffs-at-odds/article_818605b6-420b-552d-989c-148055a38cb5.html.

19. Stefania Lombardo et al., "Approaching to the Fourth Agricultural Revolution: Analysis of Needs for the Profitable Introduction of Smart Farming in Rural Areas," Proceedings of the 8th International Conference on Information and Communications Technologies in Agriculture, Food, and Environment (HAICTA 2017), Ghania, Greece, September 21–24, 2017, https://flore.unifi.it/retrieve/handle/2158/1112565/296930/360.pdf.

20. WEED-IT company website, https://www.weed-it.com/contact; "WEEDit Optical Spot Spray Technology," YouTube, https://www.youtube.com/watch?v=b-yTRpyYiRE; "Using Data to Drive Decisions," Bayer company website, https://www.cropscience.bayer.com/innovations/data-science.

21. "Using Data to Drive Decisions," Bayer company website.

22. Comments by Wolfgang Nickl, Bayer Board of Management member, Bayer Shareholders' Meeting, webcast available here: https://www.bayer.com/en/annual-stockholders-meeting-2020.aspx.

23. Bob Shapiro quoted in Joan Magretta, "Growth through Global Sustainability: An Interview with Monsanto's CEO, Robert Shapiro," *Harvard Business Review* 75, no 1 (January–February 1997): 82; "Smart Fields," Bayer company website, https://www.bayer.com/en/digital-farming-smart-fields.aspx; "Glyphosate-Based Herbicides Use in Modern Agriculture," Bayer company website, https://www.bayer.com/en/about-glyphosate-based-herbicides-and-their-role-in-agriculture.aspx.

24. See Chapter 12 of this book for an extensive summary of these studies.

25. Craig D. Osteen and Jorge Fernandez-Cornejo, "Herbicide Use Trends: A Backgrounder," *Choices* 31, no. 4 (4th Quarter 2016), 1, https://www.choicesmagazine.org/UserFiles/file/cmsarticle_544.pdf; Scott M. Swinton and Braeden Van Deynze, "Hoes to Herbicides: Economics of Evolving Weed Management in the United States," *European Journal of Development Research* 29, no. 3 (2017): 565; Craig D. Osteen and Jorge Fernandez-Cornejo,

Pest Management Science 69, no. 9 (September 2013): 1004; Sylvie Bonny, "Genetically Modified Herbicide-Tolerant Crops, Weeds and Herbicides: Overview and Impact," *Environmental Management* 57 (2016), 35, 41.

26. Bt traits could also be found in a smaller percentage of soybean farms in Latin America by the 2010s. Bruce E. Tabashnik, Thierry Brévault & Yves Carriére, "Insect Resistance to Bt Crops: Lessons from the First Billion Acres," *Nature Biotechnology* 31, no. 6 (June 2013), 510–21; Yves Carriére, et. al., "Crop Rotation Mitigates Impacts of Corn Rootworm Resistance to Transgenic Bt Corn," *Proceedings of the National Academy of Sciences of the United States of America* 117, no. 31 (August 2020): 18385–92; Charles M. Benbrook, "Impacts of Genetically Engineered Crops on Pesticide Use in the U.S.—The First Sixteen Years," *Environmental Sciences Europe* 24 (2012): 24; Daniel Hellerstein, et. al., USDA Economic Research Service, *Agricultural Resources and Environmental Indicators* (May 2019), 36, https://www.ers.usda.gov/webdocs/publications/93026/eib-208.pdf?v =2290.7; For total pesticide use in the United States, see statistics available at USDA National Agricultural Statistics Services Quick Stats website, https://quickstats.nass.usda.gov/.

27. Monsanto Company 1980 Annual Report, 26.

28. USDA Economic Research Service, *Three Decades of Consolidation in US Agriculture*, Economic Information Bulletin Number 189 (March 2018), 40.

29. Department of Homeland Security, *Threats to Precision Agriculture* (2018), 3–7, https://www.dhs.gov/sites/default/files/publications/2018%20AEP_ Threats_to_Precision_Agriculture.pdf.

30. 2020 Bayer Shareholders' Meeting, webcast available here: https://www .bayer.com/en/annual-stockholders-meeting-2020.aspx.

31. "Behind the Monsanto Deal, Doubts About the GMO Revolution," *Wall Street Journal*, September 14, 2016, https://www.wsj.com/articles/behind -the-monsanto-deal-doubts-about-the-gmo-revolution-1473880429;   "A Growing Discontent," *New York Times*, March 12, 2010, B1; For official USDA estimates of soybean and corn seed costs, see "Commodity Costs and Return," USDA Economic Research Service website: https://www.ers.usda .gov/data-products/commodity-costs-and-returns/commodity-costs-and -returns/#Recent%20Cost%20and%20Returns. Adjusted for inflation, the increase in soybean farmers' seed expenses from 1995 to 2019 was about 150 percent compared to 120 percent for corn growers.

Illustration Credits

Part Openers

Part I, p. x: Courtesy Kristofer Husted.

Part II, p. 18: Monsanto Company Records, Julian Edison Department of Special Collections, Washington University in St. Louis.

Part III, p. 76: Courtesy Jonathan Zadra.

Part IV, p. 206: Alf Ribeiro / Alamy Stock Photo.

Part V, p. 264: dpa picture alliance / Alamy Stock Photo.

Figures

Figure 1, p. 174: Graph created for this book based on data courtesy of Stephen O. Duke, USDA Agricultural Research Service.

Figures 2 and 4, pp. 186 and 225: Graphs created for this book using data from study by Dr. Ian Heap (2020), International Herbicide-Resistant Weed Database, Weedscience.org.

Figure 3, p. 219: USGS Pesticide National Synthesis Project.

Figures 5a and 5b, p. 230: Graphs created for this book using data from USDA National Agricultural Statistics Service and USGS Pesticide National Synthesis Project.

Illustration Insert

Insert p. 1: "Portrait of John Francis Queeny [1859–1933] and Dog," circa 1930. Williams Haynes Portrait Collection, Box 12, Science History Institute. Philadelphia. https://digital.sciencehistory.org/works/r207tp91k.

Insert p. 2: Left: Olga Monsanto with her children. Monsanto Company Records, Julian Edison Department of Special Collections, Washington University in St. Louis. *Below:* "Portrait of Edgar Monsanto

Queeny [1897–1968]," 1950–59. Williams Haynes Portrait Collection, Box 12, Science History Institute, Philadelphia. https://digital.sciencehistory.org/works/3n203z80v.

Insert p. 3: Right: Chemical plant in Nitro, West Virginia, 1970s. National Archives and Records Administration. *Below:* Nitro workers. Courtesy *Charleston Gazette-Mail.*

Insert p. 4: Top: Phosphate slag dumping at Monsanto's Soda Springs, Idaho, facility, 2016. Courtesy Jonathan Zadra. *Bottom:* Entryway of the Soda Springs facility. Courtesy Jonathan Zadra.

Insert p. 5: Left: "Scribbled notes from an October 1969 meeting of Monsanto's ad hoc committee." Chemical Industry Archives, a project of the Environmental Working Group. *Below:* "David Baker standing over his brother's grave." Courtesy ©Mathieu-Asselin.

Insert p. 6: Da Nang airport cleanup site, Vietnam, 2017. Courtesy Jonathan Zadra.

Insert p. 7: Top: Cotton seed bag. Courtesy Krisotofer Husted. *Bottom:* Seed label. Exhibit 1 in *Bader Farms v. Monsanto*, Case No. 1:16-CV-299-SNLJ, Doc. No.: 54-1, filed: 05/01/17.

Insert p. 8: Bayer April 2019 shareholder's meeting. dpa picture alliance/Alamy Stock Photo.

Index

A page number in italics refers to an illustration.

Abernathy v. Monsanto, 196, 199–201, 202–3
Abrams, Chris, 258
Agent Orange. *See also* dioxin contamination in Vietnam
  ecological devastation caused by, 102–4
  as Kennedy's solution, 99–100
  lawsuits by Vietnam veterans, 144, 147, 151, 255, 261
  Monsanto as largest producer, 12, 100
  protests against military use, 105–6
  toxicity to humans, 12–13, 104–5
  US compensation to veterans, 254–55
  used in Vietnam, 102–6
  VAVA assisting victims, 251, 253–54
Agent Purple, 98, 99, 101
AgrEvo, 177–78, 197
agricultural chemicals. *See also* herbicides; insecticides
  biotechnology increasing dependence on, 182
  declining farm-labor pool and, 92
  Monsanto's expansion into, 72–73, 90–91
Agricultural Division, 91, 108, 129–32, 135, 177
agriculture
  government investments of 1930s and 1940s, 71–72
  industry funding of research, 220–22, 235
Albert, Dan, 120–21
ALS inhibitors, resistance to, 185–86, 186, 187, 210

Ambien, 176, 191
Anniston, Alabama, 192–204. *See also* polychlorinated biphenyls (PCBs)
  acquisition of Swann Chemical Company in, 59, 66
  cleanup operations, 195, 196, 203–4
  government investigation of PCB pollution, 112–13
  health problems in, 195
  lawsuits by residents, 194, 196, 199–201
  Monsanto's buying up houses and buildings, 192, 194
  PCB production in, 66, 109, 117
  PCBs in residents' bodies, 193–94
  shutting down PCB production, 119, 193, 194, 201
  treated as hazardous waste dump, 193–94
Argentina, 106, 199, 241–43
Aroclors, 109, 110, 113, 114, 117
Ashe, William F., 85–86, 87–88, 90
aspirin, 24, 50, 129, 267
auxins, synthetic, 82–84

Bacillus thuringiensis. See Bt gene
Bader, Bill, 3–5, 7–8, 234–35, 249, 262
Bader Farms v. Monsanto and BASF, 3, 5, 7–8, 249, 262, 269
Bailey, Charles, 256
Baker, David, 192–93, 195–96, 200, 201–2, 203
BASF
  Bader case and, 3, 8
  chloracne at plant, 95–96
  at dawn of chemical age, 24
  in global seed market, 178, 268

BASF (*continued*)
  selling dicamba, 6, 7, 235
  World War II and, 68
Baumann, Werner, 267–70, 271–72, 278
Baur, Jacob, 31
Bayer
  acquisition of Monsanto, 4, 268–69
  aggressive marketing of GE seeds,
    212
  aspirin and, 24, 267
  Bader case and, 3
  cancer verdicts after Monsanto
    buyout, 269
  at dawn of chemical age, 24
  ethical workers in, 15–16
  haunted by Monsanto legacy,
    269–70
  herbicide in 2020 product pipeline,
    277
  Norsworthy's funding by, 235
  ongoing legal settlement talks, 272
  Queeny's visions of freeing
    Americans from, 16, 44
  rebranded as digital farming, 273–74
  seed-processing equipment of, 213
  selling Roundup Ready Xtend as
    liberating, 13
  shareholders' meeting in 2020,
    267–68, 278
  shareholders' vote of no confidence
    in 2019, 269
  stacked GE germplasm sent to
    Brazil, 250
  World War II and, 68
Bebie, Jules, 36, 44–45
Belknap, Charles, 58, 69
Bibb v. Monsanto, 272–73
Biden, Joseph R.
  EPA under, 271
Big Three German chemical firms, 24
biotechnology. *See also* genetic
  engineering
  companies competing with
    Monsanto, 177–78
  founding of companies, 159
  patent law and, 178–80
  public resistance to, 182
  Reagan era regulatory policy, 180–83
  warning by environmental groups,
    182
biotechnology at Monsanto. *See also*
  genetically engineered (GE) seeds
  bovine somatotropin (BST), 176–77
  crucial plant research, 159–61

as escape from petrochemicals,
    142–43, 276
  Shapiro's faith in, 175, 184
Birnbaum, Linda, 147
Bishop, Dan, 123
Black workers at Queeny's firm, 47–49
Boggess, James Ray
  dioxin trial and, 145–46, 155
  suffering from chloracne, 79–80,
    84–85, 96, 106
  well paid for dirty work, 89–90,
    93–94, 95, 97
Bollgard, 184, 242
Bolsonaro, Jair, 250
bovine somatotropin (BST), 176–77
Bowie, Andrew, 194, 195, 196
Bradshaw, Laura, 187
Brazil
  acquisition of seed companies in,
    246–47
  approval of Roundup Ready, 243–44
  Cerrado in, *206*, 245–49, 262, 271
  dicamba-resistant seeds in, 249–51, 262
  farmers' initial savings with
    Roundup Ready, 248
  glyphosate resistance developing,
    248–49
  illegal spread of Roundup Ready
    from Argentina, 241, 242–43
  Monsanto needing to expand into,
    199, 242
  Monsanto's investments to reduce
    imports, 244–45
  soybean varieties developed for,
    245–46
  yields with Roundup Ready seeds,
    206, 247
breast milk
  dioxin in, 145, 252
  PCBs in, 109, 123, 145
Brincefield, Tim, 168
Brown, Jesse, 254
Bt gene
  in Bollgard, 184
  competition to incorporate in
    plants, 160
  crop yields and, 237, 238, 275
  decreasing insecticide use, 275
  in Dekalb Vietnam corn seeds, 261
  insects developing resistance, 275
  misstep in sale to Pioneer, 189
  monarch butterflies and, 198–99
  Roundup Ready gene cassettes
    stacked with, 247, 250

caffeine, 34–35, 36–37, 38–41, 42–43, 68
Calgene, 159, 160–61, 178, 180, 189
Calwell, Stuart, 138–41
    Nitro case and, 141, 146, 148–53, 200, 272–73
cancer
    Baumann's settlement proposal and, 272
    of bladder caused by PAB, 139, 154
    of Calwell's client Cleo Smith, 140–41
    dioxin studies and, 146, 147–48
    glyphosate and, 10–11, 232–33
    Johnson case, 8–9, 11, 15, 233, 262, 269
Candler, Asa and John S., 39
Cantwell, John, 8
Capobianco, João Paulo, 241, 243–44, 249, 250
Carey, Boyd, 7–8
Carson, Rachel, 100–101, 106, 107, 110, 111
Carter, Jimmy, 145
Charles, prince of Wales, 198
ChemChina, 268
chemical companies, American
    changed by petrochemicals, 61–62, 68–69
    coal tar as early basis of, 50
    early dependence on Europe, 14, 25, 32
    entering synthetic organic chemistry, 24–25
    Monsanto ranking eighth in 1932, 59
    reliant on fossil fuels and mineral deposits, 14–15
chemical companies, European, 24, 33–34, 35, 44–45, 68–69
chemical warfare, 67, 82, 105
Chemstrand, 61
Chilton, Mary-Dell, 159
chloracne
    caused by PCBs, 110–11
    explosion in Italy and, 144
    in Vietnam veterans, 254
    in workers with 2,4,5-T, 79–80, 84–89, 90, 94–96, 106
chloral hydrate, 39, 44
Clapp, Richard, 148
Clean Air Act, 112
Clean Water Act, 112, 120
Cleeves, Emma, 33

coal tar
    drugs synthesized from, 24, 25
    European chemical companies and, 61, 68
    as fundamental resource, 49–50, 51
    investment in British distillery for, 63
    making intermediates from, 45–46
    replaced by petroleum, 61
    saccharin synthesized from, 29
Coca-Cola Company
    buying Monsanto caffeine, 35, 36–37, 41, 68
    buying Monsanto saccharin, 30, 34, 41
    secret formula, 35, 39–40
    trial on added caffeine, 38–40
Cochran, Johnnie L., Jr., 200
Comai, Luca, 160–61
commodity chemicals, 59, 90, 126, 133, 141–42, 143, 176
Copenhaver, John Thomas, Jr., 151, 154, 156
Corteva, 268
COVID-19, 267, 277, 278
CRISPR, 276
Cunningham, Omar, 155
Curtis, Francis J., 64

Dawson, Ray, 223
DDT, 14, 70–71, 100, 110
Dekalb Vietnam, 259, 260, 261, 262
Dekkers, Marijn, 268
Delta & Pine Land Company, 197–98
detergents, 59, 106, 110, 128, 130–32, 133
developing countries, 12, 72, 259–60.
    See also Argentina; Brazil; Vietnam
Dewine, Mike, 271
Diamond v. Chakrabarty, 178–79
dicamba
    Bader Farms case, 3, 5, 7–8, 249, 262, 269
    Bayer taking on legal cases, 270, 272
    drift problem in Brazil, 249–50, 262, 271
    drift problem known for decades, 231
    drift problem used to sell seeds, 6, 8, 13, 235–36
    drift-related damage, 4–5, 233, 234–35

dicamba (*continued*)
 formulation claimed less volatile,
  6–7, 233–34, 235
 legal uncertainties in US during
  2020 and 2021, 270–72
 not well suited to Brazilian weeds,
  249–50
dicamba-tolerant seeds. *See also*
 Roundup Ready Xtend
 in Brazil, 249–51
 development and approval, 4, 6,
  231–32, 234
 sold before EPA-approved dicamba,
  6–7
digital farming, 273–74, 277–78
dioxin contamination in Vietnam. *See
 also* Agent Orange
 cleanup at Da Nang airport, 257–58
 Ford Foundation assistance, 256
 funds for assistance with health
  problems, 258, 272
 hot spots at US air bases, 13, 251–52
 lawsuit against Monsanto and Dow,
  253–54
 Monsanto's avoidance of costs, 12,
  253–55, 258–59, 261
 Monsanto's control of information
  and, 97
 relief programs of VAVA, 251, 253
 Rieser and Leahy's work on, 255–59
 studies of human health and, 252
 US taxpayer funds for cleanup, 12,
  256–57, 258, 262, 272
dioxin toxicity. *See also* chloracne;
 Nitro plant
 birth defects and, 104, 105
 cancer and, 146, 147–48
 as cause of chloracne, 96
 controversial Monsanto-funded
  research, 146–48
 mounting evidence, 144–45
 secrets kept by Dow and Monsanto,
  104–5
 strategy for evading responsibility,
  146
Dow Chemical
 Agent Orange and, 253–54, 258
 entering synthetic auxins business, 83
 founding of, 25
 investing in biotechnology, 143–44
 merger with DuPont in 2017, 268
 Monsanto's financial rank
  compared to, 59, 68

petroleum-based chemicals and,
  61–62, 64, 65
 toxicity of 2,4,5-T impurities and,
  80, 95, 96, 104–5
Drinker, Cecil, 73–74
drought-resistant seeds, 15, 239, 277
DuBois, Gaston, 34, 35, 36, 44–45,
  58, 63
Duke, Stephen, 237, 239
DuPont
 founding of, 24–25
 investing in biotechnology, 143–44
 merger with Dow in 2017, 268
 Monsanto's financial rank
  compared to, 59, 68
 petroleum-based chemicals and, 61
 purchase of Pioneer seed company,
  197, 212
 Queeny's cocky offer to, 63–64
 winning plastic bottle race, 133

Eli Lilly, 25, 144
Embrapa, 245–46, 249, 250
energy crisis, 141–42
Enviro-Chem Systems, 108
Environmental Defense Fund (EDF),
  124, 148, 182
environmental groups, and
  biotechnology, 182, 184, 197
environmental impact statement (EIS)
 federal law and, 112, 135
 phosphate mining and, 136
Environmental Protection Agency
 (EPA). *See also* Superfund sites
 approving glyphosate in 1975, 132,
  133
 under Biden, 271
 biotechnology policy and, 180
 cleanup in Anniston and, 201–2,
  203–4
 dicamba issues under Trump and
  Biden, 235, 270–72
 dicamba with less volatility and, 6,
  7, 233, 234, 235
 finding glyphosate not carcinogenic,
  11, 232–33
 map of dioxin around Nitro plant,
  156–57
 Monsanto's phosphate mining and,
  60
 Monsanto's phosphate processing
  and, 127, 137, 161–72
 Nixon's creation of, 112

PCBs and, 119–20, 121–22, 123–26,
    196, 271
taking over pesticide regulation in
    1972, 132
Toxic Substances Control Act and,
    119, 122–24
under Trump, 11, 235, 270–72
EPSP synthase, 132, 160–61
Ex parte Hibberd, 180

Fahlberg, Constantin, 29
Farmer, Donna, 9
Federal Insecticide, Fungicide, and
    Rodenticide Act (FIFRA), 101, 108
Ferguson, Bob, 271
Fernandez-Cornejo, Jorge, 237
Fingerhut, Marilyn, 147
Foley, Jonathan, 239–40
Food and Drug Administration (FDA)
    approving BST, 177
    banning Monsanto's plastic bottles,
        133
    biotechnology policy and, 180, 183
    PCBs in food packaging, 116
    pesticide residues in food, 101
food system
    alternative future still possible, 279
    biotechnology promoted to feed the
        world, 175, 239–40
    genetic engineering used to sell
        chemicals for, 277
    transformed by Monsanto's GE
        seeds, 14, 173
    vulnerability of digital farming, 277–78
foreign competition
    changed due to petroleum, 61–62
    in commodity chemical business,
        126, 133, 141
    in early twentieth century, 33–34,
        35, 38, 44
    World War I and, 44–45, 51
    World War II and, 68–69
Forty Barrels case, 38–41
fossil fuels. *See also* coal tar;
    petroleum-based chemicals
    dependence of chemical industry
        on, 14–15, 64–65
    Monsanto's attempt to escape
        economy of, 276
Fraley, Robert "Robb," 159, 182, 188,
    189–90
Franz, John, 15, 128–32, 158
freedom, marketing based on, 13

Gaffey, Bill, 148
Geddes, Robert, 163–64
genetically engineered (GE) food
    claims about eliminating hunger, 240
    importation issue in Europe and
        Japan, 184
    mostly fed to livestock, 239–40
    public concern about, 198
genetically engineered (GE) seeds. *See
    also* Bollgard; Bt gene; Roundup
    Ready technology; Roundup
    Ready Xtend
    abandoned now by many farmers,
        278–79
    chemical dependence not reduced
        by, 275, 277
    competitors to Monsanto, 177–78
    difficulty of escaping the system,
        279
    drought resistance and, 15, 239, 277
    European backlash, 197–99, 226, 242
    by gene transfer into plant cells, 159
    launched by Monsanto in 1996, 184
    mergers and acquisitions in business
        of, 268
    patent protection for, 178–80
    quadrupled in price since 1997,
        278–79
    rapid adoption of, 173, *174*
    Reagan era policy and, 180–83
    sought to go with Roundup, 159–60
    sterility-producing technique for, 198
    in Vietnam, 12, 13, 259–63
    yield of, 236–40, 275, 276, 279
genetic engineering. *See also*
    biotechnology
    beginnings of, 158–59
    potential and problems, 276–77
glufosinate, 178, 197
glyphosate. *See also* Roundup
    Bayer's 2020 claims about, 274
    cancer risk and, 10–11, 232–33
    Chinese imports of generics to
        Brazil, 245
    disrupting microflora, 132
    efforts to deny health effects, 10–11,
        232–33
    elemental phosphorus needed for, 128
    eliminating monarch butterfly food,
        199
    EPA approval of, 132, 133
    as exit from detergent business,
        131–32

glyphosate (*continued*)
　Franz's testing of, 131
　increased use from 1992 to 2017,
　　219, 275
　manufactured in Brazil by
　　Monsanto, 244–45
　much less toxic than 2,4-D, 229
　price drop to deal with competition,
　　228
　surfactants added to, 9–10
　traces in processed foods, 13
glyphosate-resistant weeds
　appearing in Australia, 187
　developing in Brazil, 248–49
　increase in, 221–22, 224–25, *225*
　Monsanto's initial denial of, 187
　need for other herbicides and, 229,
　　230, 231, 275
Gould, Fred, 239
government regulation. *See also*
　　Environmental Protection
　　Agency (EPA); Food and Drug
　　Administration (FDA); US
　　Department of Agriculture (USDA)
　of biotech, 180–83
　changing political climate in 1970s
　　and, 112
　EPA taking over pesticide regulation
　　in 1972, 132
　final ban on US uses of 2,4,5-T, 106
　PCBs under increasing pressure,
　　112–13, 116–17, 118–19
　of phosphate detergents, 130
　Reagan era policy, 165, 180–83
　relying on corporately financed
　　studies, 97
　Sommer's fighting against, 108, 110
　by USDA and FDA in 1950s and
　　1960s, 101
Grant, Hugh, 226–29, 244, 245, 247
Green Revolution, 72, 134–35
Guararria, Leonard, 181
Gunnell family, 167–69, 170

Hamm, Philip, 130, 131
Hanley, John, 133, 142–43
Hansen, Kirk, 162, 163
Hardeman, Edwin, 269
Hays, Alastair, 148
Hein, John, 154
herbicide resistance. *See* ALS
　　inhibitors, resistance to;
　　glyphosate-resistant weeds

herbicides. *See also* Agent Orange;
　　dicamba; glyphosate; Roundup
　Bayer's digital farming and, 273–74
　Bayer's new chemical in 2020, 277
　increased use since introduction of
　　GE seeds, 229, *230*, 231, 275
　Monsanto's initial investments in,
　　72, 83, 91–92
　Monsanto's other brands beside
　　Roundup, 135
　Monsanto wiping out most
　　competitors, 218–19
Hertzberg, Vicki, 147
Heydens, William, 10
Hibberd, Ex parte, 180
Hinds, Bob, 53
Hochwalt, Carroll A., 63
Hoechst, 24, 68, 177–78
Hoover, Herbert, 69
Horsch, Robert, 159, 161
Hough, Warwick, 42, 43
Hughes, Charles Evans, 41
Hurley, Jonathan, 85–86, 88

I. G. Farben, 68
immigrants
　Edgar Queeny's disparagement of,
　　57
　as laborers at John Queeny's firm,
　　48
industrial farming, 72
influenza, 46, 50, 267
insecticides, 70–71, 72. *See also* Bt
　　gene; DDT
INTACTA 2 Xtend seed, 250

Jaworski, Ernest "Ernie," 15, 158, 159,
　　160–61, 188
Jeffers, Chester A., 80
Jenkins, Dan, 232–33
Jensen, Sören, 109–10
Johnson, Dewayne "Lee," 8–9, 11, 15,
　　233, 262, 269
Johnson, Eric, 186
Johnson, Lyndon, 105

Kelly, R. Emmet, 74, 110–11, 117, 152,
　　153
Kelton, Anna, 39
Kennedy, John F., 98–101
King, John A., 23
Kingsbury, David, 180
Kloppenburg, Jack R., Jr., 179

Langreck, Bert, 48
Leahy, Patrick, 255–56, 257, 258
Lê Kế Sơn, 257
Le Roy, Louis, 40
Liberty herbicide, 178
Liberty Link seeds, 178, 197
Limbaugh, Stephen N., Jr., 3, 5, 269
Losey, John, 198–99
Loux, Mark, 220–22, 225–26, 229, 231,
     235, 271
Love, Charles, III, 151, 153
Lula da Silva, Luiz Inácio, 241, 24243
Luu Van Tran, 260

Mahoncy, Richard "Dick," 142–43
Mandolidis case, 149, 154
Manhattan Project, 66–67, 91
Manibusan, Mary, 232
Maradona seeds, 241, 242–43
McClanahan, Ivan, 79, 85–86, 88
McKee, James, 54
McLachlin, Beverley, 224
Merck, 25, 26, 30, 44
Mexican Agricultural Program (MAP),
     72, 134
Meyer Brothers Drug Company, 26–
     29, 31–33, 35, 36
Miller, Jan, 5
monarch butterflies, 198–99
Monsanto. *See also* Queeny, Edgar;
     Queeny, John
   acquired by Bayer, 4, 268–69
   acquired by Pharmacia & Upjohn,
     200
   acquiring dozens of seed companies,
     211
   "Chemical" dropped from its name,
     106
   ethical complexity of people in, 15
   Grant chosen as CEO in 2003,
     226
   Hanley as president in 1972, 133
   income down in early 1990s, 176
   international expansion, 106, 108,
     240
   as largest seller of seeds in the
     world, 14, 211
   Mahoney as president in 1980, 142
   mergers and spinoffs of early 2000s,
     200
   needing a success in 1970s, 133
   need to break dependency on oil,
     141–42

never breaking free from chemical
     economy, 276
"new" agricultural Monsanto, 200,
     204
profiting from problems created by
     technology, 275
profits sagging in early 2000s, 226
on rebound after Anniston
     settlement, 203, 204
segregating dirtier business from
     corporate core, 92–93
Shapiro as leader in mid-1990s, 173
Sommer as president in 1960, 106
spinning off liabilities into Solutia,
     190–91
Monsanto, Olga, *18*, 22, 26, 27, 32–33
Monsanto Enviro-Chem Systems, 108
Monsanto Petroleum Chemicals
     subsidiary, 61

National Environmental Policy Act,
     112, 135
New Deal, 56, 60, 71, 189
"new" Monsanto, 200, 204
Nguyen Hong Lam, 259–60, 262
Nguyễn Thị Hồng, 102
Nitro, West Virginia, 81–82
Nitro lawsuits, 137, 140–41, 147,
     148–57, 272–73
Nitro plant. *See also* dioxin toxicity
   chloracne of workers, 79–80, 84–89,
     90, 94–96, 106
   contamination continuing in 1950s,
     93–97
   corporately financed studies and, 97
   dirtier process than competitors,
     145
   labor problems at, 93–95
   labor unions and, 86–87, 90, 97,
     140, 146
   PAB contamination in, 139, 154
   starting to make 2,4,5-T in 1948, 83
   toxic explosion, 79–80, 84–86,
     87–88, 146
   union and government accepting
     conditions, 97
   workers needing to stay, 89–90
   Workmen's Compensation denial,
     87, 96
Nixon, Richard, 112
Norsworthy, Jason, 233, 235
Nutrasweet (aspartame), 173–74, 176,
     191

oil companies, 90, 126, 133, 144
Operation Ranch Hand, 98, 100, 104,
    251
Owens case. See *Walter Owens v.*
    Monsanto

PAB, and human health, 139, 154
Papageorge, William B., 117–18, 120,
    121–22
Parry, James, 10
patent protection
    for GE seeds, 178–80
    on Monsanto's seeds, 222, 223, 224,
        242
    on Roundup, 135, 159, 217, 227–28
Patridge, Scott, 233–34
PCBs. *See* polychlorinated biphenyls
    (PCBs)
Pedroso, Rafael, 249
Perdue, Sonny, 11, 15
Pesticide Control Act of 1972, 132
pesticides. *See* herbicides; insecticides
petroleum-based chemicals, 61–62,
    68–69, 90, 141, 276
    transitioning to life sciences from,
        142–44, 174
Pfizer, 25
pharmaceuticals, 191, 197, 200, 216–
    17, 276
Pharmacia, 200, 203
Pharmacia & Upjohn, 200
phenacetin, 36, 39, 46, 50, 267
phosphate facilities of Monsanto, 60,
    127–28
    contamination near plant, 167–72
    massive energy consumption, 60,
        127
phosphate mining, 59, 60, 136, 170–71
phosphate slag, radioactive, 127,
    136–37, 162–70
phosphorus products, 59
    detergents, 59, 128, 130–31
    glyphosate, 128, 132, 170, 244, 276
    phosphoric acid, 59–60, 128
Pilliod, Albert and Alva, 269
Pioneer Hi-Bred deal, 189
Pioneer seed company, 197, 212
Piper, Ralph, 53–55
Plant B, 49, 51, 66
plastic bottles, 133
plastics, 61–62, 64, 65, 90, 106
polychlorinated biphenyls (PCBs),
    65–66, 109–26. *See also* Anniston,
    Alabama

ban on manufacture, not on use,
    122–24
Bayer's liabilities for contamination,
    271–72
continuing threat to public health,
    125–26
in dairy milk, 117
economic impact of discontinuing,
    119–20, 124, 126
Environmental Defense Fund and,
    124
EPA and, 119–20, 121–22, 123–26
in food supply, 109, 116–17, 118
Jensen's whistleblowing on, 109–10
Monsanto as sole US manufacturer,
    66, 109
Monsanto plans to deal with
    problem of, 113–16
in mother's milk, 109, 123, 145
offloading liability onto client
    companies, 118–19, 126, 194
phaseout of open applications, 117
publicity about damage due to, 111
regulatory pressure on, 112–13,
    116–17, 118–19
salespeople trained to obfuscate, 122
toxicity of, 74, 110–11, 113
uses of, 109–10
Pond, Fred, 210–15, 218
Posilac, 176–77
Powles, Stephen B., 226
Pure Food and Drug Act, 28–29, 38,
    41, 108, 181
Pursuit brand herbicide, 185
Pusztai, Arpad, 198

Queeny, Edgar
    becoming president in 1928, 51,
        55–56
    ceding presidency in 1943, 69
    Coca-Cola contract and, 35
    creating Central Research
        Department, 63
    death in 1968, 108
    early life, *18, 26, 27, 55*
    final years of ill health, 107–8
    as heavy drinker, 53–55
    as hunter, farmer, and lover of
        wildlife, 69–71, 91, 107
    investing in foreign subsidiaries, 63
    military contracts and, 68
    personal qualities, 56
    political and economic beliefs,
        56–57, 69

preferring acquisitions over
    research, 57–59, 81
preferring agricultural chemicals, 91
Queeny, John
    childhood, 22–23
    dealing with chemical
        contamination, 46–47
    death in 1933, 62
    early work life, 23–27
    family photo, 18
    fearing his son's ambitions, 57
    frugality of, 45, 52, 55
    goals in creating Monsanto, 14
    loss of sulfuric acid plant, 27
    making drugs for Spanish flu,
        267–68
    moving to synthesis of raw
        materials, 36
    postwar debt and recovery, 51–52
    quest for independence from
        German firms, 16, 268
    supporting national drug regulation,
        28–29, 108, 181
    working at Meyer Brothers, 26–29,
        31–33, 36

Raab, Edward L., 115
Randles, Bev, 3–4, 5, 269
Randles, Billy, 3–4, 5, 7–8, 269
Randox, 91–92
Reagan era regulatory policy, 165,
    180–83
Resource Conservation and Recovery
    Act, 135–36
Rieser, Tim, 255–57, 258–59
Rifkin, Jeremy, 177
Risebrough, Robert W., 111
Roberts, Cassandra, 194, 199
Robin, Marie-Monique, 146
Rogers, Steve, 159
Roosevelt, Theodore, 41–42
Roundup. See also glyphosate
    Bayer's dependence on, 279
    Bayer taking on 120,00 legal cases,
        270, 272
    beating chemical competitors,
        218–19
    creation of, 131
    early success with, 133–35, 177
    extremely profitable, 217
    lawsuit on Johnson's cancer, 8–9,
        11, 15, 233, 262, 269
    marketing of, 134–35, 216–18
    patent on, 135, 159, 217, 227–28

phosphate rock used for, 59, 76, 137,
    169
to revive Monsanto's market share, 126
after spin-off of Solutia, 191
surfactants added to, 9–10
Roundup Ready technology, 4. See
    also Brazil; genetically engineered
    (GE) seeds
    approved in Argentina in 1996,
        241–42
    approved in US in mid-1990s, 172
    in Bayer's digital farming, 274
    in competitive market, 177–78
    costs to farmers, 211–12
    creation of, 161
    in Europe, 216
    helping recovery after Anniston
        settlement, 203
    herbicide use increase and, 229, 230,
        231
    initial success with farmers, 188,
        190, 214–15, 218–20
    marketing of, 13, 174, 212–13, 215–18
    misstep in sale to Pioneer, 189
    processing of seeds by seedsmen, 213
    replanting prohibited by TUA, 190,
        211, 222–24
    solving resistance problem, 185, 210
    technology fee, 189, 215, 241–42
    in Vietnam, 12, 13, 259–63
Roundup Ready Xtend, 4. See also
    dicamba-tolerant seeds
    Bayer still deeply wedded to, 264, 268
    development and approval, 231–32,
        233
    dicamba drift in Brazil and, 249–50,
        271
    dicamba drift used to sell, 6, 8, 13,
        235–36
    used with volatile dicamba, 234–35
Rowland, Jess, 233
rubber products, 57–58, 65, 68, 81
    toxic PAB in production of, 139, 154
Rubber Services Laboratories, 57, 81
Rumsfeld, Don, 173

saccharin, 29–32
    banned by USDA in 1911, 38, 41–44, 46
    price undercut by German
        producers, 33–34
    Roosevelt's support for, 41–42
    sales overtaken by caffeine, 35, 39
    sold to Coca-Cola Company, 30,
        34, 41

saccharin (*continued*)
  sugar and, 30, 42, 68
  synthesis of, 22, 32, 36, 44, 45–46
  tariff on imports of, 30
  USDA restrictions lifted in 1925, 44
Sachs, Eric, 232–33
Saltmiras, David, 11
Sandoz, 24, 31–32
Sanford, Edward T., 40–41
Santobane, 70–71
scavenger capitalism, 50, 61, 142, 276
Schaefer, Louis, 39
Schecter, Arnold, 252, 258
Schell, Jozef "Jeff," 159
Schmeiser, Percy, 223–24
Schneiderman, Howard, 142, 143
Schwartz, Louis, 85
Searle, 173–74, 216
"seed piracy," 222–24
seedsmen, 210–13
Seymour, David, 126
Shapiro, Robert B.
  aggressive acquisitions and
    investments, 188–89
  as believer in biotechnology, 175,
    178, 183–84, 187–88, 277
  claiming GE seeds would increase
    yield, 236, 274
  environmental concerns, 175, 187–88
  expanding Monsanto's seed
    business, 211
  hoping to change the world for the
    better, 15, 173–75
  offloading toxic chemical liabilities,
    190–91, 196
  talks with Delta & Pine, 197–98
  wealth of, 197, 203
Shelby, Richard, 201
Silbergeld, Ellen, 148
Smith, Cleo, 140–41
Smith, Steve, 5–6
Soares, José, 250
Soda Springs, Idaho. *See* phosphate
    facilities of Monsanto
Soja Livre ("Free Soy" movement), 250
Solutia, 190, 191, 196, 200, 201, 202–4
Sommer, Charles H., 108, 110
Soros, George, 248
Spanish flu, 50, 267–68
Speziale, A. John, 129, 158
Steele, Jesse, 85–86, 88
Stewart, Donald, 194, 196, 199, 200,
    202

sulfuric acid, 27, 49, 50
Superfund sites
  deregulatory interests and, 165–66
  Monsanto responsible for eighty-
    nine, 176
  phosphate mines, 171
  Soda Springs plant, 76, 137, 163,
    166, 167, 170
  toxic nuclear research site, 67
Suskind, Raymond, 85–86, 87–89, 90,
    95, 96, 146–47, 148
Swann Chemical Company, 59, 66
Swinton, Scott, 229
Swiss Triumvirate, 36, 44, 45
Symms, Steve, 163
Syngenta, 268, 269
synthetic fibers, 61, 62, 90, 106

tariffs
  on Chinese glyphosate imported to
    Brazil, 245
  protecting US sugar growers, 30, 42
  Queeny's foreign competition and,
    38, 44, 51, 52
technology fee, 189, 215
  in Argentina, 241–42
technology-use agreement (TUA), 190,
    211, 222
Thomas, Charles A., 63, 66, 67, 91, 92,
    107, 108
Thomas, Gene, 89, 153, 155
Tolbert v. Monsanto, 200, 202–3
Toxic Substances Control Act (TSCA),
    119, 122–24, 136
Train, Russell, 122, 124
Trần Ngọc Tâm, 251
Trần Thị Tuyết Hạnh, 251–52
Trump, Donald, 11, 235, 250, 270–71
2,4-D, 13, 82–84, 129
  used for glyphosate resistance, 229,
    231, 275
  used in Vietnam, 13, 98, 101–2
2,4,5-T, 82–84. *See also* Agent Orange;
    dioxin toxicity
  chloracne and, 79–80, 84–89, 90,
    ·94–96, 106, 144
  closure of legal cases against
    Monsanto, 157
  final ban on US use, 106
  used in Vietnam, 101–2

Union Carbide, 59, 61, 62, 64, 65, 68, 81
Upadyayula, Narasimham, 260

US Department of Agriculture (USDA)
  biotechnology policy and, 180, 181
  Bt effect on monarch butterflies and, 198–99
  helping Monsanto in Vietnam, 260–61, 262
  industrial farming promoted by, 72
  milk stockpiling, 177
  outspent on research by industry, 221
  promoting 2,4-D and 2,4,5-T, 82–83
  regulating insecticides and herbicides, 101
  saccharin ban of 1911, 38, 41–44, 46
  studying yields of herbicide-tolerant crops, 236–37, 275
  Trump's secretary Perdue, 11, 15
  yielding pesticide regulation to EPA, 132

Vanacht, Marc, 215–18
Van Deynze Braeden, 229
VanGessel, Mark, 221
Van Montagu, Marc, 159
Van Strum, Carol, 144
Vegadex, 91–92
Veillon, Louis, 21–22, 31–37, 44–45, 47, 48
Verfaillie, Hendrik, 226
Vietnam, Monsanto's GE seeds in, 12, 13, 259–63
Vietnam War, 98–100. *See also* Agent Orange

Volz, Ed, 88
Vranes, Randy, 132

Walter Owens v. Monsanto, 194, 199–200, 201
Watson, Guy, 197
WEED-IT, 273
Weinstein, Jack B., 151, 253–54
Wheeler, Andrew, 11, 235
Wheeler, Elmer, 110–11, 112, 113, 114
Whiteside, Thomas, 102, 144–45
Wiley, Harvey, 29, 38, 39, 41–42
Willard, Paul, 85–86, 88, 90
Witherspoon, John, 40
Woodruff, Robert, 56
World War I, 44–45, 46, 48, 49–50, 81
World War II, 65–69, 70, 82
Wright, Ralph, 34

Xtend. *See* Roundup Ready Xtend
XtendiMax with VaporGrip, 6–7, 233–34, 235

yield
  claims for digital farming, 273–74
  of conventionally bred Brazilian soybeans, 246, 247, 250
  gains from conventional breeding, 239, 240, 275, 279
  of GE crops, 236–40, 275, 276, 279
  Grant's expansive claims, 247
Young, Harold, 84

Zack, Judith, 146, 148